$51.00 UCF

Image and Video Databases:
Restoration, Watermarking and Retrieval

Series Editor: J. Biemond, Delft University of Technology, The Netherlands

Volume 1　Three-Dimensional Object Recognition Systems
(edited by A. K. Jain and P. J. Flynn)
Volume 2　VLSI Implementations for Image Communications
(edited by P. Pirsch)
Volume 3　Digital Moving Pictures – Coding and Transmission on ATM Networks
(J.-P. Leduc)
Volume 4　Motion Analysis for Image Sequence Coding (G. Tziritas and C. Labit)
Volume 5　Wavelets in Image Communication (edited by M. Barlaud)
Volume 6　Subband Compression of Images: Principles and Examples
(T.A. Ramstad, S.O. Aase and J.H. Husøy)
Volume 7　Advanced Video Coding: Principles and Techniques
(K.N. Ngan, T. Meier and D. Chai)
Volume 8　Image and Video Databases: Restoration, Watermarking and Retrieval
(A. Hanjalic, G.C. Langelaar, P.M.B. van Roosmalen, J. Biemond and
R.L. Lagendijk)

ADVANCES IN IMAGE COMMUNICATION 8

Image and Video Databases:

Restoration, Watermarking and Retrieval

A. Hanjalic, G.C. Langelaar, P.M.B. van Roosmalen, J. Biemond and R.L. Lagendijk

Department of Mediamatics
Delft University of Technology
Delft, The Netherlands

2000

ELSEVIER
Amsterdam – Lausanne – New York – Oxford – Shannon – Singapore – Tokyo

ELSEVIER SCIENCE B.V.
Sara Burgerhartstraat 25
P.O. Box 211, 1000 AE Amsterdam, The Netherlands

© 2000 Elsevier Science B.V. All rights reserved.

This work is protected under copyright by Elsevier Science, and the following terms and conditions apply to its use:

Photocopying
Single photocopies of single chapters may be made for personal use as allowed by national copyright laws. Permission of the Publisher and payment of a fee is required for all other photocopying, including multiple or systematic copying, copying for advertising or promotional purposes, resale, and all forms of document delivery. Special rates are available for educational institutions that wish to make photocopies for non-profit educational classroom use.

Permissions may be sought directly from Elsevier Science Rights & Permissions Department, PO Box 800, Oxford OX5 1DX, UK; phone: (+44) 1865 843830, fax: (+44) 1865 853333, e-mail: permissions@elsevier.co.uk. You may also contact Rights & Permissions directly through Elsevier's home page (http://www.elsevier.nl), selecting first 'Customer Support', then 'General Information', then 'Permissions Query Form'.

In the USA, users may clear permissions and make payments through the Copyright Clearance Center, Inc., 222 Rosewood Drive, Danvers, MA 01923, USA; phone: (978) 7508400, fax: (978) 7504744, and in the UK through the Copyright Licensing Agency Rapid Clearance Service (CLARCS), 90 Tottenham Court Road, London W1P 0LP, UK; phone: (+44) 171 631 5555; fax: (+44) 171 631 5500. Other countries may have a local reprographic rights agency for payments.

Derivative Works
Tables of contents may be reproduced for internal circulation, but permission of Elsevier Science is required for external resale or distribution of such material.
Permission of the Publisher is required for all other derivative works, including compilations and translations.

Electronic Storage or Usage
Permission of the Publisher is required to store or use electronically any material contained in this work, including any chapter or part of a chapter.

Except as outlined above, no part of this work may be reproduced, stored in a retrieval system or transmitted in any form or by any means, electronic, mechanical, photocopying, recording or otherwise, without prior written permission of the Publisher.
Address permissions requests to: Elsevier Science Rights & Permissions Department, at the mail, fax and e-mail addresses noted above.

Notice
No responsibility is assumed by the Publisher for any injury and/or damage to persons or property as a matter of products liability, negligence or otherwise, or from any use or operation of any methods, products, instructions or ideas contained in the material herein. Because of rapid advances in the medical sciences, in particular, independent verification of diagnoses and drug dosages should be made.

First edition 2000

Library of Congress Cataloging in Publication Data
A catalog record from the Library of Congress has been applied for.

ISBN: 0 444 50502 4

⊗ The paper used in this publication meets the requirements of ANSI/NISO Z39.48-1992 (Permanence of Paper).

Printed in The Netherlands.

INTRODUCTION TO THE SERIES
"Advances in Image Communication"

Dear Colleague,

Image Communication is a rapidly evolving multidisciplinary field on the development and evaluation of efficient means for acquisition, storage, transmission, representation, manipulation and understanding of visual information. Until a few years ago, digital image communication research was still confined to universities and research laboratories of telecommunication or broadcasting companies. Nowadays, however, this field is witnessing the strong interest of a large number of industrial companies due to the advent of narrow band and broadband ISDN, GSM, the Internet, digital satellite channels, digital over-the-air transmission and digital storage media. Moreover, personal computers and workstations have become important platforms for multimedia interactive applications that advantageously use a close integration of digital compression techniques (JPEG, MPEG), Very Large Scale Integration (VLSI) technology, highly sophisticated network facilities and digital storage media.

At the same time, the scope of research of the academic environment on Image Communication has further increased to include model- and knowledge-based techniques, artificial intelligence, motion analysis, and advanced image and video processing techniques. The variety of topics on Image Communication is so large that no one can be a specialist in all the topics, and the whole area is beyond the scope of a single volume, while the requirement of up-to-date information is ever increasing.

This was the rationale for Elsevier Science Publishers to approach me to edit a book series on 'Advances in Image Communication', next to the already existing and highly successful Journal: "Signal Processing: Image Communication". The book series was to serve as a comprehensive reference work for those already active in the area of Image Communication. Each author or editor was asked to write or compile a state-of-the-art book in his/her area of expertise, including information until now scattered in many journals and proceedings. The book series therefore would help Image Communication specialists to gain a better understanding of the important issues in neighbouring areas by reading particular volumes. It would also give newcomers to the field a foothold for doing research in the Image Communication area.

In order to produce a quality book series, it was necessary to ask authorities well known in their respective fields to serve as volume editors, who would in turn attract outstanding contributors. It was a great pleasure to me that ultimately we were able to attract such an excellent team of editors and authors.

Elsevier Science and I, as Editor of the series, are delighted that this book series has already received such a positive response from the image communication community. We hope that the series will continue to be of great use to the many specialists working in this field.

Jan Biemond
Series Editor

Preface

Along with the advancement in multimedia and Internet technology, large-scale digital *image and video databases* have emerged in both the professional and consumer environments. Although digital representations have many advantages over analog representations, vast amounts of (old) film and video material are still in analog format. Re-using this material in digital format and combining it with newly produced audiovisual information is, however, only feasible if the visual quality meets the standards expected by the modern viewer, which motivates the need for automated image *restoration*. At the same time, service providers are reluctant to offer services in digital form because of the fear for unrestricted duplication and dissemination of copyrighted material. This has led to worldwide research on *watermarking* techniques to embed a secret imperceptible signal, a watermark, directly into the video data. Further, with steadily increasing information volumes stored in digital image and video databases, it is crucial to find ways for efficient information *retrieval*.

This book provides an in-depth treatment of the three aforementioned important topics related to image and video databases: *restoration, watermarking* and *retrieval*. It is an outgrowth of the participation of the Delft University of Technology in the European Union ACTS program, a pre-competitive R&D program on Advanced Communications Technologies and Services (1994-1998). In particular the book has benefited from participation in the AURORA and SMASH projects on respectively automated film and video restoration and storage for multimedia systems (watermarking & retrieval).

The research has been performed in the Information and Communication Theory Group, Department of Mediamatics, Faculty of Information Technology and Systems of the Delft University of Technology, The Netherlands, as part of the Visual Communications research program (http://www-ict.its.tudelft.nl) and has been published extensively. The restoration task was performed by P.M.B. van Roosmalen, the watermarking task by G.C. Langelaar, and the retrieval task by A. Hanjalic under the guidance of the professors J. Biemond and R.L. Lagendijk.

Delft, February 2000

Outline

In recent years, technology has reached a level where vast amounts of digital audiovisual information are available at a low price. During the same time, the performance-versus-price ratio of digital storage media has steadily increased. Because it is easy and relatively inexpensive to obtain and store digital information, while the possibilities to manipulate such information are almost unlimited, large-scale *image and video databases* in the professional and consumer environments have grown rapidly. Examples are databases of museum records, Internet image and video archives, databases available to commercial service providers and private collections of digital audiovisual information at home. All of these are characterized by a quickly increasing capacity and content variety. This book addresses three important topics related to efficient practical realization and utilization of digital image and video databases: image *restoration*, copy-protection by *watermarking* and efficient information *retrieval*.

The first topic, restoration, is addressed in Part I of this book. It considers unique records of historic, artistic, and cultural developments of every aspect of the 20th century, which are stored in huge stocks of archived moving pictures. Many of these historically significant items are in a fragile state and are in desperate need of conservation and restoration. Preservation of visual evidence of important moments in history and of our cultural past is not only of purely scientific value. Moreover, it is possible to digitize these historic records and combine them with newly produced programs for broadcasting or database-building purposes. On the one hand, huge collections of movies, soaps, documentaries, and quiz shows currently held in store provide a cheap alternative to the high costs of creating new programs. On the other hand, emerging databases in some professional spheres, such as journalism, politics or social sciences, can largely benefit from preserved and easily accessible historic records. Re-using old film and video material is, however, only feasible if the visual and audio quality meets the standards expected by the modern viewer. There is a need for an automated tool for image restoration due to the vast amounts of archived film and video and due to economical constraints. The term *automated* should be stressed because manual image restoration is a tedious and time-consuming process. At the Delft University of Technology, algorithms were developed for correcting three types of artifact common to old film and video sequences, namely intensity flicker, blotches and noise.

Intensity flicker is a common artifact in old black-and-white film sequences. It is perceived as unnatural temporal fluctuations in image intensity that do not originate from the original scene. We describe an original, effective method for correcting intensity flicker on the basis of equalizing local intensity mean and variance in a temporal sense.

Blotches are artifacts typically related to film that are caused by the loss of gelatine and dirt particles covering the film. Existing techniques for blotch detection generate many false alarms when high correct-detection rates are required. As a result, unnecessary errors that are visually more disturbing than the blotches themselves can be introduced into an image sequence by the interpolators that correct the blotches. We describe techniques to improve the quality of blotch detection results by taking into account the influence of noise on the detection process and by exploiting the spatial coherency within blotches. Additionally, a new, fast, model-based method for good quality interpolation of blotched data is developed. This method is faster than existing model-based interpolators. It is also more robust to corruption in the reference data that is used by the interpolation process.

Coring is a well-known technique for removing noise from still images. The mechanism of coring consists of transforming a signal into a frequency domain and reducing the transform coefficients by the coring function. The inverse transform of the cored coefficients gives the noise-reduced image. We develop a framework for coring image sequences. The framework is based on 3D (2D space and time) image decompositions, which allows temporal information to be exploited. This is preferable to processing each frame independently of the other frames in the image sequence. Furthermore, a method of coring can be imbedded into an MPEG2 encoder with relatively little additional complexity. The MPEG2 encoder then becomes a device for simultaneous noise reduction and image sequence compression. The adjusted encoder significantly increases the quality of the coded noisy image sequences.

Not only does image restoration improve the perceived quality of the film and video sequences, it also, generally speaking, leads to more efficient compression. This means that image restoration gives better quality at fixed bitrates, or, conversely, identical quality at lower bitrates. The latter is especially important in digital broadcasting and storage environments for which the price of broadcasting/storage is directly related to the number of bits being broadcast/stored. We investigate the influence of artifacts on the coding efficiency and evaluate how much is gained by restoring impaired film and video sequences. We show that considerable savings in bandwidth are feasible without loss of quality.

The second topic, copy-protection by watermarking, is addressed in Part II of this book. Although digital data have many advantages over analog data, service providers are reluctant to offer services in digital form because they fear unrestricted duplication and dissemination of copyrighted material. The lack of adequate protection systems for copyrighted content was for instance the reason for the delayed introduction of the DVD. Several media companies initially refused to provide DVD material until the copy protection problem had been addressed.

To provide copy protection and copyright protection for digital audio and video data, two complementary techniques are being developed: encryption and watermarking. Encryption techniques can be used to protect digital data during the transmission from the sender to the receiver. However, after the receiver has received and decrypted the data, the data is in the clear and no longer protected. Watermarking techniques can complement encryption by embedding a secret imperceptible signal, a watermark, directly into the clear data. This watermark signal is embedded in such a way that it cannot be removed without affecting the quality of the audio or video data. The watermark signal can for instance be used for copyright protection by hiding information about the author in the data. The watermark can now be used to prove ownership in court.

Another interesting application for which the watermark signal can be used is to trace the source of illegal copies by using *fingerprinting* techniques. In this case, the media provider embeds watermarks in the copies of the data with a serial number that is related to the customer's identity. If now illegal copies are found, for instance on the Internet, the intellectual property owner can easily identify customers who have broken their license agreement by supplying the data to third parties. The watermark signal can also be used to control digital recording devices by indicating whether certain data may be recorded or not. In this case, the recording devices must of course be equipped with watermark detectors. Other applications of the watermark signal include: automated monitoring systems for radio and TV broadcasting, data authentication and transmission of secret messages.

Each watermarking application has its own specific requirements. Nevertheless, the most important requirements to be met by most watermarking techniques are that the watermark is imperceptible in the data in which the watermark is hidden, that the watermark signal can contain a reasonable amount of information and that the watermark signal can not easily be removed without affecting the data in which the watermark is hidden.

In Part II of this book an extensive overview is given of different existing watermarking methods. However, the emphasis is on the particular class of watermarking techniques that is suitable for real-time embedding watermarks in and extracting watermarks from compressed video data. This class of techniques is for instance suitable for fingerprinting and copy protection systems in home-recording devices. To qualify as a real-time watermarking technique for compressed video data, a watermark technique should meet the following requirements besides the already mentioned ones. There are two reasons why the techniques for watermark embedding and extracting cannot be too complex: they are to be processed in real time, and as they are to be used in consumer products, they must be inexpensive. This means that fully decompressing the compressed data, adding a watermark and subsequently compressing the data again is not an option. It should be possible to add a watermark directly to the compressed data. Furthermore, it is important that the addition of a watermark does not influence the size of the compressed data. For instance, if the size of a compressed MPEG-video stream increases, transmission over a fixed bit rate channel can cause problems, the buffers in hardware decoders can run out of space, or the synchronization of audio and video can be disturbed. The most efficient way to reduce the complexity of real-time watermarking algorithms is to avoid computationally demanding operations by exploiting the compression format of the video data. We introduce two new watermarking concepts that directly operate on the compressed data stream, namely the least significant bit (LSB) modification concept and the Differential Energy Watermark (DEW) concept.

The end of Part II is dedicated to the evaluation of the DEW-concept. Several approaches to evaluate watermarking methods from literature are discussed and applied. Furthermore, watermark removal attacks from literature are discussed and a new watermark removal attack is proposed.

While Parts I and II are mainly related to the process of creating an image or video database in terms of providing visual material of acceptable quality (restoration) and protecting the ownership of that material (watermarking), the topic addressed in Part III of this book, information retrieval, concerns the efficiency of using image and video databases, that is, of handling large amounts of not-indexed audiovisual information stored therein.

With steadily increasing information volumes stored in image and video databases, finding efficient ways to quickly retrieve information of interest becomes crucial. Since searching manually through GBytes of unorganized stored data is tedious and time-consuming, the need grows for transferring information retrieval tasks to automated systems. Realizing this transfer in practice is, however, not trivial. The main problem is that typical retrieval tasks, such as "find me an image with a bird!", are formulated on a *cognitive level*, according to the human capability of understanding the information content and analyzing it in terms of objects, persons, sceneries, meaning of speech fragments or the context of a story in general. Opposed to this, an image or a video is analyzed at the algorithmic or *system level* in terms of *features*, such as color, texture, shape, frequency components, audio and speech characteristics, and using the algorithms operating on these features. Such algorithms are, for instance, image segmentation, detection of moving objects in video sequences, shape matching, recognition of color compositions, determination of spatio-temporal relations among different objects or analysis of the frequency spectrum of the audio or speech stream. These algorithms can be developed using the state-of-the-art in image and audio analysis and processing, computer vision, statistical signal processing, artificial intelligence, pattern recognition and other related areas. Experience has shown, however, that the parallelism between the cognition-based and feature-based information retrieval is not viable in all cases. Therefore, the development of feature-based content-analysis algorithms has not been directed to enable queries on the highest semantic level, such as the above example with a bird, but mainly towards extracting certain semantic aspects of the information that would allow for a reduction of the overall large search space. The material presented in Part III of this book is meant to contribute further to research efforts in this direction.

We first introduce a series of novel algorithms for video analysis and abstraction. These algorithms are developed to provide an overview of the video-database content and logical entry points into a video when browsing through a video database. Also a video index may be constructed based on visual features contained in the abstract, which can then be used for video queries using image retrieval techniques. On the one hand, algorithmic solutions are provided for segmenting a video into temporally homogeneous fragments called *video shots*, for condensing each of the shots into a set of characteristic frames called *key frames* and for performing a high-level analysis of a video content. This high-level analysis includes determining semantic relationships among shots in terms of their temporal characteristics and suitable features of their key frames, and identification of certain semantically interesting video shots. Examples are merging the shots of a movie into scenes or episodes or the identification of anchorperson shots in news programs. On the other hand, we develop an algorithm for automatically summarizing an arbitrary video by extracting a number of suitable key frames in such a way that the result is similar as when that video is summarized manually. One characteristic application where having such abstracts is useful is browsing through a video and searching for a scene of interest. The user only needs to check a limited amount of information contained in an abstract instead of going through the entire video in the fast-forward/rewind mode, while still having available all the characteristic information related to the video content and thus being able to understand and follow that content exclusively on the basis of the abstract.

The second contribution of Part III is the search for suitable compression methodologies which are to be applied to images and videos stored in databases. Large scientific and industrial efforts have been invested over the years in developing and improving high-quality digital image and video compression methods. Hereby, three classical optimization criteria were taken into account: (1) minimizing the resulting bit rate, (2) maximizing the quality of the reconstructed image and video and (3) minimizing the computational costs. The invested efforts have resulted in many efficient image and video compression methods, the most suitable of which were standardized (e.g. JPEG, MPEG). These methods are, however, not optimized in view of *content accessibility* which is analog to the efficiency of regaining the features of content elements being important for a given retrieval task. This means that a high computational load in reaching image and video features combined with large amount of information stored in a database, can negatively influence the efficiency of the interaction with that database.

In order to make the interaction with a database more efficient, it is necessary to develop compression methods which explicitly take into account the content accessibility of images and video, together with the classical optimization criteria. This challenge can also be formulated as to reduce the computational load in obtaining the features from a compressed image or video. As a concrete step in this direction a novel image compression methodology is presented where a good synergy among the four optimization criteria is reached.

Contents

PREFACE   vii

OUTLINE   ix

Part I: Restoration

INTRODUCTION TO RESTORATION   3

   1.1   BACKGROUND   3
   1.2   SCOPE OF PART I   3
   1.3   OVERVIEW OF PART I   7

MODELING AND CODING   9

   2.1   MODELING FOR IMAGE RESTORATION   9
      2.1.1   Model selection and parameter estimation   9
      2.1.2   Impairments in old film and video sequences   11
      2.1.3   Influence of artifacts on motion estimation   16
   2.2   IMAGE RESTORATION AND STORAGE   20
      2.2.1   Brief description of MPEG2   20
      2.2.2   Influence of artifacts on coding efficiency   23

INTENSITY FLICKER CORRECTION   27

   3.1   INTRODUCTION   27
   3.2   ESTIMATING AND CORRECTING INTENSITY FLICKER IN STATIONARY SEQUENCES   28
      3.2.1   A model for intensity flicker   28
      3.2.2   Estimating intensity-flicker parameters in stationary scenes   30
      3.2.3   Measure of reliability for the estimated model parameters   32
   3.3   INCORPORATING MOTION   34
      3.3.1   Estimating global motion with phase correlation   34
      3.3.2   Detecting the remaining local motion   35
      3.3.3   Interpolating missing parameters   37
   3.4   PRACTICAL ISSUES   41
   3.5   EXPERIMENTS AND RESULTS   42
      3.5.1   Experiments on artificial intensity flicker   43
      3.5.2   Experiments on naturally degraded film sequences   44
   3.6   CONCLUSIONS   46

BLOTCH DETECTION AND CORRECTION   51

4.1 SYSTEM FOR BLOTCH DETECTION AND CORRECTION   51
4.2 OVERVIEW OF EXISTING TECHNIQUES   54
    4.2.1   Blotch detection techniques   54
    4.2.2   Techniques for motion vector repair   60
    4.2.3   Blotch correction techniques   63
    4.2.4   Conclusions   66
4.3 IMPROVED BLOTCH DETECTION BY POSTPROCESSING   66
    4.3.1   Simplified ROD detector   68
    4.3.2   Extracting candidate blotches   68
    4.3.3   Removing false alarms due to noise   69
    4.3.4   Completing partially detected blotches   72
    4.3.5   Constrained dilation for missing details   73
    4.3.6   Experimental evaluation   74
4.4 BLOTCH DETECTION WITH INCREASED TEMPORAL APERTURE   76
4.5 FAST, GOOD QUALITY INTERPOLATION OF MISSING DATA   79
    4.5.1   Interpolating missing data with controlled pasting   81
    4.5.2   Practical implementation of controlled pasting   84
    4.5.3   Experiments with controlled pasting   87
4.6 RESULTS AND CONCLUSIONS   91

NOISE REDUCTION BY CORING   93

5.1 INTRODUCTION   93
5.2 NOISE REDUCTION TECHNIQUES   94
    5.2.1   Optimal linear filtering in the MMSE sense   94
    5.2.2   Optimal noise reduction by nonlinear filtering: coring   95
    5.2.3   Heuristic coring functions   97
5.3 CORING IMAGE SEQUENCES IN WAVELET AND SUBBAND DOMAINS   98
    5.3.1   Nondecimated wavelet transform   98
    5.3.2   Simoncelli pyramid   100
    5.3.3   An extension to three dimensions using wavelets   105
    5.3.4   Noise reduction by coring   106
    5.3.5   Perfect reconstruction   109
    5.3.6   Experiments and results   109
5.4 MPEG2 FOR NOISE REDUCTION   112
    5.4.1   Coring I, P, and B frames   113
    5.4.2   Determining the DCT coring functions   115
    5.4.3   Experiments and results   118
5.5 CONCLUSIONS   121

EVALUATION OF RESTORED IMAGE SEQUENCES   123

6.1 INTRODUCTION   123
6.2 ASSESSMENT OF RESTORED IMAGE SEQUENCES   123
    6.2.1   Influence of image restoration on the perceived image quality   123
    6.2.2   Influence of image restoration on the coding efficiency   125
6.3 EXPERIMENTS AND RESULTS   128
    6.3.1   Test sequences   129
    6.3.2   Experiments on image restoration and perceived quality   130
    6.3.3   Experiments on image restoration and coding efficiency   132
    6.3.4   Discussion of experimental results   134

APPENDIX A HIERARCHICAL MOTION ESTIMATION   135

APPENDIX B DERIVATION OF CONDITIONALS   137

    B.1   INTRODUCTION   137
    B.2   CONDITIONAL FOR AR COEFFICIENTS   138
    B.3   CONDITIONAL FOR THE PREDICTION ERROR VARIANCE   139
    B.4   CONDITIONAL FOR THE DIRECTION OF INTERPOLATION   140

APPENDIX C OPTIMAL QUANTIZERS FOR ENCODING NOISY IMAGE SEQUENCES   143

BIBLIOGRAPHY – PART I   147

Part II: Watermarking

INTRODUCTION TO WATERMARKING   157

    7.1   THE NEED FOR WATERMARKING   157
    7.2   WATERMARKING REQUIREMENTS   159
    7.3   BRIEF HISTORY OF WATERMARKING   162
    7.4   SCOPE OF PART II   164
    7.5   OVERVIEW OF PART II   167

STATE-OF-THE-ART IN IMAGE AND VIDEO WATERMARKING   171

    8.1   INTRODUCTION   171
    8.2   CORRELATION BASED WATERMARK TECHNIQUES   172
        8.2.1   Basic technique in the spatial domain   172
        8.2.2   Extensions to embed multiple bits or logos in one image   175
        8.2.3   Techniques for other than spatial domains   181
        8.2.4   Watermark energy adaptation based on HVS   187
    8.3   EXTENDED CORRELATION-BASED WATERMARK TECHNIQUES   191
        8.3.1   Anticipating lossy compression and filtering   191
        8.3.2   Anticipating geometrical transforms   194
        8.3.3   Correlation-based techniques in the compressed domain   197
    8.4   NON-CORRELATION-BASED WATERMARKING TECHNIQUES   198
        8.4.1   Least significant bit modification   198
        8.4.2   DCT Coefficient Ordering   198
        8.4.3   Salient-Point Modification   200
        8.4.4   Fractal-based Watermarking   201
    8.5   CONCLUSIONS   203

LOW COMPLEXITY WATERMARKS FOR MPEG COMPRESSED VIDEO   205

    9.1   INTRODUCTION   205
    9.2   WATERMARKING MPEG VIDEO BIT STREAMS   207
    9.3   CORRELATION-BASED TECHNIQUES IN THE COEFFICIENT DOMAIN   210
        9.3.1   DC-coefficient modification   210
        9.3.2   DC- and AC-coefficient modification with drift compensation   211
            9.3.2.1   Basic watermarking concept   211
            9.3.2.2   Drift compensation   212
            9.3.2.3   Evaluation of the correlation-based technique   213
    9.4   PARITY BIT MODIFICATION IN THE BIT DOMAIN   214
        9.4.1   Bit domain watermarking concept   214
        9.4.2   Evaluation of the bit domain watermarking algorithm   216
            9.4.2.1   Test sequence   216
            9.4.2.2   Payload of the watermark   217
            9.4.2.3   Visual impact of the watermark   218
            9.4.2.4   Drift   221

9.4.3 Robustness 222
9.5 RE-LABELING RESISTANT BIT DOMAIN WATERMARKING METHOD 223
9.6 CONCLUSIONS 224

DIFFERENTIAL ENERGY WATERMARKS (DEW) 227

10.1 INTRODUCTION 227
10.2 THE DEW CONCEPT FOR MPEG/JPEG ENCODED VIDEO 229
10.3 DETAILED DEW ALGORITHM DESCRIPTION 232
10.4 EVALUATION OF THE DEW ALGORITHM FOR MPEG VIDEO DATA 240
    10.4.1 Payload of the watermark 240
    10.4.2 Visual impact of the watermark 241
    10.4.3 Drift 245
    10.4.4 Robustness 246
10.5 EXTENSION OF THE DEW CONCEPT FOR EZW-CODED IMAGES 247
10.6 CONCLUSIONS 250

FINDING OPTIMAL PARAMETERS FOR THE DEW ALGORITHM 251

11.1 INTRODUCTION 251
11.2 MODELING THE DEW CONCEPT FOR JPEG COMPRESSED VIDEO 253
    11.2.1 PMF of the cut-off index 253
    11.2.2 Model for the DCT-based energies 256
11.3 MODEL VALIDATION WITH REAL-WORLD DATA 258
11.4 LABEL ERROR PROBABILITY 263
11.5 OPTIMAL PARAMETER SETTINGS 266
11.6 EXPERIMENTAL RESULTS 268
11.7 CONCLUSIONS 271

BENCHMARKING THE DEW WATERMARKING ALGORITHM 273

12.1 INTRODUCTION 273
12.2 BENCHMARKING METHODS 274
12.3 WATERMARK ATTACKS 276
    12.3.1 Introduction 276
    12.3.2 Geometrical transforms 277
    12.3.3 Watermark estimation 278
        12.3.3.1 Introduction 278
        12.3.3.2 Watermark estimation by non-linear filtering 280
12.4 BENCHMARKING THE DEW ALGORITHM 285
    12.4.1 Introduction 285
    12.4.2 Performance factors 286
    12.4.3 Evaluation of the DEW algorithm for MPEG compressed video 286
    12.4.4 Evaluation of the DEW algorithm for still images 291
12.5 CONCLUSIONS 296

BIBLIOGRAPHY – PART II 299

Part III: Retrieval

INFORMATION RETRIEVAL: AN INTRODUCTION 313

13.1 INFORMATION RETRIEVAL SYSTEMS: FROM NEEDS TO TECHNICAL SOLUTIONS 313
13.2 SCOPE OF PART III 317
13.3 OVERVIEW OF PART III 319

STATISTICAL FRAMEWORK FOR SHOT-BOUNDARY DETECTION 323

14.1 INTRODUCTION   323
14.2 PREVIOUS WORK ON SHOT-BOUNDARY DETECTION   328
    14.2.1   Discontinuity values from features and metrics   328
    14.2.2   Detection approaches   333
14.3 A ROBUST STATISTICAL FRAMEWORK FOR SHOT-BOUNDARY DETECTION   339
14.4 DETECTOR FOR ABRUPT SHOT BOUNDARIES   343
    14.4.1   Features and metrics   343
    14.4.2   A priori probability function   344
    14.4.3   Scalar likelihood functions   344
    14.4.4   PMI and the conditional probability functions   346
    14.4.5   Experimental validation   347
14.5 CONCLUSIONS   348

AUTOMATICALLY ABSTRACTING VIDEO USING KEY FRAMES   351

15.1 INTRODUCTION   351
15.2 PREVIOUS WORK ON KEY-FRAME EXTRACTION   355
15.3 EXTRACTING KEY FRAMES BY APPROXIMATING THE CURVE OF VISUAL-CONTENT VARIATIONS   359
    15.3.1   Modeling visual content variations along a shot   360
    15.3.2   Distributing N key frames over the sequence   362
    15.3.3   Distributing key frames within a shot   363
    15.3.4   Experimental validation   366
15.4 KEY-FRAME EXTRACTION BASED ON CLUSTER-VALIDITY ANALYSIS   368
    15.4.1   Clustering   369
    15.4.2   Cluster-validity analysis   371
    15.4.3   Key frames from clusters   375
    15.4.4   Experimental validation   375
15.5 CONCLUSIONS   377

HIGH-LEVEL VIDEO CONTENT ANALYSIS   379

16.1 INTRODUCTION   379
16.2 RELATED WORK   384
    16.2.1   Detecting different temporal events in a video   384
    16.2.2   Detecting scene boundaries in a movie   385
    16.2.3   Extracting the most characteristic movie segments   386
    16.2.4   Automated recognition of film genres   386
    16.2.5   News-program analysis   387
    16.2.6   Methods for analyzing sports programs   390
16.3 AUTOMATICALLY SEGMENTING MOVIES INTO LOGICAL STORY UNITS   391
    16.3.1   Hierarchical model of a movie structure   391
    16.3.2   Definition of LSU   393
    16.3.3   Novel approach to LSU boundary detection   396
    16.3.4   Inter-shot dissimilarity measure   398
    16.3.5   Experimental validation   400
16.4 DETECTING ANCHORPERSON SHOTS IN NEWS PROGRAMS   403
    16.4.1   Assumptions and definitions   404
    16.4.2   Finding a template   405
    16.4.3   Template matching   406
    16.4.4   Experimental validation   407
16.5 CONCLUSIONS   410

COMPRESSION TRENDS: THE "FOURTH CRITERION   413

17.1 INTRODUCTION   413
17.2 A CONCEPT OF AN ALTERNATIVE IMAGE CODEC   418

17.3 IMAGE CODEC BASED ON SIMPLIFIED VQ   422
   17.3.1   Code-book generation   422
   17.3.2   Finding the block correspondences   426
   17.3.3   Compressed image format specification   426
17.4 PERFORMANCE EVALUATION   427
   17.4.1   CODEC performance regarding classical criteria   428
   17.4.2   Increase of content accessibility   431
17.5 CONCLUSIONS   434

BIBLIOGRAPHY – PART III   437

INDEX   443

Image and Video Databases

Part I: Restoration

Chapter 1
Introduction to Restoration

1.1 Background

If one considers that archived film and video sequences will be preserved by transferring them onto new digital media, there are a number of reasons why these sequences should be restored before renewed storage. First, restoration improves the subjective quality of the film and video sequences (and it thereby increases the commercial value of the film and video documents). Second, restoration generally leads to more efficient compression, i.e., to better quality at identical bitrates, or, conversely, to identical quality at lower bitrates. The latter is especially important in digital broadcasting and storage environments for which the price of broadcasting/storage is directly related to the number of bits being broadcast/stored.

There is a need for an automated tool for image restoration due to the vast amounts of archived film and video and due to economical constraints. The term *automated* should be stressed because manual image restoration is a tedious and time-consuming process. Also, the restoration tool should operate in *real-time* in order to allow for bulk processing, and to reduce the high costs of manual labor by requiring a minimum of human intervention.

1.2 Scope of Part I

Detecting and restoring selected artifacts from archived film and video material with real-time hardware places constraints on how that material is processed and on the complexity of the algorithms used. It is stressed here that these constraints do not restrict the complexity of the methods for image restoration presented here, with the exception of the work presented in Chapter 3. Even though much of the work described here is too complex (meaning too expensive) to be implemented in hardware directly, it gives good insight into the nature of the investigated artifacts. The material presented in Chapters 1 to 6 gives an upper bound on the quality that can be achieved under relaxed constraints.

We restrict ourselves to black-and-white image sequences for two reasons. First, a large proportion of the films that require restoration is in black and white. Second, most of the algorithms can easily be extended to color, though perhaps in a suboptimal manner. An example of this would be a situation in which a color image sequence is restored by applying the restoration algorithms to the R, G, and B channels separately. Multi channel approaches [Arm98], [Ast90] could be taken from the start, at the cost of increased complexity and at the risk of achieving little significant gain compared to what single channel processing already brings.

As an inventory of impairments found in old film and video sequences, a list of over 150 entries emerged that indicates the nature of the defects and the frequency of their occurrence. From this list, the most important impairments are *noise* [Abr96], [Arc91], [Bra95], [Don94b], [Dub84], [Hir89], [Kle94], [Özk92], [Özk93], [Roo97], *blotches* [Fer96], [Goh96], [Kal97], [Kok98], [Kok95a], [Kok95b], [Mul96], [Nad97], [Roo99a], [Roo98b], [Vel88], *line scratches* [Kok98], [Mor96], *film unsteadiness* [Vla96], and *intensity flicker* [Fer96], [Mul96], [Ric95], [Roo99b], [Roo97]. Figure 1.1 shows some examples of these artifacts. This figure shows frames that are corrupted by multiple artifacts. This is often the case in practice.

Not only the quality of video has been affected by time, audio tracks often suffer degradations as well. However, restoration of audio is beyond the scope of this book.

Even though a single algorithm for restoring all the artifacts at hand in an integral manner is conceivable, a modular approach was chosen to resolve the various impairments. A divide-and-conquer strategy increases the probability of (at least partial) success. Furthermore, real-time systems for video processing require very fast hardware for the necessary computations. Modular systems allow the computational complexity to be distributed. Figure 1.2 shows a possible system for image restoration using a modular approach that was largely implemented for the purposes of this book.

The first block in Figure 1.2, *flicker correction*, removes disturbing variations in image intensity in time. Intensity flicker hampers accurate local motion estimation; therefore, it is appropriate to correct this artifact prior to applying any restoration

INTRODUCTION TO RESTORATION 5

technique that relies on local motion estimates. Next, local motion is estimated. Instead of designing (yet another) motion estimator that is robust to the various artifacts, we use a hierarchical block matcher [Bie88], [Haa92], [Tek95] with constraints on the smoothness of the motion vectors.

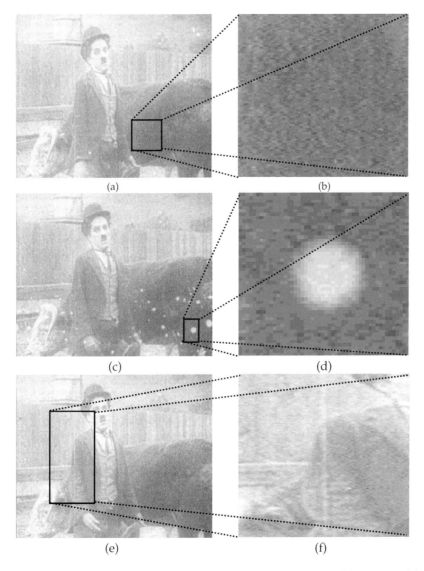

Figure 1.1: *(a,c,e) Three consecutive frames from a Charlie Chaplin film impaired by noise, blotches, and line scratches. There are also differences in intensity, which are less visible in print than on a monitor though. Zooming in on (b) noise, (d) a blotch, and (f) a scratch.*

Where the motion vectors are not reliable, due to the presence of artifacts, a strategy of vector repair is applied when necessary [Che97], [Has92], [Kok98], [Lam93], [Nar93]. Next, *blotch removal* detects and removes dark and bright spots that are often visible in film sequences. *Scratch removal*, which is not a topic of research in this book, removes vertical line scratches. *Noise reduction* reduces the amount of noise while it preserves the underlying signal as well as possible. Finally, *image stabilization* makes the sequence steadier by aligning (registering) the frames of an image sequence in a temporal sense. Image stabilization is not a topic of research in this book.

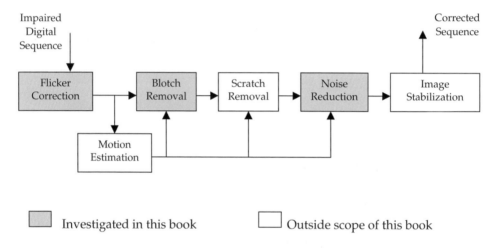

Figure 1.2: *Schematic overview of a modular system towards image restoration.*

In Figure 1.2, blotches and scratches are addressed prior to noise because they are local artifacts, corrections thereof influence the image contents only locally. Noise reduction is a global operation that affects each and every pixel in a frame. Therefore, all processes following noise reduction are affected by possible artifacts introduced by the noise reduction algorithm. Image stabilization, for which very robust algorithms exist, is placed at the back end of the system because it too affects each and every pixel by compensating for subpixel motion and by zooming in on the image. Zooming is required to avoid visible temporal artifacts near the image boundaries. As already mentioned, intensity flicker correction is an appropriate front end to the system. It is applied prior to the algorithms that require local motion estimates.

At the starting point in Figure 1.2 are digital image sequences instead of physical reels of film or tapes containing analog video. Rather than investigating the large number of formats and systems that have been used in one period or another over the last century, it is assumed that the archived material has been digitized by skilled technicians who know best how to digitize the film and video from the various sources. When the source material is film, digital image sequences are obtained by digitizing the output of the film-to-video *telecine*. It must be kept in mind that the earlier telecines have their limitations in terms of noise characteristics and resolution. Sometimes a copy on video tape obtained from an earlier telecine is all that remains of a film.

The output of the system in Figure 1.2 forms the restored image sequence. Subjective evaluations using test panels assess the improvement in perceived quality of the restored sequence with respect to the impaired input sequence.

1.3 Overview of Part I

Chapter 2 commences with general remarks on model selection, parameter estimation, and restoration. The key to an automatic restoration system lies in automatic, reliable parameter estimation. Models for noise, blotches, line scratches, film unsteadiness, and intensity flicker are reviewed. Motion estimation is an important tool in image sequence restoration, and its accuracy determines the quality of the restored sequences. For this reason, the influence of artifacts on motion estimation is investigated. It is likely that archived material selected for preservation is re-stored in a compressed format on new digital media. To appreciate the possible benefits of image restoration with respect to compression, the influence of artifacts on the coding efficiency of encoders based on the MPEG2 video compression standard is investigated.

Chapter 3 develops a method for correcting intensity flicker. This method reduces temporal fluctuations in image intensity automatically by equalizing local image means and variances in a temporal sense. The proposed method was developed to be implemented in hardware; therefore, the number of operations per frame and the complexity of these operations have been kept as low as possible. Experimental results on artificially and naturally degraded sequences prove the effectiveness of the method.

Chapter 4 investigates blotch detection and removal. Existing methods, both heuristic and model based, are reviewed. Improved methods are developed. Specifically, the performance of a blotch detector can be increased significantly by postprocessing the detection masks resulting from this detector. The postprocessing operations take into account the influence of noise on the detection process; they also exploit the spatial coherency within blotches. Where blotches corrupt the image data, the motion estimates are not reliable. Therefore, benefits of motion-vector repair are investigated. Finally, a new, relatively fast model-based method for good-quality interpolation of missing data is presented.

Chapter 5 investigates coring. Coring is a well-known technique for removing noise from images. The mechanism of coring consists of transforming a signal into a frequency domain and reducing the transform coefficients by the coring function. The inverse transform of the cored coefficients gives the noise-reduced image. This chapter develops a framework for coring image sequences. The framework is based on 3D image decompositions, which allows temporal information to be exploited. This is preferable to processing each frame independently of the other frames in the image sequence. Furthermore, this chapter shows that coring can be imbedded into an MPEG encoder with relatively little additional complexity. The adjusted encoder significantly increases the quality of the coded noisy image sequences.

Chapter 6 evaluates the image restoration tools developed in this book. First, it verifies experimentally that the perceived quality of restored image sequences is better than that of the impaired source material. Second, it verifies experimentally that, for the artifacts under consideration, image restoration leads to more efficient compression.

Chapter 2

Modeling and Coding

2.1 Modeling for image restoration

Model selection and parameter estimation are key elements in the design process of an image restoration algorithm. Section 2.1.1 reviews these key elements so that their presence can be recognized clearly in subsequent chapters. It is argued that robust automatic parameter estimation is essential to an automatic image restoration system. Section 2.1.2 models common degradations that affect old film and video sequences. These models form a basis for the restoration techniques developed in this book. They are also used for evaluation purposes. Section 2.1.3. investigates the influence of artifacts on the accuracy of motion estimation.

2.1.1 Model selection and parameter estimation

Image model. Many models that define various aspects of natural images and of image sequences are described in literature. For example, for still images, the magnitude of the Fourier spectrum has a *1/f* characteristic [Sch98], and local pixel intensities depend on each other via markov random fields [Gem84], [Ros82], [Won68], [Woo72]; for image sequences, there is a very high correlation between frames in time for image sequences [Has92].

The choice of the image model to be used depends on the problem at hand. In the case of image restoration, it is appropriate to select image models with ordinary parameter values that are affected as much as possible by the degradations under investigation. The reason for this is apparent. Suppose the model parameters of the assumed image model are not affected at all by a certain degradation. Then that image model provides no information that can be used for determining the severity of that degradation, nor does it provide any indication of how to correct the degradation.

Degradation model. Degradation models describe how data are corrupted; they imply how the model parameters for unimpaired images are altered. Models for specific degradations are obtained through a thorough analysis of the mechanisms generating the artifacts. The analysis is not always straightforward because the physical processes that underlie an impairment can be very complex and difficult to qualify. Often there is a lack of detailed knowledge on how a signal was generated. In practice, approximations and assumptions that seem reasonable have to be made. For example, in Section 2.1.2, the overall influence of the various noise sources affecting pictures in a chain of image capture, conversion, and storage is approximated by a single source instead of taking into account all the individual noise contributions explicitly.

Restoration model. Ideally, restoration would be modeled as the inverse operation of the degradation with its model parameters. Unfortunately, "the inverse" does not exist in many cases due to the singularities introduced by the degradation and due to the limited accuracy with which the model parameters are known. There are many solutions to a restoration problem that give identical observed signals when the degradation model (though be it with different parameters) is applied to them. For example, image data corrupted by blotches can be restored by a number of methods (Chapter 4), each of which gives a different solution. However, none of the solutions conflict with the degradation process and with the observed data that result from the degradation process.

The restoration problem is ill posed in the sense that no unique inverse to the degradation exists. A unique solution can be found only by reducing the space of possible solutions, by setting constraints in the form of criteria that must be fulfilled as well as is possible: the characteristics of the restored image are required to fit an image model. The goal of image restoration is to restore an image so that it resembles the original scene as closely as possible. Therefore, an often used additional criterion is that, in the spatial domain, the mean squared error between the restored image and the original, uncorrupted image must be as small as possible.

Estimating model parameters. Figure 2.1 shows how the image, degradation, and restoration models relate to each other. The central element that links the models is parameter estimation (*system identification*). The quality of a restored image sequence is determined by the quality of the estimated model parameters. Indeed, the quality of a restored image sequence can be worse than that of the degraded source material if poor choices are made for the values of the model parameters. Therefore, the level of automation for which model parameters can be estimated in a robust manner and with sufficient accuracy determines the extent to which a restoration system performs its function without user intervention. For this reason, automatic parameter estimation from the observed signals is an important part of each of the methods for image restoration presented in this book.

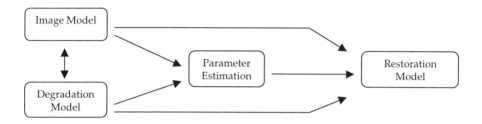

Figure 2.1: *Relationships between model selection and parameter estimation.*

Automatic parameter estimation is a non-trivial task in many cases due to the fact that insufficient numbers of data are available and due to the presence of noise. The term *noise* has a broad meaning in this context, and often it includes the signal to be restored from the observed data themselves. For example, estimating the noise variance (as a parameter for some algorithm) is hampered by the fact that it is very difficult to differentiate between noise and texture in natural images. Again, approximations and assumptions that seem reasonable have to be made.

Note that the quality of the estimated model parameters, e.g., determined by means of a direct numerical comparison to the true parameters, is not necessarily a good indication of the quality of the restoration result. This is because the quality of the restoration result varies in a different way for estimation errors in each of the parameters [Lag91].

2.1.2 Impairments in old film and video sequences

Chapter 1 mentions the most common impairments in old film and video sequences, and Figure 1.1 shows some examples of these artifacts. This subsection gives models for the various impairments. Figure 2.2 indicates the sources of the artifacts in a chain of recording, storage, conversion, and digitization.

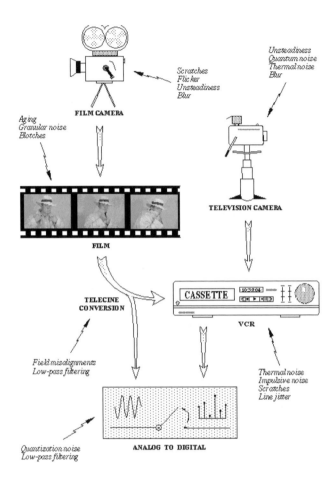

Figure 2.2: *Sources of image degradation in a chain of recording, storage, conversion and digitization.*

Noise. Any recorded signal is affected by noise, no matter how precise the recording apparatus. In the case of archived material, many noise sources can be pointed out. There is granular noise on film, a result of the finite size of the silver grains on film, that can be modeled by signal-dependent random processes [Bil75], [Jai89], [Özk93], [Pra91]. There is photon or quantum noise from plumbicon tubes and *charged coupled devices* (CCDs) that is modeled as a signal-dependent Poisson process [Dav92]. There is also thermal noise, introduced by electronic amplifiers and electronic processing, that is modeled as additive white gaussian noise [Dav91], [Pra91]. There is impulsive noise resulting from disturbances of digital signals stored on magnetic tape [Jus81].

Finally, in the case of digital signal processing, the digitizing process introduces quantization noise that is uniformly distributed [Rob87].

Many historical (and modern) film and video documents contain a combination of all the types of noise mentioned. For instance, such is the case for material originating on film that has been transferred to video. Modeling noise is often complicated by the band-limiting effects of optical systems in cameras and by the nonlinear gamma correction that makes the noise dependent on the signal [Kle94]. Quantitative analysis of the contributions of each individual noise source to a recorded image is extremely difficult, if not impossible. In practice, it is often assumed that the *Central Limit Theorem* [Leo94] applies to the various noise sources. This implies the assumption that the various noise sources generate *independent and identically distributed* (i.i.d.) noise.

Unless mentioned otherwise, it is assumed in Chapters 1 to 6 that the combined noise sources can be represented by a single i.i.d. additive gaussian noise source. Hence, an image corrupted by noise is modeled as follows. Let $y(i)$ with $i=(i, j, t)$ be an image with discrete spatial coordinates (i, j) recorded at time t. Let the noise be $\eta(i)$. The observed signal $z(i)$ is then given by:

$$z(i) = y(i) + \eta(i) \tag{2.1}$$

Many very different approaches to noise reduction are found in the literature, including optimal linear filtering techniques, (nonlinear) order statistics, scale-space representations, and bayesian restoration techniques [Abr96], [Arc91], [Bra95], [Don95], [Don94a], [Don94b], [Dub84], [Hir89], [Kle94], [Özk92], [Özk93], [Roo96].

Blotches. Blotches are artifacts that are typically related to film. In this book, the term *blotch* is used to indicate the effects that can result from two physical degradation processes of film. Both degradations lead to similar visual effects. The first degradation process is a result of dirt. Dirt particles covering the film introduce bright or dark spots on the picture (depending on whether the dirt is present on the negative or on the positive). The second degradation process is the loss of gelatin covering the film, which can be caused by mishandling and aging of the film. In this case, the image is said to be *blotched*. A model for blotches is given in [Kok98]:

$$z(i)=(1- d(i))\, y(i) + d(i)\, c(i) \tag{2.2}$$

where $z(i)$ and $y(i)$ are the observed and the original (unimpaired) data, respectively. The binary blotch detection mask $d(i)$ indicates whether each individual pixel has been corrupted: $d(i) \in \{0, 1\}$. The values at the corrupted sites are given by $c(i)$, with $c(i) \neq y(i)$. A property of blotches is that the intensity values at the corrupted sites vary smoothly; that the variance $c(i)$ within a blotch is small. Blotches seldom appear at the same location in a pair of consecutive frames. Therefore the binary mask $d(i)$ will seldom be set to one at two spatially co-sited locations in a pair of consecutive frames. However, there is spatial coherence within a blotch; if a pixel is blotched, it is likely that some of its neighbors are corrupted as well.

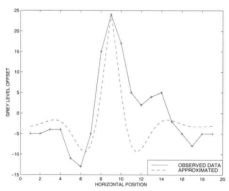

Figure 2.3: *Measured intensities (solid line) and approximated intensities (dashed line) from a cross section of the vertical scratch in Figure 1.1e*

Figure 2.4: *Example of a frame affected by a horizontal scratch on a two-inch video tape. (Photo by courtesy of the BBC).*

Films corrupted by blotches are often restored in a two-step approach. The first step detects blotches and generates binary detection masks that indicate whether each pixel is part of a blotch. The second step corrects pixels by means of spatio-temporal interpolation [Fer96], [Goh96], [Kal97], [Kok95a], [Kok95b], [Mul96], [Nar93], [Roo99a], [Roo98b], [The92]. Sometimes an additional step of motion estimation is included prior to interpolation because motion vectors are less reliable at corrupted sites. An alternative approach is presented in [Kok98], where blotches are detected and corrected simultaneously.

Line scratches. A distinction can be made between horizontal and vertical line scratches. Vertical line scratches are impairments that are typically related to film [Kok98], [Mor96]. They are caused by sharp particles scratching the film in a direction parallel to the direction of film transport within the camera. Line scratches are often visible as bright or dark vertical lines. The fact that vertical lines appear in nature frequently makes it difficult for an algorithm to distinguish between scratches and real-image structures. A one-dimensional cross-section of a scratch can be modeled by a damped sinusoid (Figure 2.3):

$$l(i) = A\,k^{|c-i|}\cos\left(\frac{|c-i|}{w}\right) + f_0 \tag{2.3}$$

where A depends on the dynamic range of the intensities over the cross-section of a scratch, k is the damping coefficient, c indicates the central position of the scratch, w indicates the width of the scratch, and f_0 is an offset determined by the local mean gray level. Once detected, line scratches can be restored by spatial or spatio-temporal interpolation.

In the case of video, horizontal scratches disturb the magnetic information stored on the tape. As a result of the helical scanning applied in video players, a horizontal scratch on the physical carrier does not necessarily give a single horizontal scratch in the demodulated image. For example, a horizontal scratch on a *two-inch* recording results in local distortions all over the demodulated image. Figure 2.4 is an example.

Film unsteadiness. Two types of film unsteadiness are defined, namely interframe and intraframe unsteadiness. The first and most important category is visible as global frame-to-frame displacements caused by mechanical tolerances in the transport system in film cameras and by unsteady fixation of the image acquisition apparatus. A model for interframe unsteadiness is:

$$z(i) = y(i - q_i(t), j - q_j(t), t) \tag{2.4}$$

Here $q_i(t)$ and $q_j(t)$ indicate the global horizontal and vertical displacement of frame t with respect to the previous frame. Intraframe unsteadiness can be caused by transfers from film to video where the field alignment is off (many older telecines used separate optical paths for the odd and even fields). This leads to interference patterns that are perceived as variances in luminance. Unsteadiness correction is estimated from the displacements and misalignments by maximizing temporal and spatial correlation, followed by resampling of the data. See [Vla96], for example.

Intensity flicker. Intensity flicker is defined as unnatural temporal fluctuations in the perceived image intensity that do not originate from the original scene. There are a great number of causes, e.g., aging of film, dust, chemical processing, copying, aliasing, and, in the case of the earlier film cameras, variations in shutter time. This book models intensity flicker as:

$$z(i) = \alpha(i)\,y(i) + \beta(i) \tag{2.5}$$

where fluctuations in image intensity variance and in intensity mean are represented by the multiplicative $\alpha(i)$ and additive $\beta(i)$. It is assumed that $\alpha(i)$ and $\beta(i)$ are spatially smooth functions. Histogram equalization has been proposed as a solution to intensity flicker [Fer96], [Mul96], [Ric95]. This book presents a more robust solution [Roo99b].

Other artifacts are line-jitter [Kok98], [Kok97], color fading, blur [Ban97], [Lag91], echoes, drop-outs and moiré effects. These are beyond the scope of this book.

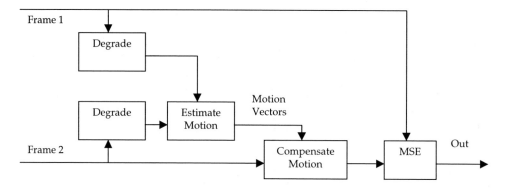

Figure 2.5: *Scheme for measuring the influence of image degradations on the accuracy of estimated motion vectors.*

2.1.3 Influence of artifacts on motion estimation

For image sequence restoration, temporal data often provide additional information that can be exploited above that which can be extracted from spatial data only. This is because natural image sequences are highly correlated in a temporal sense in stationary regions. In nonstationary regions, object motion reduces the local temporal correlation. Therefore, increasing the stationarity of the data via motion estimation and compensation is beneficial to the restoration process. Many motion estimation techniques have been developed in the context of image compression. Examples are (hierarchical) block matchers, pel-recursive estimators, phase correlators, and estimators based on bayesian techniques [Bie87], [Bie88], [Haa92], [Kon92], [Pea77], [Tek95].

This book uses a hierarchical motion estimator with integer precision and some constraints on the smoothness of the motion vectors. The constraints on smoothness are imposed by increasingly restricting the allowed deviation from the local candidate vectors passed on from lower resolution levels to higher resolution levels. Appendix A describes the details of this motion estimator. A motion-compensated frame representing a frame y(i) recorded at time t computed from a reference frame recorded at time t+k will be denoted as $y_{mc}(i,t+k)$.

The scheme depicted in Figure 2.5 was used for some experiments to get some feeling for the influence of various artifacts on the accuracy of this motion estimator. In this scheme, two consecutive frames from a sequence are degraded and the motion between the objects in the degraded frames is estimated. The *mean squared error* (MSE) between the original (unimpaired) frames is then computed. One of these frames is compensated for motion with the estimated vectors. Let N indicate the number of pixels per frame. Then, the MSE between the current frame and the motion-compensated next frame is defined as:

$$MSE(y(i), y_{mc}(i,t+1)) = \frac{1}{N} \sum_i \sum_j (y(i,j,t) - y_{mc}(i,j,t+1))^2 \qquad (2.6)$$

The rationale behind this scheme is the following. In the case that the motion estimator is not influenced much by the degradations, the correct vectors are found and the MSE is low. As the influence of the degradations on the estimated motion vectors becomes more severe, the MSE increases.

The scheme in Figure 2.5 was applied to three test sequences to which degradations of various strength are added. The first sequence, called *Tunnel*, shows a toy train driving into a tunnel. The background is steady. The second sequence, *MobCal*, has slow, subpixel motion over large image regions. The third sequence, *Manege*, shows a spinning carousel and contains a lot of motion. Table 2.1 indicates the severity of the impairments for various levels of strength. Strength zero indicates that no degradation has been added, strength four indicates an extreme level of degradation. The latter level does not occur frequently in naturally degraded image sequences.

Figure 2.6 plots the MSE for each of the test sequences as a function of the strength of the impairments. Before going into the details of the results, a few details are noted from this figure. First, in the absence of degradations, the MSE is relatively large for the *Manege* sequence. The reason for this is that the motion estimation, which was computed on a frame basis, was hampered by the strong interlacing effects. Second, the trends of the results are identical for all test sequences, i.e., the results are consistent.

Noise. Block-matching algorithms estimate motion by searching for maximal correlation between image regions in consecutive frames. If the signal-to-noise ratio is low, there is a risk that the maximum results largely from correlating noise. In the case the noise spectrum is white, hierarchical motion estimators are more robust to noise than full-search block matchers. Most of the signal energy of natural images is concentrated in the low frequencies. For a hierarchical block matcher this means that at the lower resolution levels, which are obtained by low-pass filtering the data, the signal-to-noise ratio is higher than at the higher resolution levels. Therefore, the probability of spurious matches is reduced. The influence of noise at the higher resolution levels is reduced by the constraints placed on the smoothness of the candidate motion vectors. Figure 2.6a shows the MSE computed for the three test sequences to which various amounts of white gaussian noise have been added.

	Strength 0	Strength 1	Strength 2	Strength 3	Strength 4
Noise (variance)	0	14	56	127	225
Blotches (% corrupted)	0	0.41	0.62	1.04	1.90
Number of Scratches	0	2	5	8	11
Flicker (MSE)	0	19	72	161	281

Table 2.1: *Average strength of various impairments added to test sequences. For noise the measure is the noise variance; for blotches, the measure is the percentage of pixels corrupted; for scratches, the measure is the number of scratches; and for intensity flicker, the measure is the MSE between original and corrupted frames.*

Blotches. A hierarchical block matcher will find the general direction in which data corrupted by blotches move, provided that the sizes of the contaminated areas are not too large. Because of the subsampling, the sizes of the blotches are reduced and they will have little influence on the block-matching results at the lower resolution levels. At the higher resolutions, the blotches cover larger parts of the blocks used for matching, and blotches will therefore have great influence on the matching results. However, if the number of candidate vectors is limited (e.g., in case the motion is identical in all neighboring regions) the correct motion vector may yet be found. Figure 2.6b shows the MSE computed for the three test sequences to which various numbers of blotches have been added.

Line scratches. The temporal consistency of line scratches is very good. As a result, motion estimators tend to lock onto them, especially if the contrast of the scratches is great with respect to the background. If the background motion is different from that of the line scratches, considerable errors result. Figure 2.6c shows the MSE computed for the three test sequences.

Unsteadiness. Measuring the influence of unsteadiness on motion estimates with the scheme in Figure 2.5 is not meaningful. Estimating motion between frames from an unsteady sequence is not unlike estimating motion between frames from a sequence containing camera pan. A motion estimator that performs its function well does not differentiate between global and local motion. In practice, unsteadiness (and camera pan) does have some influence. First, there are edge effects due to data moving in and out of the picture. Second, motion estimators are often intentionally biased towards zero-motion vectors. Third, the motion estimation can be influenced by aliasing if the data are not prefiltered correctly. This third effect is not of much importance because natural images have relatively little high-frequency content.

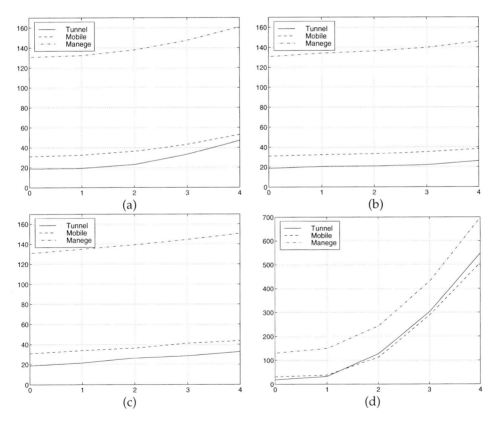

Figure 2.6: *MSE versus strength of impairment (0 = no impairment, 4 = greatly impaired): (a) noise, (b) blotches, (c) line scratches, (d) intensity flicker. Note the differences in scale*

Intensity Flicker. Many motion estimators, including the hierarchical motion estimator used in this book, assume the *constant luminance constraint* [Tek95]. This constraint, which requires that there be no variations in luminance between consecutive frames, is not met in the presence of intensity flicker. Figure 2.6d shows the MSE computed for the three test sequences to which varying amounts of intensity flicker have been added. The dramatic influence of this artifact on the quality of the estimated motion vectors compared to the other the artifacts examined becomes clear when the scale in Figure 2.6d is compared to those in Figures 2.6a-c.

In conclusion, artifacts can have a considerable impact on the accuracy of estimated motion vectors. In some cases, this leads to a chicken-and-egg problem: in order to obtain good motion estimates, the artifacts should be restored; and in order to restore the artifacts, good motion estimates are required. This problem can often be overcome by applying iterative solutions where estimates of the motion vectors and of the restored image are obtained in an alternating fashion. Alternatively, restoration methods that do not rely on motion estimates might be devised (Chapter 3) or a strategy of motion-vector repair can be applied after the severity of the impairments has been determined (Chapter 4).

2.2 Image restoration and storage

Restoration of archived film and video implies that the restored sequences will once again be archived. It is very likely that the restored documents are stored in new digital formats rather, than in analog formats similar to those from which the material originated. Most restored material will be re-archived in a compressed form due to the high costs associated with renewed storage of the vast amounts of material being held in store currently. This section investigates the effects of various impairments on the coding efficiency and uses the MPEG2 compression standard as a reference. The results of this investigation indicate the possible benefits that can be obtained by applying image restoration prior to encoding.

2.2.1 Brief description of MPEG2

The ISO/IEC MPEG2 coding standard developed by the Motion Pictures Expert Group is currently *the* industry standard used for many digital video communication and storage applications. As a result of the requirements on its versatility, it has become a very complex standard with a description that fills several volumes [IEC1], [IEC2], [IEC3]. The following describes only the basics of MPEG2 that are relevant to the restoration.

To achieve efficient compression, the MPEG2 encoding scheme exploits spatial and temporal redundancy within elementary units of pictures. Such an elementary unit is called a *group of pictures* (GOP) (Figure 2.7). MPEG2 defines three types of pictures that can be used within a GOP, namely *intra frames* (I frames), *predicted frames* (P frames), and *bi-directionally interpolated frames* (B frames). A GOP cannot consist of a random collection of I, B, and P frames. There are some rules that must be adhered to, e.g., the first encoded picture in a GOP is always an I frame. Figure 2.8 gives a schematic overview of the hybrid coding scheme that forms the heart of the MPEG2 coding system.

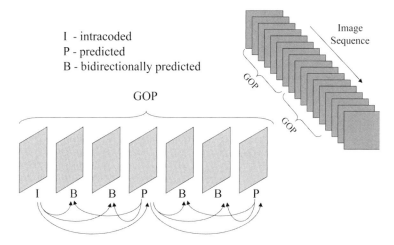

Figure 2.7: *Subdivision of an image sequence into groups of pictures (GOPs). In this example, the GOP has length 7 and it contains I, P and B frames. The arrows indicate the prediction directions.*

I frames. To encode I frames, spatial information is used only. Therefore, temporal information for decoding I frames is not required. This is important because it allows random access to the image sequence (on the level of GOPs anyhow) and it limits error propagation in the temporal direction resulting from possible bit errors in a stream of encoded data.

Efficient compression of I frames requires reduction of spatial redundancy. The MPEG2 standard reduces the spatial redundancy by subdividing I frames into 8 by 8 image blocks and applying the *discrete cosine transform* (DCT) to these blocks. The decorrelating properties of the DCT concentrate much of the signal energy of natural images in the lower-frequency DCT coefficients. A quantizer Q quantizes the transform coefficients and thereby reduces the number of representation levels and sets many coefficients to zero. Note that, as the eye is less sensitive to quantization of high frequencies, the high-frequency components can be quantized relatively coarsely. Entropy coding codes the remaining coefficients efficiently by applying *run-length coding* followed by *variable length coding* (VLC) to each 8 by 8 block of quantized DCTs. The result forms the encoder output.

The decompression of I frames is straightforward: the inverse DCT is applied to 8 by 8 blocks in which the quantized coefficients are ordered after the entropy-coded data are decoded.

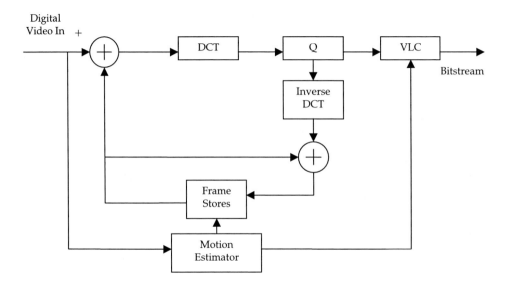

Figure 2.8: *Schematic overview of the hybrid coding scheme used in MPEG2.*

B and P frames. Efficient compression of P frames and B frames is achieved by exploiting both temporal and spatial redundancy. P frames are predicted from single I frames or P frames coded previously, for which motion estimation and compensation is often used. The prediction error signals, which contain spatial redundancy, are encoded as are the I frames, i.e., by means of the DCT and quantization. B frames are predicted from two coded I frames or P frames and are encoded like the P frames. The motion vectors are transmitted as well; these are encoded with differential coding. Note that, in the case of P frames and B frames, the encoder may well decide that it is more efficient to encode the original contents of an image region instead of encoding the prediction error signals.

Decompression consists of decoding the error signal and adding it to the motion-compensated prediction made in the decoder.

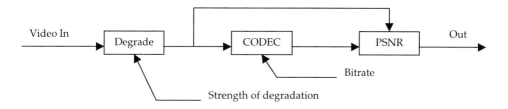

Figure 2.9: *Experimental setup for evaluating the influence of artifacts on the coding efficiency.*

2.2.2 Influence of artifacts on coding efficiency

Figure 2.9 shows an experimental setup used for evaluating the quantitative influence of artifacts on the coding efficiency of an MPEG2 encoder. Coding efficiency is defined as the amount of distortion introduced by a codec under the condition of a limited bitrate, or, vice versa, as the bitrate required by a codec under condition of limited distortion. The scheme in Figure 2.9 measures the *peak-signal-to-noise-ratio* (PSNR) of a degraded image sequence after encoding and decoding $z_c(i)$. The degraded sequence prior to encoding $z_o(i)$ serves as the reference. The PSNR is defined as:

$$PSNR[z_o(i), z_c(i)] = 10\log\left(\frac{224^2}{\frac{1}{N}\sum_i (z_o(i) - z_c(i))^2}\right) \quad (2.7)$$

The numerator in (2.7) is a result of the dynamic range of the image intensities. The allowed range of intensities is restricted here to values between 16 and 240. If the degradations have little influence on the coding efficiency, the differences $z_o(i) - z_c(i)$ will be small and the PSNR will be large. As the influence of the degradations on the coding efficiency increases, the PSNR decreases.

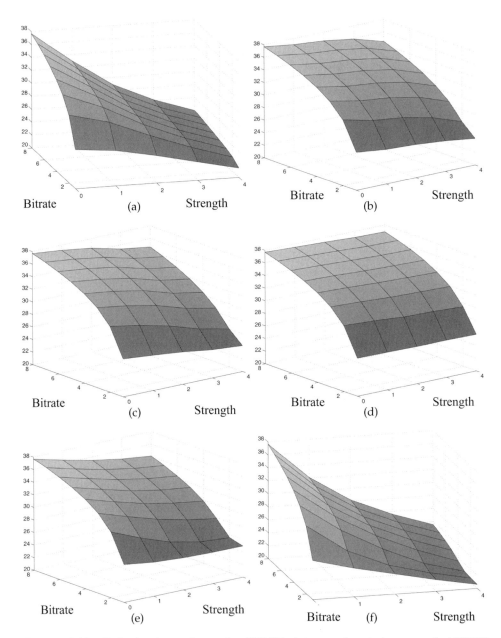

Figure 2.10: *Peak-signal-to-noise ratio (PSNR) between input image and MPEG2 encoded/decoded result as a function of bitrate and strength of artifacts: (a) noise, (b) blotches, (c) line scratches, (d) unsteadiness, (e) intensity flicker and (f) all artifacts combined.*

The degradations are introduced by applying the models for the artifacts in Section 2.1.2. Figure 2.10 plots the PSNR as a function of the bitrate of the encoder and of the strength of the impairments (Table 2.1) for the *MobCal* sequence. From this figure it can be seen that, if the strength of the impairments is held constant, the PSNR increases with increasing bitrate. This is to be expected, of course, because a signal can be encoded more accurately if more bits are available. When the bitrate is kept constant, it can been seen that the coding efficiency decreases with an increasing level of impairment. The reason for the latter is explained for each impairment in the qualitative analysis that follows.

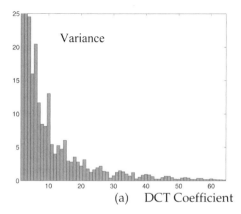 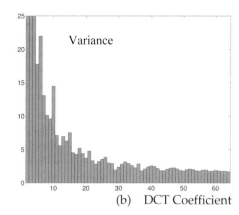

Figure 2.11: *(a) Variance of DCT coefficients (in zig-zag scan order) of a clean frame from the MobCal sequence, (b) variance of DCT coefficients from same frame but now with white gaussian noise with variance 100.*

Noise. A property of white noise is that the noise energy spreads out evenly over all the transform coefficients when an orthonormal transform is applied to it. The DCT is an orthonormal transform. Therefore, in MPEG2, the presence of additive white gaussian noise leads to fewer transform coefficients that are zero after quantization. Furthermore, on average, the amplitudes of the remaining coefficients are larger than in the noise-free case. See Figure 2.11. Both these effects lead to a decrease in coding efficiency; more coefficients must be transmitted and, on average, the codewords are longer. Similar arguments hold for the encoding of the error signals of the P frames and B frames. Note that the noise variance in the error signal is larger than that in I frames. This is so because the error signal is formed by subtracting two noisy frames. The benefits of noise reduction prior to MPEG2 encoding are shown by [Ric85], [Roo], [Roo98a].

Blotches. Blotches replace original image contents with data that have little relation to the original scene. Large prediction errors will result for P frames and B frames at spatial locations contaminated by blotches. Large prediction errors imply nonzero DCT coefficients with large amplitudes, they therefore imply a decrease in coding efficiency. The overall influence of blotches on the coding efficiency is usually less than that of noise because blotches are local phenomena that often affect only a small percentage of the total image area.

Line scratches. Scratches are image structures that, depending on their sharpness, have high energy in the frequency domain in orientations perpendicular to that of the scratch in question. For I frames this implies nonzero coefficients with large amplitudes, i.e., a decrease in coding efficiency. The situation is slightly better for P frames and B frames if the spatial locations of the scratches do not vary too much from frame to frame. In such cases, the prediction errors are small.

Unsteadiness. In principle, the influence of film unsteadiness on prediction errors for P frames and B frames is countered by motion compensation. At first glance, the overhead due to nonzero motion vectors is neglible because of the differential coding: adjacent regions affected by global motion have only zero differential motion. However, because the codeword for no motion takes fewer bits than that for zero differential motion [Erd98], unsteadiness influences the coding efficiency in a negative sense. Furthermore, near the image edges, the prediction errors can be large due to data moving in and out of the picture.

Intensity flicker. Intensity flicker decreases the coding efficiency of P frames and B frames for two reasons. First, the prediction error increases due to the fluctuations in image intensities. Thus the entropy of the error signal increases. Second, in the presence of intensity flicker the *constant luminance constraint* [Tek95] under which many motion estimators operate is violated. The result is that the motion vectors are more erratic, which leads to larger differential motion. The larger the differential motion, the more bits are required for encoding. The positive effects of reducing intensity flicker prior to compression are shown by [Ric85], [Roo].

The analysis given here shows that artifacts have a negative influence on the coding efficiency of MPEG2. Therefore removing artifacts prior to encoding is beneficial. It is difficult to quantify the benefits beforehand because they depend strongly on the nature of the unimpaired signal, the strength of the impairments, and the effectiveness of the restoration algorithms. It should be noted that not all impairments decrease the coding efficiency. For example, image blur [Ban97], [Lag91] is beneficial to compression because removes high frequency contents and thus nullifies the high-frequency transform coefficients.

Chapter 3

Intensity flicker correction

3.1 Introduction

Intensity flicker is a common artifact in old black-and-white film sequences. It is perceived as unnatural temporal fluctuations in image intensity that do not originate from the original scene. Intensity flicker has a great number of causes, e.g., aging of film, dust, chemical processing, copying, aliasing, and, in the case of the earlier film cameras, variations in shutter time. Neither equalizing the intensity histograms nor equalizing the mean frame values of consecutive frames, as suggested in [Fer96], [Mul96], [Ric95], are general solutions to the problem. These methods do not take changes in scene contents into account, and they do not appreciate the fact that intensity flicker can be a spatially localized effect. This chapter describes a method for equalizing local intensity means and variances in a temporal sense to reduce the undesirable temporal fluctuations in image intensities [Roo99b].

Section 3.2 models the effects of intensity flicker, and derives a solution to this problem for stationary sequences that is robust to the wide range of causes of this artifact. The derived solution is optimal in a *linear mean square error* sense. The sensitivity to errors in estimated model parameters and the reliability of those parameters are analyzed. Section 3.3 extends the applicability of the method to include nonstationary sequences by incorporating motion. In the presence of intensity flicker, it is difficult to compensate for motion of local objects in order to satisfy the requirement of temporal stationarity. A strategy of compensating for global motion (camera pan) in combination with a method for detecting the remaining local object motion is applied. The model parameters are interpolated where local motion is detected. Section 3.4 shows the overall system of intensity-flicker correction and discusses some practical aspects. Section 3.5 describes experiments and results. Conclusions relevant to this chapter are given in Section 3.6.

3.2 Estimating and correcting intensity flicker in stationary sequences

3.2.1 A model for intensity flicker

It is not practical to find explicit physical models for each of the mechanisms mentioned that cause intensity flicker. Instead, the approach taken here models the effects of this phenomenon on the basis of the observation that intensity flicker causes temporal fluctuations in local intensity mean and variance. Since noise is unavoidable in the various phases of digital image formation, a noise term is included in the model:

$$z(i) = \alpha(i) \, y(i) + \beta(i) + \eta(i) \tag{3.1}$$

The multiplicative and additive intensity-flicker parameters are denoted by $\alpha(i)$ and $\beta(i)$. In the ideal case, when no intensity flicker is present, $\alpha(i) = 1$ and $\beta(i) = 0$ for all i. It is assumed that $\alpha(i)$ and $\beta(i)$ are spatially smooth functions. Note that $y(i)$ does not necessarily need to represent the original scene intensities; it may represent a signal that, prior to the introduction of intensity flicker, may already have been distorted. The distortion could be due to signal-dependent additive granular noise that is characteristic of film [Bil75], [Özk93], for example.

The intensity-flicker-independent noise, denoted by $\eta(i)$, models the noise that has been added to the signal after the introduction of intensity flicker. It is assumed that this noise term is uncorrelated with the original image intensities. It is also assumed that $\eta(i)$ is a zero-mean signal with known variance. Examples are quantization noise and thermal noise originating from electronic studio equipment (VCR, amplifiers, etc.).

Correcting intensity flicker means estimating the original intensity for each pixel from the observed intensities. Based on the degradation model in (3.1), the following choice for a linear estimator for estimating $y(i)$ is obvious:

$$\hat{y}(i) = a(i) \, z(i) + b(i) \tag{3.2}$$

If the error between the original image intensity and the estimated original image intensity is defined as:

$$\varepsilon(i) = y(i) - \hat{y}(i) \qquad (3.3)$$

then it can easily be determined that, given $\alpha(i)$ and $\beta(i)$, the optimal values for $a(i)$ and $b(i)$ in a *linear minimum mean square error* (LMMSE) sense are given by:

$$a(i) = \frac{\text{var}[z(i)] - \text{var}[\eta(i)]}{\text{var}[z(i)]} \frac{1}{\alpha(i)} \qquad (3.4)$$

$$b(i) = -\frac{\beta(i)}{\alpha(i)} + \frac{\text{var}[\eta(i)]}{\text{var}[z(i)]} \frac{E[z(i)]}{\alpha(i)} \qquad (3.5)$$

where $E[.]$ stands for the expectation operator and $\text{var}[.]$ indicates the variance. It is interesting that it follows from (3.4) and (3.5) that $a(i) = 1/\alpha(i)$ and $b(i) = -\beta(i)/\alpha(i)$ in the absence of noise. In such a case, it follows from (3.1) and (3.2) that $\hat{y}(i) = y(i)$. That is to say, the estimated intensities are exactly equal to the original intensities. In the extreme case that the observed signal variance equals the noise variance, we find that $a(i) = 0$ and $\hat{y}(i) = b(i) = E[y(i)]$; the estimated intensities equal the expected values of the original intensities.

In practical situations, the true values for $\alpha(i)$ and $\beta(i)$ are not known and estimates $\hat{\alpha}(i)$ and $\hat{\beta}(i)$ are made from the observed data (this is the topic of Section 3.2.2). Because these estimates will never be perfect, the effects of errors in $\hat{\alpha}(i)$ and $\hat{\beta}(i)$ on $\hat{y}(i)$ is investigated. To simplify the analysis, the influence of noise is discarded. For ease of notation, the following analysis leaves out the spatial and temporal indices. Let $\hat{\alpha} = \alpha + \Delta\alpha$ and $\hat{\beta} = \beta + \Delta\beta$. The reconstruction error Δy is then given by:

$$\Delta y = y - \hat{y} = \frac{\Delta\alpha}{\alpha + \Delta\alpha} y + \frac{\Delta\beta}{\alpha + \Delta\alpha} \qquad (3.6)$$

Figure 3.1 plots the reconstruction error as a function of $\Delta\alpha$ and $\Delta\beta$ with $\alpha = 1$, $\beta = 0$ and $y=100$. Now, if $|\Delta\alpha| \ll \alpha$, then it can be seen that the sensitivity of Δy to errors in $\hat{\alpha}(i)$ is linear in y, and that the sensitivity of Δy to errors in $\hat{\beta}(i)$ is constant:

$$\frac{d\Delta y}{d\Delta\alpha} = y \quad \text{and} \quad \frac{d\Delta y}{d\Delta\beta} = 1 \qquad (3.7)$$

Equation (3.7) shows that Δy is much more sensitive to errors in $\hat{\alpha}(i)$ than to errors in $\hat{\beta}(i)$. It also shows that the sensitivity due to errors $\Delta\alpha$ can be minimized in absolute terms by centering the range of image intensities around 0. For example, consider a noiseless case in which $\hat{\alpha} = \alpha + 0.1$ and $\hat{\beta} = \beta$. If y ranges between 0 and 255 with $\alpha = 1$ and $\beta = 0$, then it can be seen from (3.6) that Δy is maximally 23.2. After the range of image intensities is centered around 0, y ranges between -127 and 128. The maximal absolute error is halved and, unlike the previous case, the sensitivity to errors in $\hat{\alpha}(i)$ for the mid-gray values is relatively small.

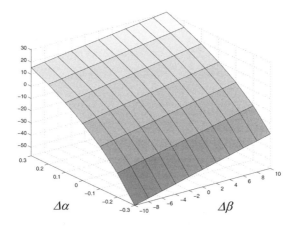

Figure 3.1: *Error Δy in a reconstructed image as a function of errors $\Delta\alpha$ and $\Delta\beta$ computed for y=100*

3.2.2 Estimating intensity-flicker parameters in stationary scenes

In the previous section, a LMMSE solution to intensity flicker is derived on the assumption that the intensity-flicker parameters $\alpha(i)$ and $\beta(i)$ are known. This is not the case in most practical situations, and these parameters will have to be estimated from the observed data. This section determines how the intensity-flicker parameters can be estimated from temporally stationary image sequences, i.e., image sequences that do not contain motion. It was already assumed that $\alpha(i)$ and $\beta(i)$ are spatially smooth functions. For practical purposes it is now also assumed that the intensity-flicker parameters are constant locally:

$$\begin{cases} \alpha(i,j,t) = \alpha_{m,n}(t) \\ \beta(i,j,t) = \beta_{m,n}(t) \end{cases} \forall i,j \in \Omega_{m,n} \tag{3.8}$$

INTENSITY FLICKER CORRECTION

where $\Omega_{m,n}$ indicates a small image region. The image regions $\Omega_{m,n}$ can, in principle, have any shape, but they are rectangular blocks in practice, and m, n indicate their horizontal and vertical spatial locations. The $\alpha_{m,n}(t)$ and $\beta_{m,n}(t)$ correspondig to $\Omega_{m,n}$ are considered frame-dependent matrix entries at m, n. The size $M \times N$ of the matrix depends on the total number of blocks in the horizontal and vertical directions.

Keep in mind the assumption that the zero-mean noise $\eta(i)$ is signal independent. The expected value and variance of $z(i)$ taken from (3.1) in a spatial sense for $i, j \in \Omega_{m,n}$ is given by:

$$E[z(i)] = \alpha_{m,n}(t) E[y(i)] + \beta_{m,n}(t) \tag{3.9}$$

$$\text{var}[z(i)] = \alpha_{m,n}^2(t) \text{var}[y(i)] + \text{var}[\eta(i)] \tag{3.10}$$

Rewriting (3.9) and (3.10) gives exact analytical expressions for $\alpha_{m,n}(t)$ and $\beta_{m,n}(t)$ for $i, j \in \Omega_{m,n}$:

$$\beta_{m,n}(t) = E[z(i)] - \alpha_{m,n}(t) E[y(i)] \tag{3.11}$$

$$\alpha_{m,n}(t) = \sqrt{\frac{\text{var}[z(i)] - \text{var}[\eta(i)]}{\text{var}[y(i)]}} \tag{3.12}$$

Equations (3.11) and (3.12) must now be solved in a practical situation. The means and variances of $z(i)$ can be estimated directly from the observed data of regions $\Omega_{m,n}$. The noise variance is assumed to be known or estimated. What remains to be estimated are the expected values and variances of $y(i)$ in the various regions $\Omega_{m,n}$.

Two methods for estimating the mean and variance of $y(i)$ for $i, j \in \Omega_{m,n}$ are discussed here. The first method estimates $y(i)$ by averaging the observed data in a temporal sense. In this case the underlying assumption is that the effects of flicker will be averaged out:

$$E[y(i,j,t)] = \frac{1}{p+q+1} \sum_{l=-p}^{q} E[z(i,j,t+l)] \tag{3.13}$$

$$\text{var}[y(i,j,t)] = \frac{1}{p+q+1} \sum_{l=-p}^{q} \text{var}[z(i,j,t+l)] \tag{3.14}$$

The second method takes the frame corrected previously as a reference:

$$E[y(i,j,t)] = E[\hat{y}(i,j,t-1)] \tag{3.15}$$

$$\text{var}[y(i,j,t)] = \text{var}[\hat{y}(i,j,t-1)] \tag{3.16}$$

The latter approach is adopted here because it has the significant advantage that the requirement of temporal stationarity is more likely to be fulfilled when a single reference frame, rather than multiple reference frames, is used. This approach is also more attractive in terms of computational load and memory requirements. Hence, for $i,j \in \Omega_{m,n}$, the estimated intensity-flicker parameters are given by:

$$\hat{\beta}_{m,n}(t) = E[z(i,j,t)] - \hat{\alpha}_{m,n}(t) E[\hat{y}(i,j,t-1)] \tag{3.17}$$

$$\hat{\alpha}_{m,n}(t) = \sqrt{\frac{\text{var}[z(i,j,t)] - \text{var}[\eta(i,j,t)]}{\text{var}[\hat{y}(i,j,t-1)]}} \tag{3.18}$$

3.2.3 Measure of reliability for the estimated model parameters

Note that, by using (3.15) and (3.16), recursion is introduced into the method for flicker correction. As a result, there is a risk of error propagation leading to considerable distortions in a corrected sequence. A source of errors lies in the estimated model parameters $\hat{\alpha}_{m,n}(t)$ and $\hat{\beta}_{m,n}(t)$, which may not be exact. Therefore, it is useful to have a measure of reliability for $\hat{\alpha}_{m,n}(t)$ and $\hat{\beta}_{m,n}(t)$ that can be used to control the correction process by means of weighting and smoothing the estimated model parameters as is done in Section 3.3.3.

The $\hat{\alpha}_{m,n}(t)$ and $\hat{\beta}_{m,n}(t)$ are not very reliable in a number of cases. The first case is that of uniform image intensities. For any original image intensity in a uniform region, there are infinite combinations of $\alpha(i)$ and $\beta(i)$ that lead to the same observed intensity. The second case in which $\hat{\alpha}_{m,n}(t)$ and $\hat{\beta}_{m,n}(t)$ are potentially unreliable is caused by the fact that (3.15) and (3.16) discard the noise in $\hat{y}(i)$ originating from $\eta(i)$. This leads to values for $\hat{\alpha}_{m,n}(t)$ that are too small. Considerable errors result in regions $\Omega_{m,n}$ in which the signal variance is smaller than the noise variance.

The signal-to-noise ratio, defined as $\text{var}(y)/\text{var}(\eta)$, determines the variance of the errors in the estimated model parameters. Figure 3.2 illustrates this by plotting the reciprocal values of the error variances $\sigma^2_{\Delta\alpha}$ and $\sigma^2_{\Delta\beta}$ as a function of signal-to-noise ratio. These values were obtained experimentally by synthesizing 100.000

INTENSITY FLICKER CORRECTION

textured areas of 30x30 pixels with a 2D autoregressive model to which gaussian noise and flicker were added. The flicker parameters were then determined with (3.11) and (3.12). Figure 3.2 shows that the variance in the estimated model parameters is inversely proportional to the signal-to-noise ratio.

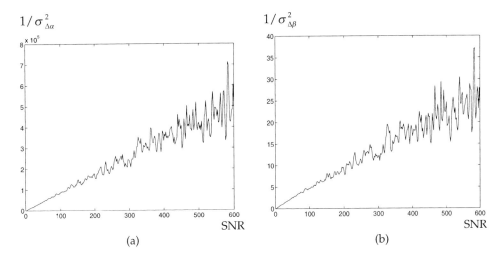

Figure 3.2: (a) Plot of $1/\sigma^2_{\Delta\alpha}$ vs. signal-to-noise ratio, (b) plot $1/\sigma^2_{\Delta\beta}$ vs. signal-to-noise ratio. Note that the relationships are linear.

In Section 3.3.3, the model parameters that are estimated over an image are smoothed and weighted using a 2D polynomial fit. The weighted least-squares estimate of the polynomial coefficients is optimal if the weights are proportional to $1/\sigma_{\Delta\alpha}$ and $1/\sigma_{\Delta\beta}$ [Str88], i.e., if the weights are proportional to the squared root of the signal-to-noise ratio. Hence, the following measure of reliability $W_{m,n}(t)$, for $i, j \in \Omega_{m,n}$, is defined:

$$W_{m,n}(t) = \begin{cases} 0, & \forall \ \text{var}[z(i)] < T_n \\ \sqrt{\dfrac{\text{var}[z(i)] - T_n}{T_n}}, & \text{otherwise} \end{cases} \qquad (3.19)$$

where T_n is a threshold depending on the variance of $\eta(i)$. Large values for $W_{m,n}(t)$ indicate reliable estimates; small values indicate unreliable estimates.

3.3 Incorporating motion

The previous sections model the effects of intensity flicker and derive a solution for temporally stationary sequences. The necessity of temporal stationarity is reflected by (3.15) and (3.16), which assume that the mean and variance of $\hat{y}(i,j,t)$ and $\hat{y}(i,j,t-1)$ are identical. Real sequences, of course, are seldom temporally stationary. Measures will have to be taken to avoid estimates of $\alpha(i)$ and $\beta(i)$ that are incorrect due to motion. Compensating motion between $z(i,j,t)$ and $\hat{y}(i,j,t-1)$ helps satisfy the assumption of temporal stationarity. This requires motion estimation.

Robust methods for estimating global motion (camera pan) that are relatively insensitive to fluctuations in image intensities exist. Unfortunately, the presence of intensity flicker hampers the estimation of local motion (motion in small image regions) because local motion estimators usually have a constant luminance constraint. This includes pel-recursive methods and all motion estimators that make use of block matching in one stage or another [Tek95]. Even if motion can be well compensated, a strategy is required for correcting flicker in previously occluded regions that have become uncovered.

For these reasons, the strategy presented here for estimating the intensity-flicker parameters in temporally nonstationary scenes is based on local motion detection. First, a pair of frames are registered to compensate for global motion (Section 3.3.1). Then the intensity-flicker parameters are estimated as outlined in Section 3.2.2. With these parameters, the remaining local motions is detected (Section 3.3.2). Finally, the missing model parameters in the temporally nonstationary regions are spatially interpolated from surrounding regions without local motion (Section 3.3.3).

3.3.1 Estimating global motion with phase correlation

In sequences with camera pan, applying global motion compensation helps satisfy the requirement of stationarity. Let the global displacement vector be $(q_i, q_j)^T$. Global motion compensation can be applied to the model parameter estimation by replacing (3.17) and (3.18) with:

$$\hat{\beta}_{m,n}(t) = E[z(i,j,t)] - \hat{\alpha}_{m,n}(t)\, E[\hat{y}(i-q_i, j-q_j, t-1)] \tag{3.20}$$

$$\hat{\alpha}_{m,n}(t) = \sqrt{\frac{\mathrm{var}[z(i,j,t)] - \mathrm{var}[\eta(i,j,t)]}{\mathrm{var}[\hat{y}(i-q_i, j-q_j, t-1)]}} \tag{3.21}$$

Global motion compensation is only useful if the global motion vectors (one vector to each frame) are accurate: i.e., if the global motion estimator is robust against intensity flicker. A global motion estimator that meets this requirement is one that is based on the phase correlation method applied to high-pass-filtered versions of the images [Pea77], [Tek95].

The phase correlation method estimates motion by measuring phase shifts in the Fourier domain. This method is relatively insensitive to fluctuations in image intensity because it uses Fourier coefficients that are normalized by their magnitude. The direction of changes in intensity over edges and textured regions is preserved in the presence of intensity flicker because the amount of intensity flicker was assumed to vary smoothly in a spatial sense. This means that the phases of the higher-frequency components will not be affected by intensity flicker. However, the local mean intensities can vary considerably from frame to frame, and this gives rise to random variations in the phase of the low-frequency components. These random variations are disturbing factors in the motion estimation process that can be avoided by removing the low-pass frequency components from the input images.

The phase correlation technique estimates phase shifts in the Fourier domain as follows:

$$C_{t,t-1}(w_1,w_2) = \frac{Z_t(w_1,w_2) Z^*_{t-1}(w_1,w_2)}{\|Z_t(w_1,w_2) Z^*_{t-1}(w_1,w_2)\|} \qquad (3.22)$$

where $Z_t(w_1,w_2)$ stands for the 2D Fourier transform of $z(i, j, t)$, and * denotes the complex conjugate. If $z(i, j, t)$ and $z(i, j, t-1)$ are spatially shifted, but otherwise identical images, the inverse transform of (3.22) produces a delta pulse in the 2D correlation function. Its location yields the global displacement vector $(q_i, q_j)^T$.

3.3.2 Detecting the remaining local motion

It is important to detect the remaining local motion after compensating for global motion. Local motion causes changes in local image statistics that are not due to intensity flicker. This leads to incorrect estimates of $\alpha(i)$ and $\beta(i)$; to visible artifacts in the corrected image sequence. First, two obvious approaches to motion detection are discussed. It is concluded that these are not appropriate. Next, a robust alternative strategy is described.

Two methods for detecting local motion are (1) detecting large local frame differences between the corrected current and previous frames and (2) comparing the estimated intensity-flicker parameters $\hat{\alpha}_{m,n}(t)$ and $\hat{\beta}_{m,n}(t)$ to threshold values and detect motion when these thresholds are exceeded. These methods have disadvantages that limit their usefulness. The first method is very sensitive to film unsteadiness; slight movements of textured areas and edges lead to large frame differences and thus to "false" detections of motion. The second method requires threshold values that detect motion accurately without generating too many false alarms. Good thresholds are difficult to find because they depend on the amount of intensity flicker and the amount of local motion in the sequence.

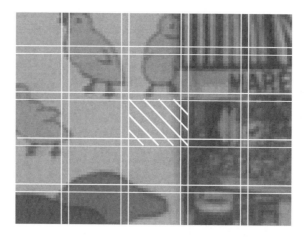

Figure 3.3: *Example of part of a frame subdivided in blocks $\Omega_{m,n}$ that overlap each other by one pixel.*

To overcome problems resulting from small motion and hard thresholds, a robust motion-detection algorithm that relies on the current frame only is developed here. The underlying assumption of the method is that motion should only be detected if visible artifacts would otherwise be introduced. First, the observed image is subdivided into blocks $\Omega_{m,n}$ that overlap their neighbors both horizontally and vertically (Figure 3.3). The overlapping boundary regions form sets of reference intensities. The intensity-flicker parameters are estimated for each block by (3.20) and (3.21). These parameters are used with (3.2), (3.4), and (3.5) for correcting the intensities in the boundary regions. Then, for each pair of overlapping blocks, the common pixels that are assigned significantly different values are counted:

$$n_{q,r} = \sum_{i \in S_{q,r}} boolean[\,|\hat{y}_q(i) - \hat{y}_r(i)| > T_d\,] \tag{3.23}$$

Here q and r indicate two adjacent image blocks, $S_{q,r}$ indicates the set of boundary pixels, T_d is a threshold above which pixels are considered to be significantly different and *boolean*[.] is a boolean function that is one if its argument is true and is zero otherwise. Motion is flagged in both regions q and r if too many pixels are significantly different, that is, if:

$$n_{q,r} > D_{max} \tag{3.24}$$

where D_{max} is a constant.

3.3.3 Interpolating missing parameters

Due to noise and motion, the estimated intensity-flicker parameters are unreliable in some cases. These parameters are referred to as *missing*. The other parameters are referred to as *known*. The goal is to find estimates of the missing parameters by means of interpolation. It is also necessary to smooth the known parameters to avoid sudden changes in local intensity in the corrected sequence. The interpolation and smoothing functions should meet the following requirements. First, the system of intensity-flicker correction should switch itself off when the correctness of the interpolated values is less certain. This means that the interpolator should incorporate biases for $\hat{\alpha}_{m,n}(t)$ and $\hat{\beta}_{m,n}(t)$ towards unity and zero, respectively, that grow as the smallest distance to a region with known parameters becomes larger. Second, the reliability of the known parameters should be taken into account.

Three methods that meet these requirements are investigated. Each of these methods uses the $W_{m,n}(t)$ determined by the measure of reliability as defined in (3.19). The interpolation and smoothing algorithms are described for the case of the multiplicative parameters $\hat{\alpha}_{m,n}(t)$. The procedures for the $\hat{\beta}_{m,n}(t)$ are similar and are not described here.

Interpolation by dilation. With each iteration of this iterative dilation approach, regions of known parameters grow at the boundaries of regions with missing parameters. Consider the matrix containing the known $\hat{\alpha}_{m,n}(t)$ corresponding to the regions $\Omega_{m,n}$ for a frame t. Figure 3.4a graphically depicts such a matrix that can be divided into two areas: the black region indicates the matrix entries for which the multiplicative parameters are known, and the white region indicates the missing entries. Each missing $\hat{\alpha}_{m,n}(t)$ and its corresponding weight $W_{m,n}(t)$ at the boundary of the two regions is interpolated by:

$$\hat{\alpha}_{m,n}(t) = \frac{\sum_{\{q,r\}\in S_{m,n}} W_{q,r}(t)\hat{\alpha}_{q,r}(t)}{\sum_{\{q,r\}\in S_{m,n}} W_{q,r}(t)} \rho^k + 1 - \rho^k \tag{3.25}$$

$$W_{m,n}(t) = \frac{\sum_{\{q,r\}\in S_{m,n}} W_{q,r}(t)}{|S_{m,n}|} \tag{3.26}$$

where $S_{m,n}$ indicates the set of known parameters adjacent to the missing parameter being interpolated, ρ (with $0 \leq \rho \leq 1$) determines the trade-off between the interpolated value and the bias value as a function of iteration number k. After the first iteration, Figure 3.4b results. Repeating this process assigns estimates for $\hat{\alpha}_{m,n}(t)$ to all missing parameters (Figure 3.4c,d).

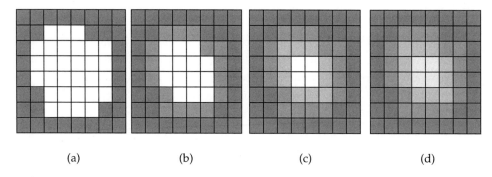

(a) (b) (c) (d)

Figure 3.4: *Interpolation process using dilation: (a) initial situation, (b), (c), (d) results after 1, 2 and 3 iterations.*

Next, a postprocessing step smooths all the matrix entries with a 5x5 gaussian kernel. Figure 3.5(a,b) shows respectively, an original set of known and missing parameters and the interpolated, smoothed parameters.

Interpolation by successive overrelaxation (SOR). SOR is a well-known iterative method based on repeated low-pass filtering [Pre92]. Unlike the dilation technique, this method interpolates the missing parameters and smooths the known parameters simultaneously. SOR starts out with an initial approximation $\hat{\alpha}^0_{m,n}(t)$. At each iteration k, the new solution $\hat{\alpha}^{k+1}_{m,n}(t)$ is computed for all (m, n) by computing a residual term $r^{k+1}_{m,n}$ and subtracting this from the current solution:

$$r_{m,n,t}^{k+1} = W_{m,n}(t)\,(\alpha_{m,n}^{k}(t) - \alpha_{m,n}^{0}(t)) + \\ \lambda\,(4\alpha_{m,n}^{k}(t) - \alpha_{m-1,n}^{k}(t) - \alpha_{m+1,n}^{k}(t) - \alpha_{m,n-1}^{k}(t) - \alpha_{m,n+1}^{k}(t))$$ (3.27)

$$\alpha_{m,n}^{k+1}(t) = \alpha_{m,n}^{k}(t) - \overline{w}\,\frac{r_{m,n,t}^{k+1}}{W_{m,n}(t) + 4\lambda}$$ (3.28)

Here $W_{m,n}(t)$ are the weights, λ determines the smoothness of the solution, and $\overline{w}$ is the so-called overrelaxation parameter that determines the rate of convergence. The $\hat{\alpha}_{m,n}^{0}(t)$ are initialized to the known multiplicative intensity-flicker parameters at (m, n), and to the bias value for the missing parameters.

The first term in (3.27) weighs the difference between the current solution and the original estimate, and the second term measures the smoothness. The solution is updated in (3.28) so that where the weights $W_{m,n}(t)$ are great, the original estimates $\hat{\alpha}_{m,n}^{0}(t)$ are emphasized. In contrast, when the measurements are deemed less reliable, i.e., when $\lambda \gg W_{m,n}(t)$, emphasis is laid on achieving a smooth solution. This allows the generation of complete parameter fields where the known parameters, depending on their accuracy, are weighted and smoothed. Figure 3.5c shows results of this method.

Interpolation by 2D polynomial fitting. By fitting a 2D polynomial $P(m, n, t)$ to the known parameters, the missing parameters can be interpolated and the known parameters are smoothed simultaneously. The 2D polynomial is given by [Hay67]:

$$P(m,n,t) = \sum_{k=0}^{D_c} \sum_{l=0}^{D_r} c_{k,l,t}\, m^k n^l$$ (3.29)

where D_r and D_c determine the degree of the polynomial surface and the coefficients $c_{k,l,t}$ shape the function. Polynomial fitting entails finding the coefficients $c_{k,l,t}$ so that the weighted mean squared difference of $P(m, n, t)$ and $\hat{\alpha}_{m,n}(t)$ is minimized for a given t:

$$\min_{c_{k,l,t}} \left(\sum_{m,n} W_{m,n}(t)\,(P(m,n,t) - \hat{\alpha}_{m,n}(t))^2 \right)$$ (3.30)

The complexity of solving (3.30) is typical of a weighted least squares problem that requires computation of a pseudo inverse of a square matrix [Str88]. The number of columns (and rows) depends on the order of the polynomial and is determined by the number of coefficients $c_{k,l,t}$ at an instant t.

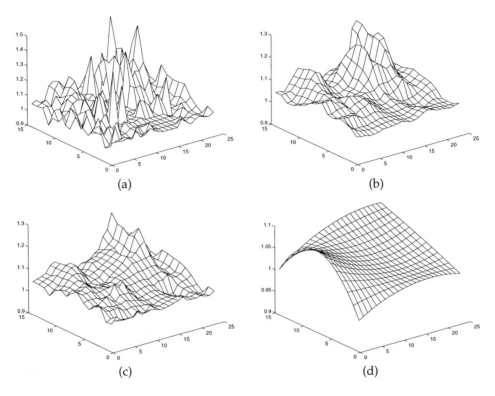

Figure 3.5: *(a) Set of original measurements with variable accuracy; the missing measurements have been set to 1, (b) parameters interpolated and smoothed by repeated dilation, (c) parameters interpolated and smoothed by SOR (250 iterations), (d) parameters interpolated and smoothed by polynomial fitting ($D_r = D_c = 2$). Note the differences in scale.*

Biases are applied by setting the missing parameters to their bias value; the weights corresponding to these parameters are set to a fraction (e.g., one tenth) of the largest weight found for the known parameters. This will have little effect on the shape of the polynomial surface if only a few parameters are missing locally. Where many parameters are missing, the combined influence of the biased parameters will shape the polynomial locally towards the bias value.

The range of the results obtained by the dilation and SOR interpolation methods is limited to the range of the data. This is not the case for 2D polynomial fitting. The higher-order terms cause spurious oscillations if the order of the polynomial is taken too high, which leads to incorrect values for the interpolated

INTENSITY FLICKER CORRECTION

and smoothed parameters. In practice, taking $D_r = D_c = 2$ gives the best results. Figure 3.5d shows a result of this interpolation and smoothing method.

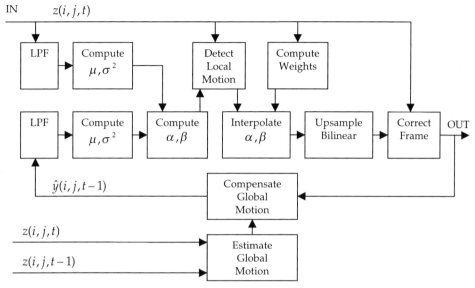

Figure 3.6: *Global structure of the intensity-flicker correction system.*

3.4 Practical issues

Figure 3.6 shows the overall structure of the system of intensity-flicker correction. Some operations have been added in this figure that have not yet been mentioned. These operations improve the system's behavior. First, the current input and the previous system output (with global motion compensation) are low-pass filtered with a 5x5 gaussian kernel. Prefiltering suppresses the influence of high-frequency noise and the effects of small motion. Then, local means μ and variances σ^2 are computed to be used for estimating the intensity-flicker parameters. The estimated model parameters and the current input are used to detect local motions. Next, the missing parameters are interpolated and the known parameters are smoothed. Bilinear interpolation is used for upsampling the estimated parameters to full spatial resolution. The latter avoids the introduction of blocking artifacts in the correction stage that follows.

As mentioned in Section 3.2.3, the fact that a recursive structure is used for the overall system of intensity-flicker correction introduces the possibility of error propagation. Errors certainly do occur, for example, as a result of the need to approximate the expectation operator and from model mismatches. Therefore, it is useful to bias corrected intensities towards the contents of the current frame to avoid possible drift due to error accumulation. For this purpose, (3.2) is replaced by:

$$\hat{y}(i) = \kappa \, (\alpha(i) \, z(i) + b(i)) + (1-\kappa) \, z(i) \qquad (3.31)$$

where κ is the forgetting factor. If $\kappa = 1$, the system relies completely on the frame corrected previously, and it tries to achieve the maximal reduction in intensity flicker. If $\kappa = 0$, we find that the system is switched off. A practical value for κ is 0.85.

3.5 Experiments and results

This section applies the system of intensity-flicker correction both to sequences containing artificially added intensity flicker and to sequences with real (non-synthetic) intensity flicker. This first set of experiments takes place in a controlled environment and evaluates the performance of the correction system under extreme conditions. The second set of experiments verifies the practical effectiveness of the system and forms a verification of the underlying assumptions of the approach presented in this chapter. The same settings for the system of intensity-flicker correction were used for all experiments to demonstrate the robustness of the approach (see Table 3.1).

Image Blocks	Motion Detection	2D Polynomial	Successive OverRelaxation	Miscellaneous
Size: 30x20 Overlap: 1 pixel	$T_d = 5$ $D_{max} = 5$	$D_r = D_c = 2$	$\overline{w} = 1$ $\lambda = 5$	$\kappa = 0.85$ var$[\eta(x,y,t)] = 5$ $T_n = 25$

Table 3.1: *Parameter settings of intensity-flicker correction system for the experiments.*

Some thought should be given to what criteria are to be used to determine the effectiveness of the proposed algorithm. If the algorithm functions well and the image contents does not change significantly, then the equalized frame means and variances should be similar from frame to frame. Indeed, the converse need not be true, but visual inspection helps to verify the results. Therefore, the temporal smoothness of frame means and frame variances measures the effectiveness of intensity-flicker correction system.

A sliding window approach is adopted here: the variance in frame mean and frame variance is computed locally over 24 frames (which corresponds to 1 second of film) and the estimated local variances are averaged over the whole sequence. There is a reason for using this sliding window. If the variation in frame means and variances are computed over long sequences, there are two components that determine the result: (1) variations due to flicker, and (2) variations due to changes in scene content. We are only interested in the first component which can be isolated by computing the variations over short segments.

3.5.1 Experiments on artificial intensity flicker

For the first set of experiments the *Mobile* sequence (40 frames), containing moving objects and camera panning (0.8 pixels/frame), is used. Artificial intensity flicker was added to this sequence according to (3.1). The intensity-flicker parameters were artificially created from 2D polynomials, defined by (3.29), with degree $D_r = D_c = 2$. The coefficients $c_{k,l,t}$ are drawn from the normal distribution $N(0, 0.1)$, and from $N(1, 0.1)$ for $c_{0,0,t}$, to generate the $\alpha(i)$ and from $N(0, 10)$ to generate the $\beta(i)$. Visually speaking, this leads to a severe amount of intensity flicker (Figure 3.7).

	MobCal		Soldier		Mine		Charlie	
	Mean	Var.	Mean	Var.	Mean	Var.	Mean	Var.
Degraded	19.8	501	2.7	44	2.3	61	8.5	435
Dilation	5.5	110	0.8	29	1.0	37	5.6	319
SOR	5.2	86	0.8	31	1.0	40	4.9	235
2D Polynomial	5.8	105	0.9	27	1.2	41	6.3	333

Table 3.2: *Standard deviation of averaged frame mean and frame variance of degraded and sequences corrected by various interpolators in the intensity flicker correction system.*

The degraded sequence is corrected three times, and each time a different interpolation and smoothing algorithm is used, as described in Section 3.3.3. Figure 3.7 shows some corrected frames. Figure 3.9 plots the frame means and the frame variances of original, degraded and corrected sequences. It can be seen from these graphs that the variations in frame mean and variance have been strongly reduced. Visual inspection confirms that the amount of intensity flicker has been reduced significantly. However, residues of local intensity flicker are clearly visible when the dilation interpolation method is used. The SOR interpolation method gives the best visual results.

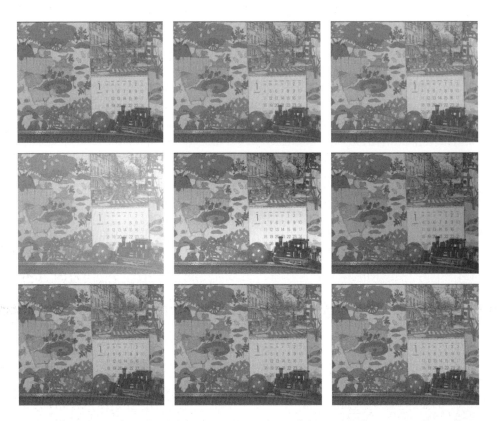

Figure 3.7: *Top row: original frames 16, 17, and 18 of the MobCal sequence. Central row: degraded frames. Bottom row: frames corrected by the intensity-flicker correction system with successive overrelaxation.*

Table 3.2 lists the standard deviation of the frame means and frame variances computed over short segments by the sliding window approach and averaged as mentioned before. This table shows that the artificial intensity flicker severely degraded the sequence. It also shows that the intensity-flicker correction system strongly reduces fluctuations in frame mean and frame variance. The SOR interpolation method gives the best numerical results.

3.5.2 Experiments on naturally degraded film sequences

Three sequences from film archives were used for the second set of experiments. Table 3.2 lists the results. The first sequence, called *Soldier*, is 226 frames long. It shows a soldier entering the scene through a tunnel. There is some camera unsteadiness during the first 120 frames, then the camera pans to the right and up. There is film-grain noise and a considerable amount of intensity flicker in this sequence. The total noise variance was estimated to be 8.9 by the method described in [Mar95]. Figure 3.8 shows three frames from this sequence, original and corrected. Figure 3.10 indicates that the fluctuation in frame means and variances have significantly been reduced by the intensity-flicker correction system. Visual inspection shows that all three methods significantly reduce the intensity flicker without introducing visible new artifacts. The best visual results are obtained with the SOR interpolation method.

Figure 3.8: *Top: frames 13, 14, and 15 of the naturally degraded Soldier sequence. Bottom: frames corrected by the intensity-flicker correction system using the 2D polynomial interpolation method.*

The second naturally degraded sequence, called *Mine*, consists of 404 frames. This sequence depicts people in a mine. It contains camera pan, some zoom, and it is quite noisy (estimated noise variance 30.7). The intensity flicker is not as severe as in the *Soldier* sequence. Figure 3.11 shows the frame means and variances of the degraded and the corrected sequences. Visually, the results obtained from the dilation interpolation method show some flickering patterns. The 2D polynomial interpolation leaves some flicker near the edges of the picture. The SOR method shows good results.

The third sequence is a clip of 48 frames from a Charlie Chaplin film, called *Charlie*. Some frames have so much intensity flicker that it looks as if the film has been overexposed and the texture is lost completely in some regions. Besides intensity flicker, this sequence is characterized by typical artifacts occurring in old films, such as blotches, scratches, and noise (estimated variance 5.0). Figure 3.12 shows that the fluctuations in frame means and variances have diminished. Again, from a subjective point of view, the SOR interpolation technique gives the best result, but a slight loss of contrast is noted in the corrected sequence.

Table 3.2 indicates that the intensity-flicker correction system significantly reduces the fluctuations in frame mean and frame variance of all the test sequences. The SOR interpolation method gives the best numerical results: in all cases it gives the largest reduction in variation of the mean image intensity and it gives a reduction in variation of image variance that is similar or better than that obtained by the other interpolation methods.

3.6 Conclusions

This chapter introduced a novel method for removing intensity flicker from image sequences that significantly reduces the temporal fluctuations in local image mean and variance. The system is based on simple block-based operations and motion detection. Therefore the complexity of the system is limited. This is advantageous for real-time implementation in hardware.

Improvements to the system are certainly conceivable. For instance, effort could be put into reducing the sizes of the image regions for which estimated flicker parameters are discarded due to local motion. In the current scheme, data in whole image blocks are discarded even though large parts of those blocks may not have been affected by motion. Alternatively, instead of detecting motion, an approach that incorporates robust motion estimation into the flicker correction system could be developed. This would result in a system for simultaneous motion and parameter estimation and intensity-flicker correction.

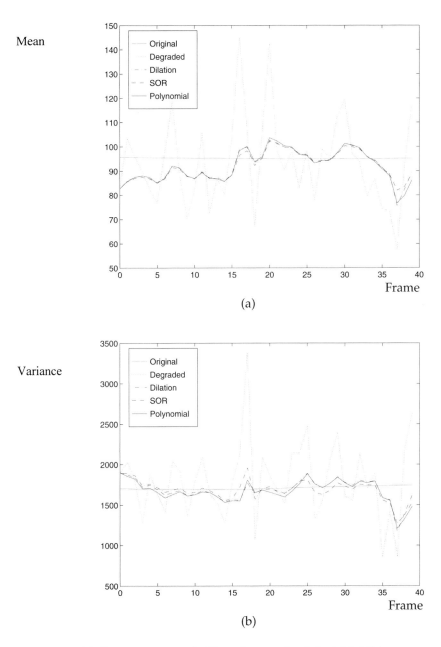

Figure 3.9: *(a) Frame means and (b) variances of original MobCal sequence, MobCal sequence with artificial intensity flicker, and sequences corrected by various interpolation and smoothing methods within the system for intensity-flicker correction.*

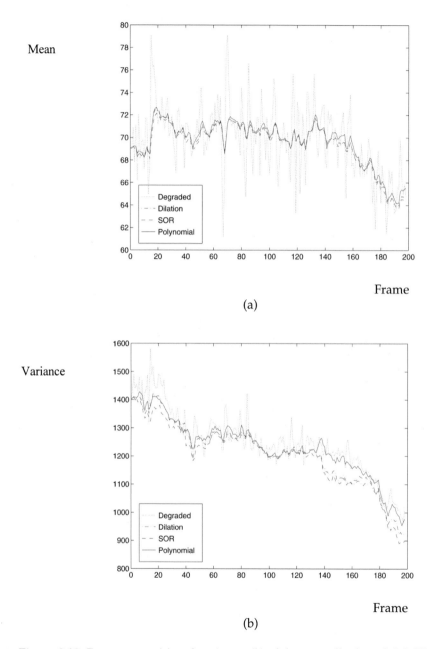

Figure 3.10: *Frame means (a) and variances (b) of the naturally degraded Soldier sequence and sequences corrected by the system for intensity-flicker correction with various interpolation and smoothing methods.*

Mean

Variance

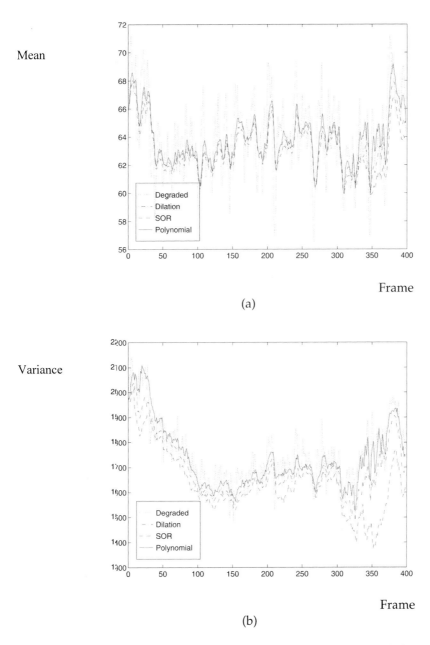

(a)

(b)

Figure 3.11: *Frame means (a) and variances (b) of the naturally degraded Mine sequence and sequences corrected by the system for intensity-flicker correction with various interpolation and smoothing methods.*

Mean

Variance

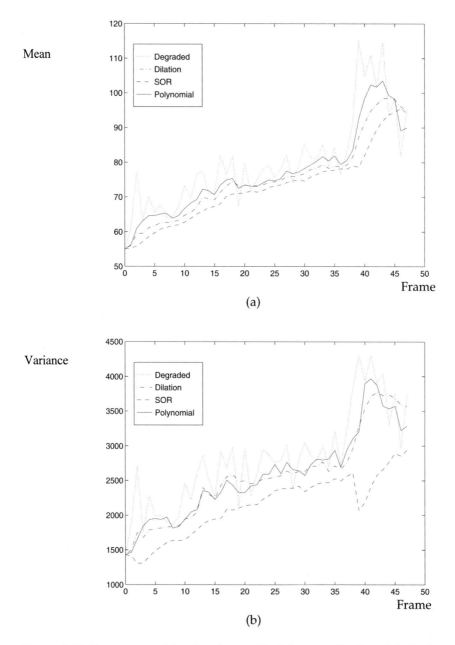

Figure 3.12: *Frame means (a) and variances (b) of the naturally degraded Charlie sequence and sequences corrected by the system for intensity-flicker correction with various interpolation and smoothing methods.*

Chapter 4

Blotch detection and correction

4.1 System for blotch detection and correction

Blotches are artifacts typically related to film. The loss of gelatin and dirt particles covering the film cause blotches. The original intensities corrupted by blotches are lost and will be referred to as *missing data*. Correcting blotches entails detecting the blotches and interpolating the missing data from data that surround the corrupted image region. The use of temporal information often improves the quality of the results produced by the interpolation process. This means that reference data from which the missing data are interpolated, need to be extracted from frames preceeding and/or following the frame currently being restored. Motion estimation and compensation is required to obtain optimal interpolation results.

The methods for blotch detection presented in this chapter assume the degradation model from (2.2), either implicitly or explicitly [Kok98]:

$$z(i) = (1 - d(i))\, y(i) + d(i)\, c(i) \qquad (4.1)$$

where $z(i)$ and $y(i)$ are the observed and the original (unimpaired) data, respectively. The binary blotch detection mask $d(i)$ indicates whether each pixel has been corrupted: $d(i) \in \{0, 1\}$. The values at the corrupted sites are given by $c(i)$, with $c(i) \neq y(i)$. One property of blotches is the smooth variation in intensity values at the corrupted sites; the variance $c(i)$ within a blotch is small. Blotches seldom appear at the same location in a pair of consecutive frames. Therefore the binary mask $d(i)$ will seldom be set to one at two spatially co-sited locations for a pair of consecutive frames. However, there is spatial coherence within a blotch; if a pixel is blotched, it is likely that some of its neighbors are corrupted as well, i.e., if $d(i) = 1$ it is likely that some other $d(i \pm 1, j \pm 1, t) = 1$ also.

The following sections use various models for the original, uncorrupted image data. The common element is that these models do not allow large temporal discontinuities in image intensity along the motion trajectories. This constraint results from the fact that $c(i) \neq y(i)$ in the degradation models, which implies that blotches introduce temporal discontinuities in image intensity. Temporal discontinuities in image intensity are also caused by moving objects that cover and uncover the background. There is a difference between the effects of blotches and the effects of motions. Motion tends to cause temporal discontinuities in either the forward or the backward temporal direction, but not in both directions at the same time. Blotches cause discontinuities simultaneously in both temporal directions.

The estimated motion vectors are unreliable at image locations corrupted by blotches because they are determined with incorrect, corrupted data. Models for motion vector repair and for blotch correction assume a relationship between the original image data at the corrupted sites and the data surrounding those sites (temporally and/or spatially). For example, for motion vector repair, this relationship can be smoothness of the motion vector field. For blotch correction, this relationship can be defined by *autoregressive* (AR) image models.

Figure 4.1 illustrates two possible approaches for detecting and correcting blotches. The first approach computes the locations of the blotches, the motion vectors, and the corrected intensities simultaneously within a single bayesian framework. *Maximum a posteriori* (MAP) estimates for the true image intensities, $\hat{y}(z)$, the motion vectors $v(i)$, the blotch detection mask $d(i)$ and the intensities of the blotches $c(i)$ are computed from the observed images $z(i)$:

$$\arg\max_{\hat{y}(i), v(i), d(i), c(i)} P[\hat{y}(i), v(i), d(i), c(i) \mid z(i)] \tag{4.2}$$

This is an elegant framework because it defines an optimal solution that takes dependencies between the various parameters into account. It was applied successfully in [Kok98]. A disadvantage of this method, besides its great computational complexity, is the difficulty of determining what influence the individual assumptions for the likelihood functions and priors have on the final outcome of the overall system. Hence, it is difficult to determine whether the assumed priors and likelihood functions give optimal results.

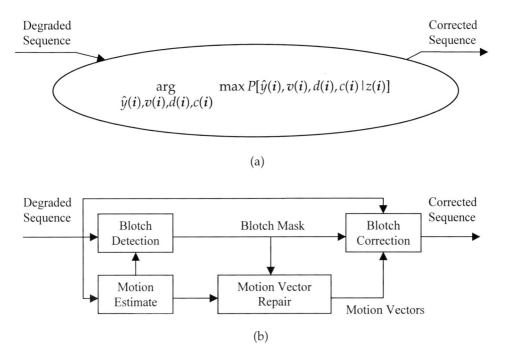

Figure 4.1: *(a) Simultaneous approach for blotch detection and correction vs. (b) modular approach.*

The second approach towards resolving blotches is a modular approach, as shown in Figure 4.1b. The *motion estimate* module estimates motion between consecutive frames in the forward and backward directions (from t to $t+1$ and from t to $t-1$ respectively). On the basis of motion estimates and the incoming degraded data, the *blotch detection* module detects blotches. The *motion vector repair* module corrects faulty motion vectors. Finally, the *blotch correction* module corrects blotches using the corrected motion vectors, the binary blotch detection mask, and the degraded image sequence.

This chapter concentrates on the modular approach for blotch detection and correction. This approach has the advantage that the modules can be designed and evaluated independently of each other. Furthermore, the modular approach has the advantage of being computationally much less demanding than the simultaneous bayesian approach.

This chapter is structured as follows. Section 4.2 reviews existing techniques for blotch detection, motion vector repair and blotch correction. Section 4.3 introduces a new technique for improving the detection results by postprocessing blotch detection masks. The postprocessing operations significantly reduce the number of false alarms that are inherent to any detection problem. Section 4.4 shows that increasing the temporal aperture of a detector gives significant gains in some cases. Section 4.5 presents a new, fast model-based method for excellent quality of missing data interpolation. Section 4.6 evaluates the performance of the complete blotch removal system and concludes this chapter.

4.2 Overview of existing techniques

4.2.1 Blotch detection techniques

The parameter estimation problem for the degradation model consists of determining the binary blotch detection mask $d(i)$ for each frame. If required, $c(i)$ can easily be found once $d(i)$ is known. The blotch detectors presented in this section all apply the same principle: they check whether the observed data $z(i)$ fit an image model for $y(i)$. If this is not the case, the image is assumed to be corrupted and a blotch is flagged.

SDIa detector. The *spike detection index-a* (SDIa) is a simple heuristic method for detecting temporal discontinuities in image intensity [Kok98], [Kok95a]. It compares each pixel intensity of the current frame $z(i)$ to the corresponding intensities in the forward and backward temporal directions by computing the minimum squared difference $SDIa(i)$:

$$SDIa(i) = \min[(z(i) - z_{mc}(i, t+1))^2, (z(i) - z_{mc}(i, t-1))^2] \qquad (4.3)$$

Large values for $SDIa(i)$ indicate discontinuities in image intensity in both the forward and backward temporal directions. A blotch is detected if $SDIa(i)$ exceeds a threshold T_1:

$$d_{SDIa}(i) = \begin{cases} 1 & \text{if } SDIa(i) > T_1 \\ 0 & \text{otherwise} \end{cases} \quad \text{with } T_1 \geq 0 \qquad (4.4)$$

where T_1 is a threshold selected by the user. If a small value is chosen for this threshold, the detector is very sensitive and will detect a large percentage of the blotches corrupting an image. However, due to the great sensitivity, many false alarms will result as well. Increasing the value of T_1 reduces the sensitivity; it reduces both the number of false alarms and the number of correct detections.

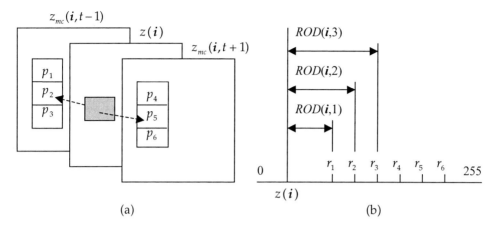

Figure 4.2: *(a) Selection of reference pixels p_k from motion compensated previous and next frames, (b) computation of ROD(i,l) based on pixels p_k ordered by rank: r_m.*

A variation on the SDIa detector is the SDIp detector. SDIp has an additional constraint that requires the signs of $z(i) - z_{mc}(i, t+1)$ and $z(i) - z_{mc}(i, t-1)$ to be identical before a blotch can be detected. This constraint reduces the number of false alarms resulting from erroneous motion estimates. In the case of correct motion estimation, the reference pixels in the previous and next frames are assumed to be identical, and therefore the intensity differences with the corrupted data in the current frame should have the same polarity. Note that this is not necessarily true in the case of occlusion and noisy data.

ROD detector. The *rank-ordered differences* (ROD) detector is a heuristic detector based on *order statistics* (OS) [Nad97]. Let p_k with k=1, 2, ..., 6 be a set of reference pixels relative to a pixel from $z(i)$. These reference pixels are taken from the motion compensated previous and next frames at locations spatially co-sited with pixel $z(i)$ and its two closest vertical neighbors (see Figure 4.2a). Let r_m be the reference pixels p_k ordered by rank with $r_1 \leq r_2 \leq r_3 \leq r_4 \leq r_5 \leq r_6$. The rank order mean r_{mean} and rank-order differences ROD(i,l) with l = 1, 2, 3 are defined by (see Figure 4.2b):

$$r_{mean} = \frac{r_3 + r_4}{2} \qquad (4.5)$$

$$ROD(i,1) = \begin{cases} r_l - z(i) & \text{if } z(i) \leq r_{mean} \\ z(i) - r_{7-l} & \text{if } z(i) > r_{mean} \end{cases} \quad \text{with} \quad l = 1, 2, 3. \qquad (4.6)$$

A blotch is detected if at least one of the rank-order differences exceeds a specific threshold T_l. The T_l are set by the user and determine the detector's sensitivity:

$$d_{ROD}(i) = \begin{cases} 1 & \text{if } ROD(i,l) > T_l \\ 0 & \text{else} \end{cases} \quad \text{with } 0 \leq T_1 \leq T_2 \leq T_3 \text{ and } l = 1, 2, 3. \tag{4.7}$$

MRF detector. In [Kok95a] an a posteriori probability for a binary occlusion map, given the current frame and a motion-compensated reference frame, is defined. The occlusion map indicates whether objects in the current frame are also visible in a reference frame. The *probability mass function* (pmf) for the a posteriori probability of the occlusion map is given by:

$$P[d_k(i) \mid z(i), z_{mc}(i, t+k)] \propto P[z(i) \mid d_k(i), z_{mc}(i, t+k)] \, P[d_k(i)] \tag{4.8}$$

where the symbol $\propto$ means *is proportional to*, and k indicates which reference frame is used. Maximizing (4.8) gives the MAP estimate for an occlusion mask.

Blotches are detected where occlusions are detected both in forward and backward temporal directions; $k = 1$ and $k = -1$:

$$d_{MRF}(i) = \begin{cases} 1 & \text{if } (d_1(i,l) = 1) \wedge (d_{-1}(i) = 1) \\ 0 & \text{otherwise} \end{cases} \tag{4.9}$$

The likelihood function in (4.8) is defined by:

$$P[z(i) \mid d_k(i), z_{mc}(i, t+k)] \propto \exp\left(-\sum_{i \in S}[(1 - d_k(i))(z(i) - z_{mc}(i, t+k))^2]\right) \tag{4.10}$$

where S indicates the set of all spatial locations within a frame. This likelihood function indicates that, in the absence of occlusion, $d_k(i) = 0$, the squared difference between the current pixel i and the corresponding pixel from the motion-compensated reference frame is likely to be small. The prior in (4.8) is given by:

$$P[d_k(i)] \propto \exp\left(-\sum_{i \in S}[\beta_1 \, f(d_k(i)) + \beta_2 \, d_k(r)]\right) \quad \text{with } \beta_1, \beta_2 \geq 0 \tag{4.11}$$

where the function $f(d_k(i))$ counts the number of neighbors of $d_k(i)$ that are different from $d_k(i)$. The term $\beta_1 f(d_k(i))$ in (4.11) constrains the occlusion map to be consistent locally. If an occlusion mask is locally inconsistent, $\beta_1 f(d_k(i))$ is large and the probability of $P[d_k(i)]$ is made smaller. The term $\beta_2 d_k(r)$ in (4.11) is a penalty term that suggests that it is unlikely that many pixels are occluded. The user controls the strength of the self-organization and the sensitivity of the detector by selecting values for β_1 and β_2.

Combining (4.8), (4.10), and (4.11) gives:

$$P[d_k(i)\,|\,z(i),z_{mc}(i,t+k)] \propto \exp\left(-\sum_{i \in S}[(1-d_k(i))(z(i)-z_{mc}(i,t+k))^2 + \beta_1 f(d_k(i)) + \beta_2 d_k(i)]\right) \quad (4.12)$$

Equation (4.12) can be maximized with *simulated annealing* (SA) [Gem84]. It is maximized once for $k = 1$ and once for $k = -1$. The resulting occlusion masks are combined by (4.9) to give the binary blotch detection mask $d_{MRF}(i)$.

AR detector. The assumptions that underlie the AR detector are that uncorrupted images follow AR models and that the images can be predicted well from the motion compensated preceeding and/or following frames [Kok95a]. If the motion-compensated frame at $t + k$ is used as a reference, the observed current frame $z(i)$ is given by:

$$z(i) = \sum_{l=1}^{n} a_l z_{mc}(i+q_l, t+k) + e(i,t+k) = \hat{z}(i) + e(i,t+k) \quad (4.13)$$

where the a_l are the n AR model coefficients estimated from the observed data (see, for example, [The92]), q_k give the relative positions of the reference pixels with respect to the current pixel and $e(i,t+k)$ denotes the prediction error.

In the absence of blotches and occlusion, the prediction errors $e(i,t+k)$ are small. A blotch is detected if the squared prediction error exceeds a user defined threshold T_1 in both the forward ($k = 1$) and backward ($k = -1$) directions:

$$d_{AR}(i) = \begin{cases} 1 & \text{if } (e^2(i,t+1) > T_1) \wedge (e^2(i,t-1) > T_1) \\ 0 & \text{otherwise} \end{cases} \quad \text{with} \quad T_1 \geq 0 \quad (4.14)$$

Evaluation. To compare the effectiveness of the detectors described in this section, Figure 4.3 plots their *receiver operator characteristics* (ROCs) for four test sequences. An ROC plots the false alarm rate (P_L) versus the correct detection rate (P_D) of a detector. Ideally, the ratio of correct detections to false alarms is large. For the SDIa, ROD, and AR detectors, the curves were obtained by letting T_1 vary so that $1 \leq T_1 \leq 35$ (for the ROD detector, $T_2 = 39$ and $T_3 = 55$ were used). For the AR detector, the image was subdivided into blocks of 28x28 pixels, and a set of AR coefficients was computed for each block. The support consisted of five pixels as in [Kok95a] (see Figure 4.4). For the MRF detector, $3 \leq \beta_1 \leq 8$ and $9 \leq \beta_2 \leq 1369$ were used.

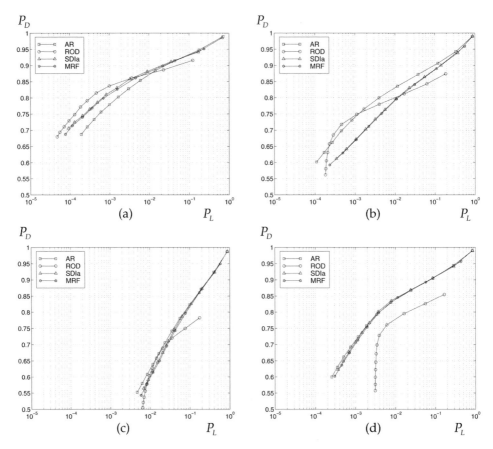

Figure 4.3: *Receiver operator characteristics for various blotch detectors for (a) Western sequence, (b) MobCal sequence, (c) Manege sequence, (d) Tunnel sequence.*

The detectors were applied to four test sequences, namely *Western*, which was also used in [Kok95a], *Mobcal, Manege,* and *Tunnel*. To avoid problems caused by the combination of interlacing and fast motion, only the odd fields from the last two sequences were used[1].

---

[1]This is reasonable because blotches are artifacts that are typically related to film with no interlacing.

All sequences were degraded by adding artificial blotches. Each artificial blotch had a fixed gray value that was drawn uniformly between 16 and 240, which is here the allowed range for pixel intensities. The *Western* sequence originates from film and therefore contains granular noise. The *MobCal*, *Manege*, and *Tunnel* sequences, which were recorded by modern cameras, have little noise. To let them resemble real film data more closely, white gaussian noise with variance 10 was added after the blotches were added. Therefore, for these sequences, unlike for the *Western* sequence, the blotches are no longer completely smooth. Motion was estimated by an hierarchical motion estimator (Appendix A).

Figure 4.3 shows that the performance of the detectors strongly depends on the sensitivity to which they are set and on the sequences themselves. The best detection results are obtained for the *Western* sequence, which has relatively low local contrasts. The poorest results are obtained for the *Manege*, sequence which contains fast motion and sharp local contrasts.

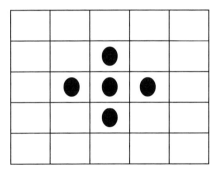

Figure 4.4 *Support (circles) from reference frame at t+k used for AR detector. The center of the support is aligned with pixel being processed in the current frame t*

The experiments show that no detector consistently outperforms any other. In some instances, the AR detector shows the best performance; in other instances, it shows the poorest performance. It can been seen from the ROCs that greater complexity does not necessarily lead to better results. The SDIa detector requires only a fraction of the number of computations required by the MRF detector, and both give a similar performance for all sequences. The ROD detector performs well for most sequences. However, it breaks down in the *Tunnel* sequence. This is because many false alarms are generated in this sequence as a result of the fixed settings chosen for T_2 and T_3.

4.2.2 Techniques for motion vector repair

Estimated motion vectors are less reliable when an image is blotched. Hence, the reference data extracted from the motion-compensated reference frames and used for interpolating the missing data may be erroneous. Motion vector repair can improve the likelihood of obtaining correct reference data. This repair has been investigated in the context of error concealment in (compressed) digital video transmission where each 8x8 or 16x16 image block gets one motion vector assigned.

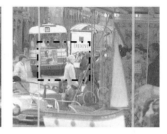

Figure 4.5: *Three frames from the Manege sequence with a single blotch (black) in the central frame and a bounding box. The regions within the dashed boxes in the outermost frames indicate the search region for the block matcher.*

Two basic approaches to motion vector repair are found in literature. The first approach re-estimates the unreliable motion vectors by interpolating them from the surrounding reliable motion vectors. In [Has92], [Nar93], median filtering and averaging are proposed for this purpose. The second approach re-estimates the motion vector on basis of the image intensities. The methods in [Che97], [Lam93] exploit the correlation between pixels along the boundaries between adjacent image blocks. An erroneous motion vector is replaced by a new vector so that the mean squared difference in image intensity over the boundaries with the neighboring blocks is minimized. The approach in [Kok98], which was developed in the context of blotch correction, re-estimates the motion of corrupted image blocks. The motion estimation process discards the corrupted pixels and constrains the smoothness of the motion vectors.

This section gives an indication of how well either approach can be expected to perform. Two algorithms are evaluated for this purpose. The first algorithm interpolates the unreliable motion estimates by applying the dilation interpolation technique described in Chapter 3 to the horizontal and vertical components of the motion vector fields independently. All weights $W_{m,n}$ are set to one, and no biases are applied.

The second algorithm re-estimates motion vectors on the basis of the observed image intensities, as illustrated by Figure 4.5. First, a bounding box is computed around each blotch with an additional (small) horizontal and vertical margin. Motion is estimated between the region contained by the box and the previous/next frames by block matching. To avoid biases resulting from blotched data, the block matcher discards the corrupted pixels in the current frame and in the reference frames. To limit the computational effort, the search range is limited to ±20 pixels for both horizontal and vertical directions.

Additionally, a third algorithm is evaluated. This algorithm simply replaces the motion vectors at blotched sites with vectors that indicate *zero motion*. Large parts of images tend to be temporally stationary and, therefore, assuming no motion is correct in a large number of cases.

The effectiveness of the three methods is evaluated by applying the scheme in Figure 4.6 to the same four test sequences used in the previous section. In this scheme, the motion vectors are estimated between pairs of consecutive frames that are corrupted by artificial blotches. Next, the motion vectors are repaired at locations indicated by the blotch detection masks. The blotch masks are made available to the motion vector repair block in Figure 4.6, though this is not shown explicitly in the figure. Then, one of the original, uncorrupted input frames is compensated for motion, and the MSE with the other original, uncorrupted input frame is computed. The MSE is computed only over the locations indicated by the blotch mask. If the motion vectors are accurate, the MSE will be small.

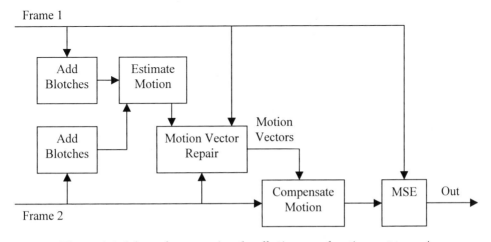

Figure 4.6: *Scheme for measuring the effectiveness of motion vector repair.*

Two sets of experiments are carried out. The first set uses the true locations of the pixels corrupted by blotches. The second set uses detection masks resulting from the SDIa detector set to a detection rate of approximately 70%. The first set of experiments shows the improvements that are obtained under ideal circumstances. The second set of experiments shows the improvements obtained under realistic circumstances where false alarms influence the results. In this second set of experiments, 30% of the blotched pixels are not detected. These are the so-called *misses*. Motion vector repair does not influence the motion vectors assigned to misses because the motion vectors are only re-estimated at locations when blotches are detected.

Blotch Mask	Vector Repair	Western (MSE)	Mobcal (MSE)	Manege (MSE)	Tunnel (MSE)
Exact	None	94	338	1027	396
Exact	Block Matching	32	58	215	82
Exact	Dilation	62	218	656	122
Exact	Zero Motion	63	190	748	97
Estimated	None	157	721	2136	1044
Estimated	Block Matching	103	538	2138	693
Estimated	Dilation	140	594	2658	908
Estimated	Zero Motion	126	496	2841	847

Table 4.1: *Evaluation of quality of motion vectors before and after motion vector repair. The MSE is computed only at sites indicated by the blotch mask. Both the true blotch mask and an estimated blotch mask, estimated with the SDIa detector set to a detection rate of approximately 70%, are used.*

Blotch Mask	Western (MSE)	Mobcal (MSE)	Manege (MSE)	Tunnel (MSE)
Exact	21	35	113	18
Estimated	72	343	1925	689

Table 4.2: *MSE computed with the motion vectors estimated from the original, unimpaired image sequences. The MSE is computed only at sites indicated by the blotch mask.*

Table 4.1 gives the experimental results. This table indicates that applying vector repair significantly increases the accuracy of the corrupted motion vectors if the locations of the corrupted sites are known exactly. When the estimated blotch mask is used, the MSE increases and the gains are smaller. This is not surprising. Because of false alarms, motion vectors are re-estimated at locations that are not corrupted.

The new motion estimates for the false alarms are suboptimal because correct image data are discarded in the motion estimation process.

The lowest MSEs are obtained with motion vectors repaired by the block matching technique, and, therefore, this method is to be preferred to the other methods for vector repair. The *zero motion* technique shows good results for those test sequences that contain large areas without motion, i.e., all test sequences except the *Manege* sequence. The dilation method has a relatively poor performance, yet it is to be preferred to no vector repair at all.

Table 4.2 shows the MSEs obtained from motion vectors computed from the original, unimpaired test sequences. These form the lower bound for the MSEs that can ideally be achieved. The conclusion is that the block-matching vector-repair technique bridges the gap between the MSE obtained from the corrupted vectors and the "true" vectors to a large extent.

4.2.3 Blotch correction techniques

MMF interpolator. A *multistage median filter* (MMF) is a concatenation of median filtering operations. The ML3Dex MMF is a heuristic method for interpolating missing data [Kok95b]. ML3Dex first applies five subfilters centered around the pixel being processed. Figure 4.7 shows the subfilter masks. In this figure, the top plane of each subfilter refers to data in the motion-compensated next frame, the center plane refers to data in the current frame, and the bottom plane refers to data in the motion-compensated previous frame. Next, the output of all the subfilters are combined and give the interpolated value according to:

$$m_l = median[W_l] \quad \text{with} \quad 1 \leq l \leq 5 \tag{4.15}$$

$$ML3Dex = median[m_1, m_2, m_3, m_4, m_5] \tag{4.16}$$

Note that ML3Dex does not necessarily fulfill any of the image models used by the detectors described in Section 4.2.1. In other words, if a detector is applied again to a corrected image, the corrected data may well be flagged as being blotched. In such instances, there is no objective reason to prefer the corrected data to the observed data and, from an engineering point of view, it may actually be better to stick to the observed data. This reduces the risk of introducing corruption at locations at which blotches were mistakenly detected.

MRF interpolator. A MRF formulation towards interpolating missing data is given in [Kok95b]. This approach tries to find the MAP estimate of the missing data, $\hat{y}(i)$, given the locations of the corrupted sites and the observed (motion-compensated) previous, current and next frames by maximizing:

$$P[\hat{y}(i)\,|\,d(i),z_{mc}(i,t-1),z(i),z_{mc}(i,t+1)] \propto$$
$$\exp\left(-\sum_{i:d(i)=1}\left(\sum_{s\in S_S(i)}(\hat{y}(i)-\hat{y}(s))^2 + \sum_{s\in S_T(i)}\lambda[(\hat{y}(i)-z_{mc}(s,t-1))^2 + (\hat{y}(i)-z_{mc}(s,t+1))^2]\right)\right) \quad (4.17)$$

where S_S and S_T indicate the spatial and temporal neighborhoods, and λ is the relative weight for the temporal neighborhood. Equation (4.17) is optimized only over blotched image locations. The term $(\hat{y}(i)-\hat{y}(s))^2$ on the right hand side of (4.17) indicates the assumption that the interpolated values are likely to be smooth spatially. The other quadratic terms indicate the assumption that it is unlikely that the interpolated values introduce temporal discontinuities in image intensity along the motion trajectories. Equation (4.17) can be maximized with SA [Gem84].

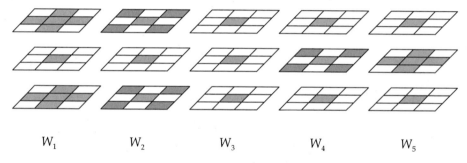

Figure 4.7: *Subfilter masks for ML3Dex. Gray elements indicate the included data, white elements indicate excluded data.*

AR interpolator. A method for interpolating missing data based on a 3D AR model is described in [Kok95b]. For each region of an image with missing data a set of AR parameters is determined. It is assumed that the data in this region are stationary. The AR parameters are computed from data of the (motion-compensated) previous, current, and next frames. Note that the blotched data in the current frame t are discarded so that they do not bias the estimates of the AR parameters. Next, the missing data are interpolated so that the linear-mean-squared-prediction-error, computed with the estimated AR parameters, is minimized.

Consider the data to be ordered in a lexicographic fashion [Pra75]. Let e indicate a vector of prediction errors, let z_+ indicate a vector containing the observed data from the current frame plus that from the motion-compensated previous and next frames, and let A be a matrix with the AR coefficients placed at suitable locations. The prediction errors are denoted compactly by:

$$e = Az_+ \tag{4.18}$$

The prediction errors consist of two parts. One part depends on the product of the known data z_{k+} (data that are not to be interpolated) in z_+ with a number of columns from A, these columns will be denoted by A_k. The other part consists of the product of unknown data z_{u+} (data that are to be interpolated) in z_+ and the remaining columns of A, which will be denoted by A_u:

$$e = A_k z_{k+} + A_u z_{u+} \tag{4.19}$$

The unknown data are interpolated so that the mean-squared-prediction-error $e^T e$ is minimized. Taking the derivative of $e^T e$ with respect to z_{u+}, setting it to zero and solving for z_{u+} gives the required result:

$$z_{u+} = -[A_u^T A_u]^{-1} A_u^T A_k z_{k+} \tag{4.20}$$

Variations on this 3D AR method are described in [Goh96], [Kal97]. In [Goh96] it is pointed out that the assumption of stationarity is not met for occluded regions that have become uncovered (and vice versa). The authors suggest estimating the AR model parameters and interpolating the missing data with two frames only. One frame is the current frame that contains the missing data. The other frame is either the preceeding or the following frame. This depends on which (motion-compensated) frame gives the smallest mean squared difference with the current frame in the region of the missing data. This method is referred to as the B3DAR method. In [Kal97] this approach is refined by subdividing regions with missing data into multiple regions and interpolating the missing data for each region. This is done because a single set of AR coefficients may not be able to model a block of pixels adequately when the missing data cover a large region.

Drawbacks. There are a number of drawbacks to the methods for interpolating missing data described in this section. The multistage median filter has no model for the corrected image. Therefore the interpolation results are not necessarily consistent, either with the data surrounding the corrupted region or within the corrected region itself. The MRF interpolator gives an overly smooth result because it interpolates the data so that the differences between an interpolated pixel and its spatio-temporal neighbors are minimized. The MRF interpolator takes no measures for resolving the effects resulting from occlusion.

The AR interpolators can also smooth the data, and therefore the fidelity of the interpolated data in textured regions and in noisy film sequences is not that of their surroundings. As mentioned before, the problem of occlusion can, in principle, be solved with the method in [Goh96]. However, unlike the method described in [Goh96], the direction of interpolation should be determined pixelwise instead of blockwise. Because occlusion can vary on a pixel-by-pixel basis, the optimal direction of interpolation should be allowed to vary on a pixel-by-pixel basis. Furthermore, by subdividing missing data into a number of regions, as suggested in [Kal97], mismatches may well occur within the interpolated results near the region boundaries.

Finally, all the approaches described in this section assume that the reference regions in the motion-compensated previous/next frames do not contain missing data in the regions of interest. This assumption is not always correct and can lead to incorrect interpolated data, as will be shown in Section 4.5.

4.2.4 Conclusions

Existing techniques for blotch detection show good performance, though even better performance is desirable in an automated environment for image restoration. For example, consider the ROC curves in Figure 4.3. These indicate that the false alarm rate varies between 0.5 and 15% for a correct detection rate of 85%. With other words, not only are many blotches removed, which is good, but also two thousand to sixty thousand pixels are also interpolated unnecessarily for each frame of a PAL image, which has a resolution of 720x576 pixels. Because the interpolators are fallible, false alarms can lead to artifacts in the corrected sequence that are visually more disturbing than the blotches themselves.

The development of improved methods for blotch detection and correction is the topic of the remaining sections of this chapter. Sections 4.3 and 4.4 investigate how to improve the detectors. Section 4.5 develops an interpolator for correcting blotches that is robust to errors in the reference data obtained from motion-compensated frames.

4.3 Improved blotch detection by postprocessing

The goal of this section is to improve the ratio of correct detections to false alarms of existing blotch detectors. The approach taken here is not one of designing yet another detector. Instead, a strategy of postprocessing that removes possible false alarms and that finds parts of blotches missed by the detector is developed. Figure 4.8a shows how postprocessing fits in the scheme of Figure 4.1b. Figure 4.8b shows the proposed set of postprocessing techniques.

What is the idea behind the postprocessing operations? Blotches are not just random sets of individual pixels, but that they are spatially coherent regions and can be manipulated as such. How these regions can be extracted from the blotch detection masks is discussed in Section 4.3.2. Because it is not certain at this point that the extracted regions are true blotches, rather than something that resulted from false alarms made by the detector, the term *candidate blotches* is used to refer to the extracted regions.

Section 4.3.3 dllows a probabilistic approach towards identifying and eliminating candidate blotches as a result of false alarms due to noise. The other candidate blotches, resulting from correct detections, have been detected only partially. Applying techniques called *hysteresis thresholding* and *constrained dilation* can make the detections more complete. These techniques are explained in Section 4.3.4 and Section 4.3.5. Section 4.3.6 concludes with experimental evaluations that demonstrate the effectiveness of the postprocessing approach applied to a simplified version of the ROD detector, which is described next.

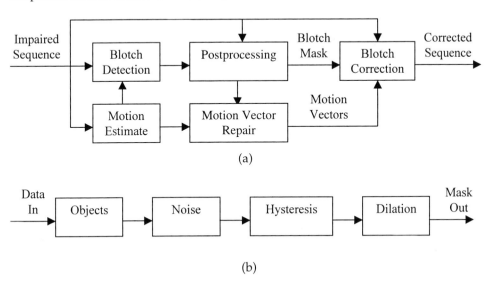

Figure 4.8: *(a) Place of postprocessing in a system of blotch detection and correction, (b) chain of postprocessing operations for increasing the ratio between correct detections and false alarms.*

4.3.1 Simplified ROD detector

By letting $T_2 \to \infty$ and $T_3 \to \infty$, the output of the ROD detector is completely determined by T_1. In this case, T_2 and T_3 can removed from the equations and a *simplified ROD* (SROD) detector results. The SROD detector is computationally much more efficient than the ROD detector because it no longer requires the reference pixels to be ordered by rank:

$$SROD(i) = \begin{cases} \min(p_k) - z(i) & \text{if } \min(p_k) - z(i) > 0 \\ z(i) - \max(p_k) & \text{if } z(i) - \max(p_k) > 0 \quad \text{with} \quad k = 1, \ldots, 6 \\ 0 & \text{otherwise} \end{cases} \quad (4.21)$$

A blotch is detected if:

$$d_{SROD}(i) = \begin{cases} 1 & \text{if } SROD(i) > T_1 \\ 0 & \text{otherwise} \end{cases} \quad \text{with} \quad T_1 \geq 0 \quad (4.22)$$

The SROD detector looks at a range of pixel intensities obtained from motion-compensated frames and compares this range to the pixel intensity under investigation. Blotches are detected if the current pixel intensity lies far enough outside the range. What is considered "far enough" is determined by T_1.

4.3.2 Extracting candidate blotches

The SROD detector is a pixel based detector. If the spatial coherence within blotches is to be exploited, regions consisting of pixels with similar properties will have to be extracted from the available data. Adjacent pixels within a blotch tend to have similar intensities. A pair of pixels are considered to be similar if their difference is smaller than twice the standard deviation of the noise. This means at least 96% of the pixels will be labeled as belonging to the same candidate blotch if additive white gaussian noise is assumed to be corrupting the image.

Therefore, adjacent pixels with similar intensities that have been flagged by the blotch detector are considered to be part of the same candidate blotch. To differentiate between the various candidate blotches, a unique label is assigned to each of them.

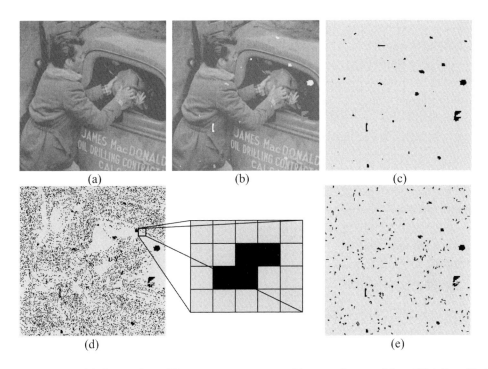

Figure 4.9: *(a) Frame from Western test sequence, (b) same frame with artificially added blotches, (c) true blotch mask, (d) blotch mask estimated with the SROD detector with $T_1 = 0$ and a zoom in on a candidate blotch, (e) estimated blotch mask after possible false alarms due to noise are removed.*

4.3.3 Removing false alarms due to noise

After the labelling procedure, a candidate blotch is an object with spatial support S and it consists of K pixels, each of which has a specific detector output $SROD(i)$. By selecting a small value for T_1, the detector is set to a great degree of sensitivity. In this case, it is not only sensitive to blotches, but also to noise. An example of this is given in Figure 4.9, which shows a frame from the original *Western* test sequence, the same frame degraded with artificial blotches, and the blotch mask used for adding the artificial blotches. The estimated blotch mask, estimated with the SROD detector with $T_1 = 0$, shows many false alarms.

Figure 4.9d also zooms in on a candidate blotch. The question for this candidate blotch is whether it is likely that it was detected purely as a result of false alarms due to noise. If so, the complete candidate blotch should be removed from the blotch detection mask. Figure 4.9e shows a result of this approach, for which the details follow, applied to Figure 4.9d. Many false alarms have been removed.

The probability of a candidate blotch being detected purely due to false alarms is equal to the probability of the detector giving specific set of values $SROD(i)$, all of which are larger than T_1. This probability can be computed in two steps. The first step determines the probability of a specific detector response for an individual pixel under the influence of noise. The second step determines the probability that a collection of such pixels belong to a single object. The details of these two steps are given now.

For the first step, it is assumed that the reference pixels p_k and the current pixel $z(i)$ are identical except for the additive noise in the absence of blotches, i.e., $z(i) = y + \eta_i$ and $p_k = y + \eta_k$, where η_i and η_k indicate a specific noise realization. It is also assumed that the noise is i.i.d., has zero mean, and is symmetrically distributed around the mean. The probability that the SROD detector generates a false alarm due to noise is:

$$\begin{aligned}
P(SROD(i) > T_1) &= P[z(i) - \max(p_k) > 0, z(i) - \max(p_k) > T_1] + \\
&\quad P[\min(p_k) - z(i) > 0, \min(p_k) - z(i) > T_1] \\
&= P[z(i) - \max(p_k) > 0 | z(i) - \max(p_k) > T_1] P[z(i) - \max(p_k) > T_1] + \\
&\quad P[\min(p_k) - z(i) > 0 | \min(p_k) - z(i) > T_1] P[\min(p_k) - z(i) > T_1] \quad (4.23)\\
&= P[z(i) - \max(p_k) > T_1] + P[\min(p_k) - z(i) > T_1] \\
&= 2\, P[z(i) - \max(p_k) > T_1] \\
&= 2\, P[\eta_i - \max(\eta_k) > T_1]
\end{aligned}$$

where the last but one line follows from symmetry. Using the fact that $\eta_i - \max(\eta_k) > T_1$ requires that $\eta_i - \eta_k > T_1$ for all k gives:

$$\begin{aligned}
P(SROD(i) > T_1) &= 2\, P[\eta_i - \eta_1 > T_1, \eta_i - \eta_2 > T_1, \ldots, \eta_i - \eta_{61} > T_1] \\
&= 2 \int_{-\infty}^{\infty} P[\eta_i - \eta_1 > T_1, \eta_i - \eta_2 > T_1, \ldots, \eta_i - \eta_6 > T_1 | \eta_i]\, P[\eta_i]\, d\eta_i \\
&= 2 \int_{-\infty}^{\infty} \prod_k P[\eta_i - \eta_k > T_1 | \eta_i]\, P[\eta_i]\, d\eta_i \quad (4.24)\\
&= 2 \int_{-\infty}^{\infty} P^6[\eta_i - \eta_k > T_1 | \eta_i]\, P[\eta_i]\, d\eta_i \\
&= 2 \int_{-\infty}^{\infty} \left(\int_{-\infty}^{\eta(i) - T_1} P[\eta]\, d\eta \right)^6 P[\eta_i]\, d\eta_i
\end{aligned}$$

BLOTCH DETECTION AND CORRECTION

The step from the second to the third line in (4.24) is obtained by applying the theorem on total probability [Leo94]. The fourth and the fifth lines in (4.24) are obtained by considering that the η_k are independent of each other. This is indicated by dropping index k from η_k. Equation (4.24) gives the probability that the SROD detector generates a false alarm for an individual pixel due to noise and can be evaluated numerically once the parameters of the noise have been determined.

In the case that the pixels of an image sequence are represented by integer values, the output of SROD also consists of integer values. The probability $P_{mass}[SROD(i) = x]$ that the SROD detector gives a specific response x, with $x \geq 0$, for an individual pixel is given by:

$$P_{mass}[SROD(i) = x] = P[SROD(i) > x - 0.5] - P[SROD(i) > x + 0.5] \qquad (4.25)$$

Table 4.3 lists the computed probabilities of specific detector responses in the case of white gaussian noise with a variance of 9.6. (The method described in [Mar95] was used to estimate a noise variance of 9.6 for the *Western* sequence).

	Probability		Probability
1	0.091921	7	0.002224
2	0.060748	8	0.000892
3	0.036622	9	0.000304
4	0.020492	10	0.000108
5	0.010353	11	0.000028
6	0.005168	12	0.000007

Table 4.3: *Probability of a specific detector response SROD(i) computed for a constant signal corrupted by additive white gaussian noise with variance 9.6.*

For the second step, it is assumed that the individual pixels within a blotch are flagged independently of their neighbors. Strictly speaking this assumption is incorrect because, depending on the motion vectors, sets of reference pixels p_k can overlap. The effects of correlation are ignored here. Let H_0 denote the hypothesis that an object is purely the result of false alarms and that each of the sets of reference pixels were identical to the true image intensity $y(i)$ except for the noise. $P[H_0]$ is then the probability that a collection of K individual pixels are flagged by the SROD detector, each of which with a specific response $x(i)$:

$$P[H_0] = \prod_{i \in S} P_{mass}[SROD(i) = x(i)] \qquad (4.26)$$

where S is the spatial support of the candidate blotch. Those objects for which the probability that they are solely the result of noise exceeds a risk R are removed from the detection mask:

$$P[H_0] > R \qquad (4.27)$$

The result of this approach, as mentioned before, is indicated in Figure 4.9e.

4.3.4 Completing partially detected blotches

The technique for removing possible false alarms due to noise can be applied to any value of T_1. When a blotch detector is set to a low detection rate, not much gain is to be expected from this technique because the detector is insensitive to noise. A second method for improving the ratio of correct detections to false alarms is described here.

Many blotches are not detected at all and others are detected only partially at lower detection rates. The strategy is now to make the partial detections more complete. This is achieved by noting from Figure 4.3 that the probability of false alarms decreases rapidly as the correct detection rate is lowered. Therefore, detections resulting from a blotch detector set to a low detection rate are more likely to be correct and can thus be used to validate the detections by the same detector set to a high detection rate.

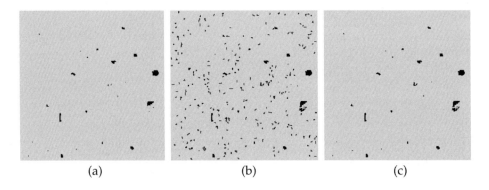

(a) (b) (c)

Figure 4.10: *Example of hysteresis thresholding. Detection masks from (a) detector set to low sensitivity ($T_1 = 30$) with removal of possible false alarms due to noise, (b) detector set to high sensitivity ($T_1 = 0$) with removal of possible false alarms due to noise, (c) hysteresis thresholding.*

The validation can be implemented by applying hysteresis thresholding [Can86]; see Figure 4.10. The first stage computes and labels the set of candidate blotches with a user-defined setting for T_1. Possible false alarms due to noise are removed as already described. The second stage sets the detector to a very high detection rate, i.e., $T_1 = 0$, and again a set of candidate blotches is computed and labeled. Candidate objects from the second set can now be validated; they are preserved if corresponding candidate objects in the first set exist. The other candidate blotches in the second set, which are more likely to have resulted from false alarms, are discarded. Effectively blotches detected with the operator settings are preserved and are made more complete.

4.3.5 Constrained dilation for missing details

There is always a probability that a detector fails to detect elements of a blotch, even when it is set to its most sensitive setting. For example, the large blotch on the right hand side in Figure 4.9c is not completely detected in Figure 4.9d. In this final postprocessing step, the detected blotches are refined by removing small *holes* in the candidate blotches and by adding parts of the blotches that may have been missed near the edges.

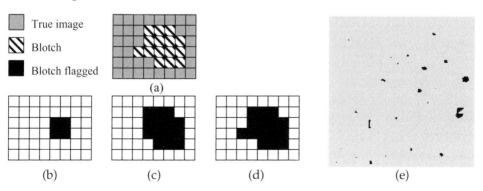

Figure 4.11: *Example of constrained dilation: (a) image with blotches, (b) initial detection mask, (c) detection mask after one iteration, (d) detection mask after two iterations, (e) result of constrained dilation applied to Figure 4.10(c).*

For this purpose, a constrained dilation operation is suggested here. Dilation is a well known technique in morphological image processing [Mar98]. The constrained dilation presented here applies the following rule: if a pixel's neighbor is flagged as being blotched and its intensity difference with that neighbor is small (e.g., less than twice the standard deviation of the noise) then that pixel should also be flagged as being blotched. The constraint on the differences in intensity reduces the probability that uncorrupted pixels surrounding a corrupted region are mistakenly flagged as being blotched. It uses the fact that blotches tend to have gray values that are significantly different from their surroundings. Figure 4.11a-c illustrates the procedure, Figure 4.11e shows the result of this method when applied to the blotch mask in Figure 4.10c.

It is important not to apply too many iterations of the constrained dilation operation because it is always possible that the contrast between a candidate blotch and its surrounding is small. The result would be that the candidate blotch grows completely out of its bounds and many false alarms occur. In practice, if the detector is set to a great sensitivity, applying two iterations favorably increases the ratio of the number of correct detections to false alarms. When the detector is set to less sensitivity, the constrained dilation is less successful and should not be applied. In the latter case, the blotches that are initially detected by the SROD detector must have sharp contrast with respect to the reference data. Because of the sharp contrast, the blotches are made fairly complete by the hysteresis thresholding. The dilation therefore adds little to the number of correct detections, yet it significantly increases the number of false alarms.

4.3.6 Experimental evaluation

Figure 4.12 summarizes the effects of the consecutive postprocessing operations. Visually speaking, the final result in this figure compares well to the true blotch mask in Figure 4.9c. Now the effectiveness of the postprocessing operations is evaluated objectively.

Figure 4.13 plots the ROCs for ROD detector, SROD detector, and the SROD detector with postprocessing. The results from either the MRF or the AR detector, depending on which showed the best results in Figure 4.3, are also plotted for comparison. Figure 4.13 makes it clear that the SROD detector has a performance similar to that of the ROD detector for small values of T_1 (high detection rates). When set to a lesser sensitivity, the SROD detector shows performance either similarly to or better than the ROD detector. This is explained by the fact that detection mask of the SROD detector is a subset of the detection mask of the ROD detector; each detection made by the SROD detector is also made by the ROD detector. However, the SROD detector generates not only fewer correct detections, but also (significantly) fewer false alarms.

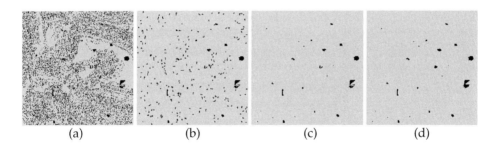

Figure 4.12: *Summary of postprocessing: (a) initial detection, (b) result after removal of false alarms, (c) result after hysteresis thresholding, (d) final result after constrained dilation.*

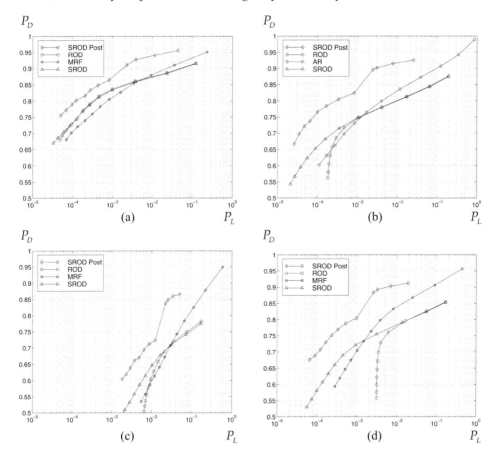

Figure 4.13: *Receiver operator characteristics for (a) Western sequence, (b) MobCal sequence, (c) Manege sequence, (d) Tunnel sequence (P_L - false alarm rate, P_D - correct detection rate)*

The postprocessing applied to the detection masks obtained from the SROD detector improves the performance considerably over the whole range of operation of the detector. Note that the constrained dilation operation was not applied for $T_1 > 12$. This explains the sometimes large change in trend between the fourth and fifth measuring point of the SROD ROCs. The postprocessed results are significantly better than any results from the detectors without postprocessing. For instance, before postprocessing, a correct detection rate of 85% corresponds with a false alarm rate between 0.5 and 15%. After postprocessing a correct detection rate of 85% corresponds with a false alarm rate between 0.05 and 3%.

4.4 Blotch detection with increased temporal aperture

Objects for which the motion cannot be tracked accurately from frame to frame pose severe problems to blotch detectors. Incorrect motion vectors lead to incorrect sets of reference pixels and hence to false alarms. An obvious solution to this problem would be to use a "robust" motion estimator. Though techniques that are more robust to complex motion than the hierarchical block matcher used here do exist, e.g., motion estimators that use affine motion models [Odo95], [Wan], it is questionable whether the increase in performance justifies the increase in complexity. Motion in natural image sequences often involves objects of which shape, texture, illumination, and size vary in time. No motion estimation algorithm is truly capable of dealing with this type of motion.

An alternative way to reduce the number of false alarms is to incorporate more temporal information. False alarms result from the fact that object motion cannot be tracked to any of the reference frames. Increasing the number of reference frames increases the probability that good correspondence to at least one of the reference frames is found. Once good correspondence is found for an object, it is assumed that this object is not a blotch. Therefore, increasing the temporal aperture of a blotch detector reduces the number of false alarms. However, increasing the temporal aperture also increases the probability that blotches are mistakenly matched to other blotches or to some part of the image contents. This decreases the correct detection rate. Obviously there is a trade-off.

The SROD detector can easily be extended to use four reference frames by taking into account three extra reference pixels from each of the frames at $t - 2$ and at $t + 2$. The extended SROD detector is denoted by SRODex. The postprocessing operations can be applied as before, all that is necessary is to recompute the probability of false alarms due to noise (taking into account that there are now twelve reference pixels instead of six).

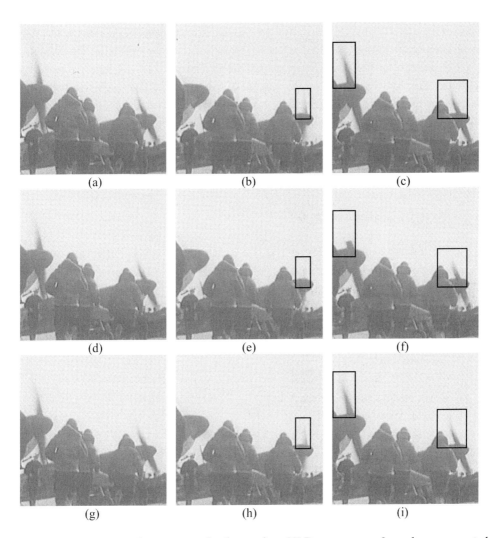

Figure 4.14: *Top row: three consecutive frames from VJ Day sequence. Second row: corrected frames using SROD with postprocessing and ML3Dex. Note the distortion of the propellers in the boxed regions. Third row: corrected frames after combining the SROD detection results with SRODex results. (Original photos courtesy by the BBC).*

Consider two sets of candidate blotches detected by the SROD detector and the SRODex detector, respectively. The SRODex detections form a subset of the SROD detections; the SROD detector finds blotches everywhere the SRODex detector does, and more. The blotches detected by the SROD detector are more complete than those detected by the SRODex detector, but the SRODex detections are less prone to false alarms. As in Section 4.3.5, hysteresis thresholding can be applied. The reliable, possibly incomplete SRODex detected blotches can be used to validate less reliable, but more complete SROD detections. In case of true blotches, the shapes and sizes of the regions flagged by both detectors should be similar. If this is not the case, it is likely that the detections are a result of false alarms due to complex motion. Hence, preserving SROD-detected candidate blotches that are similar to corresponding SRODex-detected blotches reduces the probability of false alarms. The other SROD-detected candidate blotches are discarded. Two candidate blotches and are considered to be similar if the ratio of their sizes is smaller than some constant ζ:

$$\frac{\text{Size of blotch in A}}{\text{Size of blotch in B}} < \zeta \tag{4.28}$$

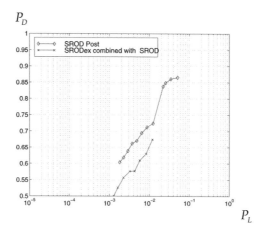

Figure 4.15: *Receiver operator characteristics for the SROD detector and for the SROD detector combined with the SRODex detector (all with postprocessing) computed for the MobCal sequence (P_L - false alarm rate, P_D - correct detection rate).*

Figure 4.14a-c show frames 27-29 from the *VJ Day* sequence. Besides blotches, this sequence contains a lot of action in the form running men and rotating propellers. Some of the propellers are not visible at all in some of the frames. Figure 4.14d-f shows data restored from the SROD detector ($T_1 = 10$) with postprocessing and ML3Dex for interpolation. The blotches have been removed very efficiently, but, as an unwanted side effect, parts of the propellers have been removed as well. Figure 4.14g-i shows restored data, but now the proposed combination of SROD and

SRODex has been used with $T_1 = 10$ and $\zeta = 2$. Most of the blotches have been removed and, very importantly, the propellers have been preserved.

The proposed algorithm was very successful for the *VJ Day* sequence because it is capable of dealing with the periodic presence of the propellers. However, increasing the temporal aperture does not necessarily always increase the performance, as can be observed from the ROC curves for the *Manege* sequence in Figure 4.15. In this case, the SRODex detector misses too many correctly SROD-detected are discarded. Whether increasing the temporal aperture is beneficial to the restoration process depends on the particular image sequence. In practice, it is up to an operator to decide which detector is most appropriate.

4.5 Fast, good quality interpolation of missing data

Section 4.2.1 showed that model-based interpolation of missing data can be done with 3D AR processes. This method gives good-quality interpolation results and its performance in resynthesizing textures of missing data is superior to that of other interpolators. Equation (4.20) gives a closed form solution to the 3D AR interpolation method. Unfortunately, there are a number of drawbacks to this method. First, it is very expensive in computational terms. For example, resynthesizing the texture for a region with a blotch of 20x20 pixels requires inverting a matrix with 400x400 elements. Second, the method as described in Section 4.2.3 assumes that the data in the reference frames are always correct. This assumption is not always true. Incorrect reference data can result from erroneous motion estimates, occlusions, and corruptions due to blotches. Third, AR interpolations can be overly smooth if the interpolated regions are large.

It is important to realize that full 3D AR restoration is not necessary in most cases. The most common differences between the frames of an image sequence can be characterized by a rearrangement of the object location. Therefore, it is likely that missing data in one frame can be restored by pasting (copying) pixels from corresponding regions in a reference frame. Reliable motion estimates must be available for pasting. In fact, pasting can be viewed as a one-tap AR interpolator with a coefficient of 1.0.

This section investigates the concept of interpolating missing data by pasting. Each pixel of the missing data in a blotched frame is replaced by a pixel from the corresponding location in either the motion-compensated previous frame or the motion-compensated next frame. A strategy for determining the direction of interpolation (i.e., pasting from the previous frame or pasting from the next frame) is required. The strategy used here constrains the interpolated data to fit in well with the region surrounding the missing data. Hence, the data surrounding the missing data define a set of boundary conditions to the solution of the interpolation problem. This constraint is enforced by requiring corrected image regions to follow 2D AR processes as well as possible.

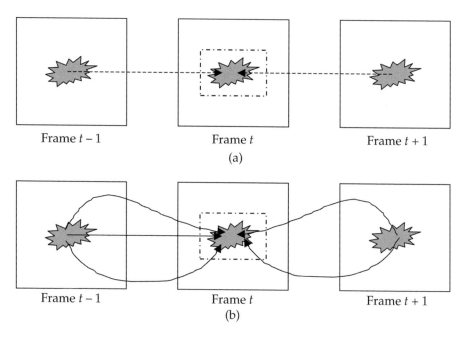

Figure 4.16: *(a) Region-based pasting: a region from either the previous or next frame (motion-compensated) is pasted into the current frame, (b) pixel-based pasting: pixels are pasted from either of the reference frames. In both cases (a) and (b), the pasting is done so that the corrected region, indicated by the dashed box, can fit a 2D AR model as well as possible.*

The question is now how to decide which reference frame should supply the pixels for pasting. One approach is to paste complete regions from either the previous or the next frame, depending on which result fits in better according to the 2D AR process (Figure 4.16a). To get good visual results with this approach, the motion-compensated reference data must represent the missing data at all locations. This requirement is less likely to be fulfilled as the size of the region to be pasted increases. The probability of some of the missing data being unavailable is proportional to the size of the region due possible to occlusions, blotches, and erroneous motion estimates.

A better approach is to determine the direction of interpolation for each individual pixel, as illustrated by Figure 4.16b. The pixel intensity from the motion-compensated reference frame with the smallest prediction error is pasted into the current frame. The advantage of pasting single pixels from either the previous or next reference frame is evident: if the reference data in one reference frame are inconsistent with the 2D AR model for the corrected frame, large prediction errors will result. In which case, data can be pasted from the other reference frame. This mechanism requires no explicit knowledge about errors in the reference data. Hence, corruptions in one of the reference frames do not influence the interpolated result negatively if the data in the other reference frame are correct.

At this point, the direction of interpolation in the pasting method described can vary erratically from pixel to pixel. Everything depends on which reference frame provides the pixel closest to the value predicted by the AR model. This can lead to two possible side effects. First, AR predictors tend to give overly smooth prediction results. Because the reference pixels closest to the AR predictions are selected, the pasted result can be overly smooth. Second, if the textures in reference frames are different (e.g., due to uncovering/occlusion), the pasted result might be a mixture of textures. In this case, the result is different from the true texture that underlies the missing data. These effects can be avoided by constraining the direction of interpolation to be consistent locally. For this purpose, a markov random field is applied.

4.5.1 Interpolating missing data with controlled pasting

This section formulates the ideas mentioned in mathematical terms. Because the aim is to paste pixels from either the previous or the next motion-compensated frame, a binary *direction mask o(i)* is introduced. This mask indicates for each spatial location which of the motion-compensated reference frames is most appropriate to serve as a reference for pasting, e.g., "0" for $z_{mc}(i, t-1)$ and "1" for $z_{mc}(i, t+1)$.

At this point it is assumed that the binary blotch detection mask $d(i)$ has already been determined. This could be done by any of the methods in the previous sections. The corrected frame $\hat{y}(i)$, which is an estimate of the true data $y(i)$, is given by:

$$\hat{y}(i) = \begin{cases} z_{mc}(i, t-1) & \text{if } d(i) = 1, o(i) = 0 \\ z_{mc}(i, t+1) & \text{if } d(i) = 1, o(i) = 0 \\ z(i) & \text{otherwise} \end{cases} \quad (4.29)$$

Now, the aim is to find $o(i)$. The reconstructed image $\hat{y}(i)$ follows through this variable. The image data model underlying the corrected image $\hat{y}(i)$ is assumed to be a 2D AR model of order n with coefficients a_l, with $l = 1, \ldots, n$. The prediction error $e(i)$ is a gaussian random variable with zero mean and variance σ_e^2:

$$\hat{y}(i) = \sum_{l=1}^{n} a_l \, \hat{y}(i+q_l) + e(i) \tag{4.30}$$

The binary field $o(i)$ must be found so that, on one hand, the corrected image $\hat{y}(i)$ fits the image model in (4.30) as well as possible, i.e., so that the prediction error variance σ_e^2 is as low as possible. On the other hand, as already explained, the direction of interpolation must be a consistent one locally. Note that not only must $o(i)$ be found, but also the parameters that define the AR process, namely the AR coefficients a_l and the prediction error variance σ_e^2.

To come to a tractable solution, the number of computations must be kept as low as possible. Therefore $o(i)$ is not computed for the complete frame. Instead, $o(i)$ is computed only for regions that contain missing data. The image regions are selected so that, at most, 20% of the area consists of missing data. Each region is modeled by a single set of AR model parameters and a single prediction error variance σ_e^2.

Proceeding in a probabilistic fashion, these requirements translate to finding the maximum of $P[o(i), a_1, \ldots, a_n, \sigma_e^2 | z_{mc}(i, t-1), z(i), z_{mc}(i, t+1), d(i), O]$. Here O indicates the direction of interpolation for the pixels in the local region surrounding $o(i)$. With Bayes' rule, this can be seen to be proportional to:

$$P[o(i), a, \sigma_e^2 | z_+(i), d(i), O] \propto P[z_+(i) | o(i), a, \sigma_e^2, d(i)] \, P[o(i) | O] \, P[a] \, P[\sigma_e^2] \tag{4.31}$$

where the terms $a_1, a_2, \ldots, a_n$ have been grouped together into a and $z_{mc}(i, t-1)$, $z(i)$ and $z_{mc}(i, t+1)$ have been grouped together in $z_+(i)$ for convenience.

The first term on the right hand side of (4.31), $P[z_+(i) | o(i), a, \sigma_e^2, d(i)]$, indicates the likelihood of observing the data $z_+(i)$, given the direction of interpolation, the AR model parameters, and the blotch mask. Let $AR(\hat{y}, a, i)$ be the prediction of the corrected image $\hat{y}$ at location i. $AR(\hat{y}, a, i)$ is determined completely by $z_+(i)$, $o(i)$, a, σ_e^2 and $d(i)$. The likelihood can then be defined by (4.32).

The second line in (4.32) states that at locations at which no blotches have been detected ($d(i) = 0$), the likelihood of observing a specific pixel intensity in the current frame $z(i)$ is proportional to the squared AR prediction error weighted by the prediction error variance. The third line in (4.32) states that, at locations where a pixel from $z_{mc}(i,t+1)$ was pasted, the likelihood of that pixel intensity being observed is proportional to the weighted squared AR prediction error of the restored frame. The fourth line of (4.32) makes a statement somewhat similar to that in the third line, but then for $z_{mc}(i,t-1)$.

$$P[z_+(i)\,|\,o(i),a,\sigma_e^2,d(i)] \propto \exp(-[(1-d(i))\frac{(z(i)-AR(\hat{y},a,i))^2}{2\sigma_e^2} +$$

$$d(i)\,o(i)\frac{(z_{mc}(i,t+1)-AR(\hat{y},a,i))^2}{2\sigma_e^2} + \qquad (4.32)$$

$$d(i)\,(1-o(i))\frac{(z_{mc}(i,t-1)-AR(\hat{y},a,i))^2}{2\sigma_e^2}])$$

Equation (4.32) can be simplified to:

$$P[z_+(i)\,|\,o(i),a,\sigma_e^2,d(i)]$$

$$\propto \exp\left(-\frac{(1-d(i))\,z(i)+d(i)\,(o(i)\,z_{mc}(i,t+1)+(1-o(i))\,z_{mc}(i,t-1))-AR(\hat{y},a,i))^2}{2\sigma_e^2}\right)$$

$$\propto \exp\left(-\frac{(\hat{y}(i)-AR(\hat{y},a,i))^2}{2\sigma_e^2}\right) \qquad (4.33)$$

$$\propto \frac{1}{\sqrt{2\pi\sigma_e^2}}\exp\left(-\frac{e(i)^2}{2\sigma_e^2}\right)$$

This means that likelihood function of the observed data $P[z_+(i)\,|\,...]$ is proportional to probability of the prediction error $e(i)$ of the restored frame as defined by (4.30).

The other three terms in (4.31) describe a priori knowledge related to the model parameters. To achieve local consistency in the direction mask $o(i)$, the following prior is assumed:

$$P[o(i)\,|\,O] \propto \exp\left(-\sum_k \beta |o(i)-o(i+q_k)|\right) \qquad (4.34)$$

where β is a constant that defines the strength of the self-organization. The eight-connected neighbors of $o(i)$ are indicated by $o(i+q_k)$, with $k=1, ..., 8$. Equation (4.34) simply states that the direction of interpolation for a pixel is likely to be similar to that of the majority of its neighbors.

Following [Kok98], a uniform prior is assigned to a, and a Jeffreys' prior [Rua96] is assigned to the prediction error variance σ_e^2:

$$P[\sigma_e^2] \propto \frac{1}{\sigma_e^2} \tag{4.35}$$

Equation (4.31) is completely defined now. The next section describes the practical implementation for correcting blotched image sequences on the basis of maximizing (4.31) jointly for all $o(i)$ in a region with missing data.

4.5.2 Practical implementation of controlled pasting

The MAP estimate for (4.31) jointly for all $o(i)$ can be found with SA [Gem84]. SA as described here involves two elements. The first element is a global control parameter T called *temperature*, which is used to shape the probability functions in (4.31). The second element is a mechanism for drawing random samples from conditionals, called a *Gibbs sampler*. SA can be summarized by four steps:

1. Initialize temperature: $T = T_{begin}$,
2. Sample the unknowns with the Gibbs sampler,
3. Repeat step 2 until convergence is obtained,
4. Lower T according to a cooling schedule and go to 2 if $T > T_{final}$.

In [Gem84] it is proved that if T_{begin} is sufficiently large and that if a logarithmic cooling schedule is applied, the algorithm converges to the MAP solution. The most involved part of the SA scheme is the Gibbs sampler. The Gibbs sampler operates iteratively by drawing random samples for the unknowns in turn, which are derived in Appendix B:

$$\begin{aligned} a &\sim P[a|\sigma_e^2, o(i), z_+, d] \\ \sigma_e^2 &\sim P[\sigma_e^2|a, o(i), z_+, d] \\ o(i) &\sim P[o(i)|a, \sigma_e^2, z_+(i), d(i), O] \end{aligned} \tag{4.36}$$

One might argue that using such heavy machinery as SA just to determine the direction of interpolation for a set of pixels is slightly overdoing things. The goal of this section is to simplify this machinery somewhat and to come to an efficient implementation.

The number of computations has to be kept small for an efficient implementation. As mentioned in the previous section, the controlled pasting scheme is not applied to the complete image, but only to image regions containing missing data. The image regions are selected so that, at most, 20% of the area consists of missing data. A single set of three AR model parameters a_l is computed for each region. A quarter plane prediction model is used (see Figure 4.17).

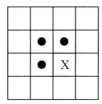

Figure 4.17: *Support (dots) used for AR prediction (cross).*

Strictly speaking, all unknowns should be sampled in the SA scheme, and this includes the sampling the AR coefficients a and the error variance σ_e^2 from the probability functions derived in Appendix B. Drawing samples from these distributions is costly in terms of computational complexity, and it is noted here that good results are obtained by just using the least squares estimate for the AR coefficients instead of sampling them. (In fact, this is equivalent to sampling from (B.10) with zero variance). Similarly, it is not necessary to sample for σ_e^2 to get good results. Hence:

$$a = R_{\hat{y}\hat{y}}^{-1} r_{\hat{y}\hat{y}} \qquad (4.37)$$

Here $R_{\hat{y}\hat{y}}$ and $r_{\hat{y}\hat{y}}$ are the autocorrelation matrix and autocorrelation vector that are required for solving the normal equations [Lag94], [The92]. What remains are the samples to be drawn for $o(i)$ from (B.16):

$$\begin{aligned}
&P[o(i)\,|\,a,\sigma_e^2,z_+(i),d(i),O] \\
&\propto \exp(-\frac{1}{T}[(1-d(i))\,(z(i)-AR(\hat{y},a,i))^2 + \\
&\qquad d(i)\,(o(i)\,z_{mc}(i,t+1)+(1-o(i))\,z_{mc}(i,t+1)-AR(\hat{y},a,i))^2 + \\
&\qquad \sum_k \beta|o(i)-o(i+q_k)\,|])
\end{aligned} \qquad (4.38)$$

where $AR(\hat{y},a,i)$ indicates the spatial AR prediction of $\hat{y}(i)$ from its surroundings. The reconstructed image $\hat{y}$, required for the AR predictions is obtained via (4.29). Drawing samples from (4.38) with the Gibbs sampler is very easy. It involves evaluating (4.38) at a specific site i for $o(i) = 0$ and for $o(i) = 1$, while keeping the other values for the direction mask and the $\hat{y}(i)$ fixed. The results are assigned to c_1 and c_2, respectively. Next a value for $o(i)$ (and thereby the corresponding $\hat{y}(i)$) is chosen at random, with a probability $c_1/(c_1+c_2)$ that $o(i) = 0$ and with a probability $c_2/(c_1+c_2)$ that $o(i) = 1$. A single update of an image region consists of applying the Gibbs sampler to each site in that region in turn, using, for instance, a checkerboard scanning pattern.

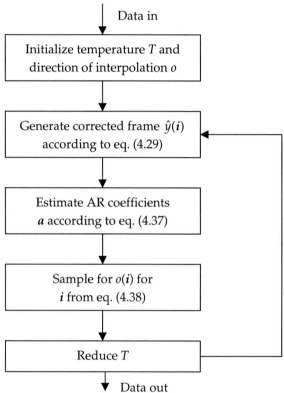

Figure 4.18: *Overview of a practical implementation of the CP scheme.*

Figure 4.18 summarizes the practical *controlled pasting* (CP) scheme that results. The data put into the system consist of the current frame and the motion-compensated previous and next frames. The blotch detection mask, which indicates for each pixel whether it is considered to be part of a blotch, also belongs to the input data. Initially, the direction field $o(i)$ is assigned binary values at random, and an initial temperature T is chosen. The main loop is as follows. First a corrected frame $\hat{y}(i)$ is generated. Next, a set of AR coefficients a is estimated for each missing region. This is used for predicting the corrected image intensities. Next, the direction of interpolation is updated by sampling from (4.38) as already described. The main loop is repeated at each temperature level T, until the solution has converged or until a fixed number of iterations have been done. The temperature is lowered with an exponential cooling schedule:

$$T_k = \gamma^k T_{begin} \tag{4.39}$$

where γ controls the rate of decrease and k indicates the k-th temperature level. The main loop is iterated again until the final temperature has been reached.

4.5.3 Experiments with controlled pasting

The scheme in Figure 4.18 is ready to be applied now. The result it yields is the joint distribution of the $o(i)$ within an image region S_r, as is given by (4.40). The term defined by the summation in (4.40) is known as the *potential function*. Lower potential functions indicate better solutions.

$$\begin{aligned} &P[o|a,\sigma_e^2,z_+(i),d(i)] \\ &\propto \exp(-\frac{1}{T}\sum_{i \in S_r} [(1-d(i))\,(z(i)-AR(\hat{y},a,i))^2 + \\ &\qquad d(i)\,(o(i)\,z_{mc}(i,t+1)+(1-o(r))\,z_{mc}(i,t+1)-AR(\hat{y},a,i))^2 + \\ &\qquad \sum_k \beta|o(i)-o(i+q_k)|]) \end{aligned} \tag{4.40}$$

To get some idea about what sensible values are for T_{begin}, T_{final}, and κ, two experiments are carried out on a blotched frame from the *Western* test sequence. For the first experiment $T_{begin} = 100.0$, $T_{final} = 1.0$ and $\gamma = 0.9$ is chosen. At each temperature level, 30 iterations are applied. For the second experiment, only one temperature level $T_{begin} = T_{final} = 1$ is assumed. Again, 30 iterations of the Gibbs sampler are applied. Figure 4.19 plots the potential functions for both experiments as a function of the number of iterations.

Figure 4.19 shows that both experiments converge. The solution found in the full SA scheme converged to a lower potential (final potential 483) than the solution found with the Gibbs sampler only (final potential 1307). The difference is, however, that the first experiment required about 625 iterations to reach its optimum, whereas the second experiment required only 25 iterations. Visually, the corrected results are not noticeably different. The conclusion is that it is not necessary to apply an elaborate cooling schedule and that sufficiently good results can be obtained in relatively few iterations. It must be emphasized that the result obtained by applying the Gibbs sampler only (without a cooling schedule) does not in general result in a MAP estimate. The reason why it is so successful here is probably because the distributions from which the samples are drawn are very compact; there is not a lot of ambiguity in drawing a sample.

The top row in Figure 4.20 shows three frames from the *Western* test sequence: (motion-compensated) previous, current, and next. The second row shows three corrections of the current frame, made with the *3DAR* and the *ML3Dex* methods, described in Section 4.2.3, and with the CP method described in the previous section. The results from the CP method were obtained by using just 30 iterations of the Gibbs sampler.

All the corrected frames show a great improvement over the corrupted frame. However, the *3DAR* and the *ML3Dex* methods fail where the motion-compensated frames are corrupted (see the highlighted boxes in the figures). These methods fail because they always incorporate data from both motion-compensated frames, regardless of the fact that some of those data may be corrupted. The *B3DAR* method, of which the results are not shown, also fails in this particular case because a block-based approach is used to determine the direction of interpolation, regardless on the validity of the data within the block. Figure 4.20g-i zooms in on the boxed regions. Clearly, the proposed CP method outperforms the other methods in terms of visual quality.

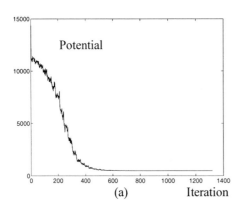
(a)
Iteration

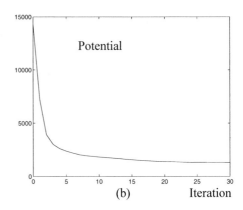
(b)
Iteration

Figure 4.19: *Potential function as function of iteration number: (a) for $T_{begin} = 100.0$, $T_{final} = 1.0$ and $\gamma = 0.9$, (b) for $T_{begin} = T_{final} = 1$.*

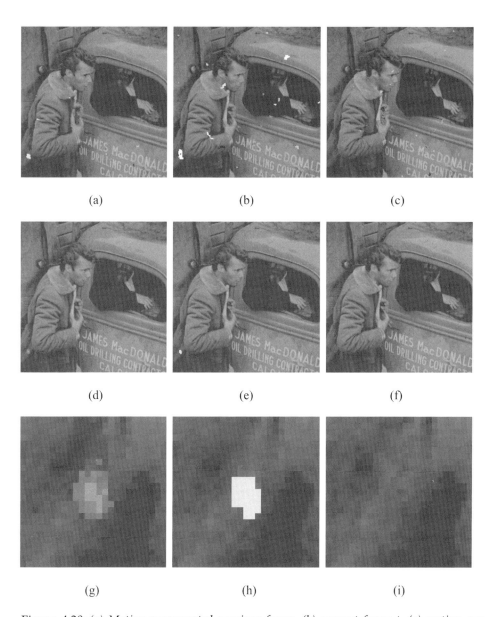

Figure 4.20: *(a) Motion-compensated previous frame, (b) current frame t, (c) motion-compensated next frame, (d), (e), (f) restored frame t by the 3DAR, ML3Dex, and CP schemes, respectively. Note the differences within the boxed regions. (g), (h), (i) Zoom-in to the boxed regions of panels (d), (e), and (f), respectively.*

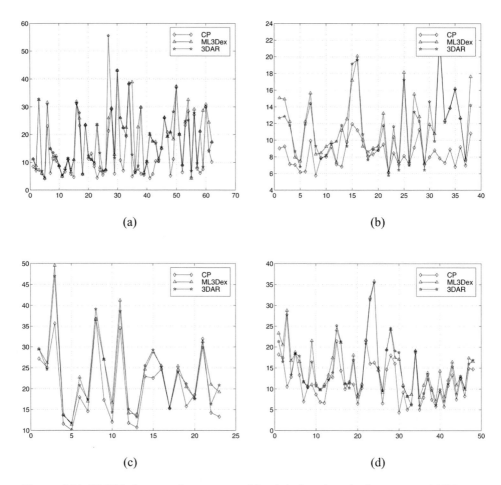

Figure 4.21: *RMSE of corrected sequences with original, unimpaired sequences: (a) Western, (b) MobCal, (c) Manege, and (d) Tunnel.*

Interpolator	Western (RMSE)	Mobcal (RMSE)	Manege (RMSE)	Tunnel (RMSE)
None	113.2	81.4	86.7	90.5
ML3Dex	20.8	12.6	25.2	16.7
3DAR	20.9	12.1	24.8	15.9
CP	16.1	8.5	22.1	12.4

Table 4.4: *RMSE computed between the corrected and original, unimpaired sequences.*

4.6 Results and conclusions

This section evaluates the complete chain of blotch detection, postprocessing, motion vector repair, and interpolation as depicted in Figure 4.8a. All experiments apply the SROD detector with postprocessing because this gives the highest ratio of correct detections to false alarms. The motion vector repair uses the block matching technique described in Section 4.2.2. Three interpolators are evaluated, namely, the ML3Dex, the 3DAR, and the CP method (using 30 iterations per frame).

Figure 4.21 shows the *root mean squared error* (RMSE), which is defined as the squared root out of the MSE, for the test sequences as a function of frame number. For each sequence the SROD detector with postprocessing was set to an overall correct detection rate of about 85%. The RMSE was computed only at locations at which the true blotch mask or the estimated blotch mask indicate corruptions (i.e., at locations where the original image data was altered by blotches or by interpolating false alarms). Figure 4.21 indicates that the CP interpolation method has the best performance. Whether the ML3Dex performs better than the 3DAR method is difficult to determine from this figure. Table 4.4 lists RMSE computed over all frames. It can be seen from this table that the interpolation considerably decreases the average errors. These data confirm that the CP method gives the best performance. Furthermore, it can be seen that the ML3Dex method, on average, performs slightly better than the AR method.

In terms of computational load, the CP method is to be preferred to the 3D AR method. The 3D AR method requires a matrix to be inverted, see (4.20), the size of which increases with increasing blotch size. Therefore, the number of computations for this method grows exponentially (order 3) [Pre92], [Str88] with increasing blotch size. There is also a risk that the system in (4.20) is singular and that no unique solution exists. In such cases, singular value decomposition [Pre92], [Str88] is useful. The ML3Dex interpolator is, computationally speaking, the most efficient: it is a non-iterative method that has to be evaluated only at the locations containing missing data, and it can be implemented efficiently with fast sorting algorithms [Pre92].

The methods for blotch detection and correction introduced in this chapter give significantly better results than those obtained by existing methods. However, as can be seen from the ROCs in Figure 4.13, the ratio of false alarms to correct detection remains relatively high for some sequences. There is room for further improvements. Nonetheless, even though too many false alarms are generated in some cases, the methods described in this chapter are very useful and can be applied efficiently in practical situations. Visually disturbing artifacts introduced into a corrected sequence due to false alarms can be removed by manual intervention. Removing regions of false alarms and undoing erroneous interpolations by single mouse clicks is much more efficient than having an operator mark and correct blotches in image sequences manually.

Chapter 5
Noise reduction by coring

5.1 Introduction

As computers with memories sufficiently large to store images and even short image sequences became widespread some 25 years ago, many researchers began to investigate digital algorithms for noise reduction. The well-known theories developed by Wiener and Kalman for optimal linear filtering were applied in the digital domain on a large scale. New types of nonlinear filters, such as order statistics filters and switching filters, were developed. Nowadays many very different approaches towards noise reduction are found in the literature [Abr96], [Arc91], [Bra95], [Don95], [Don94b], [Hir89], [Kle94], [Özk92], [Özk 93], [Roo96]. One such approach that has gained great popularity in recent years and that has proven to be very successful for denoising 2D images is *coring*. This chapter investigates this method for noise reduction and extends its application to image sequences.

Coring is a technique in which each frequency component of an observed signal is adjusted according to a certain characteristic, the so-called coring function. Originally coring was developed as a heuristic technique. It was first applied in 1951 for removing spurious oscillations in the luminance signal that were caused by a system designed to make television pictures more crisp [Gol51]. In 1968 it was recognized that this technique could also be used for removing imperfections such as noise from signals [McM68]. In the 1970s and the early 1980s coring was applied in the digital domain for noise reduction [Ade84], [Pow82], [Ros78]. The technique of *thresholding* or *coring* received a lot of attention after Donoho and Johnstone applied it successfully in the wavelet transform domain [Don95], [Don94b] in 1994.

Section 5.2 describes techniques for optimal filtering in a *minimum-mean-squared-error* (MMSE) sense. An example of such an optimal filter, the Wiener filter, is derived. The Wiener filter is a linear filter. If the constraint of linearity is dropped, more general nonlinear filters result. The filter characteristics of these nonlinear filters are represented by coring functions.

The domain in which coring is applied determines the effectiveness of coring for noise reduction. Section 5.3 describes two spatial signal transforms. One is a bi-orthogonal wavelet transform, and the other is a directionally sensitive subband decomposition. It is shown how to extend these 2D transforms to include the temporal dimension. The spatio-temporal decomposition provides a good basis for coring image sequences.

Noise-reduced signals are often stored or broadcast in a digital format. Section 5.4 investigates how noise can be reduced and compressed simultaneously within an MPEG2 encoder by coring the DCT coefficients. Section 5.5 contains some conclusions which are relevant to this chapter.

5.2 Noise reduction techniques

5.2.1 Optimal linear filtering in the MMSE sense

Any recorded signal is affected by noise, no matter how accurate the recording equipment. In this chapter noise is modeled by a additive white gaussian source. Let $y(i)$ be an original, unimpaired frame and let the noise be $\eta(i)$. The observed frame $z(i)$ is given by:

$$z(i) = y(i) + \eta(i) \tag{5.1}$$

A class of linear filters are the *finite impulse response* (FIR) filters, which are defined by:

$$\hat{y}(i) = \sum_{k=1}^{n} h_k z(i + q_k) \tag{5.2}$$

Here h_k, with $k = 1, \ldots, n$, are the n filter coefficients and the q_k define the support of the filter. The optimal filtering coefficients in MMSE sense can be found by:

$$\arg\min_{h_1,\ldots,h_k} E[(y(i) - \hat{y}(i))^2] \tag{5.3}$$

The filter that results is known as the Wiener filter. The Wiener filter can be implemented efficiently via the Fourier domain [Lag94], [The92]. Let Fourier transform of (5.1) be given by:

$$Z(w) = Y(w) + N(w) \tag{5.4}$$

The estimates $\hat{Y}(w)$ are given by:

$$\hat{Y}(w) = \frac{S_{yy}(w)}{S_{yy}(w) + S_{\eta\eta}(w)} Z(w) \tag{5.5}$$

Here $S_{yy}(w)$ and $S_{\eta\eta}(w)$ indicate the *power spectral density* (PSD) functions of the unimpaired signal and the noise. From (5.5) it can be seen that each frequency component of the observed data is weighted depending on the spectral power densities of the original, unimpaired signal and noise.

5.2.2 Optimal noise reduction by nonlinear filtering: coring

The Wiener filter imposes a FIR structure onto the solution of the MMSE problem. The optimal solution to the MMSE problem that is obtained when no constraints are placed on the filter structure is often a nonlinear function. Let $\hat{Y}(w)$ be a general function of the observed data $Z(w)$. The optimal estimate $\hat{Y}(w)$, given a single observation $Z(w)$, is found with the conditional expectation [Leo94]:

$$\begin{aligned} E[(Y(w) - \hat{Y}(w))^2] &= E[E[(Y(w) - \hat{Y}(w))^2 | Z(w)]] \\ &= \int_{-\infty}^{\infty} E[(Y(w) - \hat{Y}(w))^2 | Z(w)] \, P[Z(w)] \, dZ(w) \end{aligned} \tag{5.6}$$

The integrand in (5.6) is positive for all $Z(w)$; therefore, the integral is minimized by minimizing $E[(Y(w) - \hat{Y}(w))^2 | Z(w)]$ for each w. This minimum is given by:

$$\hat{Y}(w) = E[Y(w) | Z(w)] \tag{5.7}$$

The general solution given by (5.7) yields the smallest possible mean square error for estimating $\hat{Y}(w)$, given a single observation $Z(w)$. In general, the Wiener solution will have larger mean square errors. Further development of (5.7) gives:

$$\hat{Y}(w) = E[Y(w)|Z(w)] = \int_{-\infty}^{\infty} Y(w)\, P_{Y(w)|Z(w)}[Y(w)|Z(w)]\, dY(w) \qquad (5.8)$$

Here $P_{A|B}[A|B]$ indicates the pdf of A, given B. If the distributions of $Y(w)$ and $N(w)$ are known, then $P_{Y(w)|Z(w)}[Y(w)|Z(w)]$ can be determined via Bayes' rule:

$$\begin{aligned}
P_{Y(w)|Z(w)}[Y(w)|Z(w)] &= \frac{P_{Z(w)|Y(w)}[Z(w)|Y(w)]\, P_{Y(w)}[Y(w)]}{P_{Z(w)}[Z(w)]}\\
&= \frac{P_{N(w)}[Z(w)-Y(w)]\, P_{Y(w)}[Y(w)]}{\int_{-\infty}^{\infty} P_{N(w)}[Z(w)-Y(w)]\, P_{Y(w)}[Y(w)]\, dY(w)}
\end{aligned} \qquad (5.9)$$

In (5.7), (5.8), and (5.9), the interpretation given to w is that of frequency. Note that this frequency need not necessarily be obtained by applying a Fourier transform to a signal. Other transforms, such as the DCT, wavelet transforms, and subband transforms, may well be used.

Figure 5.1a shows a typical characteristic that results from (5.8). This characteristic is called a coring function. Sometimes this characteristic is also referred to as *Bayesian optimal coring* because of the relationship in (5.9) [Sim96]. In general, coring functions leave transform coefficients with high amplitudes unaltered, and the coefficients with low amplitudes are shrunk towards zero. Intuitively speaking, this is appealing. Coefficients with high amplitudes are reliable because they are influenced relatively little by noise. These coefficients should not be altered. Coefficients with low amplitudes carry relatively little information and are easily influenced by noise. Therefore, these coefficients are unreliable, and their contribution to the observed data should be reduced.

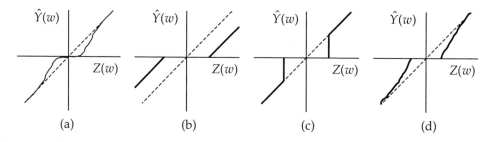

Figure 5.1: *Coring functions: (a) Bayesian optimal coring, (b) soft thresholding, (c) hard thresholding, (d) piecewise linear coring.*

5.2.3 Heuristic coring functions

Originally, coring was developed as a heuristic technique for removing noise. Three well-known heuristic coring functions are described here.

Soft thresholding. Soft thresholding is defined by [Don95], [Pow82]:

$$\hat{Y}(w) = \begin{cases} \text{sgn}(Z(w))(|Z(w)|-T) & \text{if } |Z(w)| > T \\ 0 & \text{otherwise} \end{cases} \quad (5.10)$$

where the $\text{sgn}(Z(w))$ gives the sign (or phase) of $Z(w)$. Figure 5.1b plots this coring function.

Natural signals tend to have weak high-frequency components. Therefore, soft thresholding nullifies the high-frequency transform coefficients obtained from a signal. The result is that, besides the noise being removed, the slopes of edges are reduced and their rise time increases. For images this is perceived as blurring of edges in images. Soft thresholding has another effect, namely, it reduces contrast because it shrinks the magnitudes of all AC transform coefficients indiscriminately.

Hard thresholding. Hard thresholding is defined by [Don95], [Pow82]:

$$\hat{Y}(w) = \begin{cases} Z(w) & \text{if } |Z(w)| > T \\ 0 & \text{otherwise} \end{cases} \quad (5.11)$$

Figure 5.1c plots this coring function. A disadvantage of hard thresholding is that it introduces spurious oscillations or so-called ringing patterns. These occur because hard thresholding not only removes noise energy at selected frequencies, but also signal energy. The removal of signal energy can be viewed as adding impulses to the original, unimpaired signal. The amplitudes of these impulses are equal to those of the original signal contents, but the signs are opposite. In the synthesis stage, where the signal is transformed back from the frequency domain to the spatial domain, the impulse responses of the synthesis filters are superimposed on the result. These superimposed filter responses are perceived as ringing.

Piecewise linear coring. A compromise between soft thresholding and hard thresholding is piecewise linear coring:

$$\hat{Y}(w) = \begin{cases} Z(w) & \text{if } |Z(w)| > T_1 \\ \dfrac{|Z(w)|-T_0}{T_1-T_0} T_1 \, \text{sgn}(Z(w)) & \text{if } T_0 \leq |Z(w)| \leq T_1 \\ 0 & \text{otherwise} \end{cases} \quad (5.12)$$

Figure 5.1d plots this function. Piecewise linear coring is intended to reduce the ringing artifacts resulting from hard thresholding on one hand and to preserve low-contrast picture detail, which is lost by soft thresholding, on the other hand [Pow82].

5.3 Coring image sequences in wavelet and subband domains

The frequency domain implementation of the Wiener filter as described in Section 5.2.1 can be viewed as an implementation of coring: each observed frequency component is adjusted according to a characteristic that is determined by the PSDs of signal and noise. However, the use of the Fourier transform as a decorrelating transform has the disadvantage of forfeiting knowledge of the spatial locations of dominant signal components. This implies that the cored signal is not adapted to local statistics, but depends on global statistics only. Clearly, this is suboptimal because local statistics can be very different from the global statistics.

The objective of transforming data prior to coring is to separate the signal from the noise as well as possible. To get optimal separation of the signal and the noise, it is advantageous to use transforms that compact the signal energy as much as possible [Don94b], [Nat95]. Unlike the Fourier transform, scale-space representations [Bur83], [Mal89], [Wan95] allow local signal characteristics at different scales to be taken into account. In the case of noise-reducing image sequences, adaptation to local statistics is advantageous due to the nonstationary, scale-dependent nature of natural images.

This section describes two 2D scale-space decompositions. The first is a nondecimated wavelet transform known as the *algorithm à trous* [Hol89], [Vet95]. The second is a subband decomposition based on directionally sensitive filters that is known as the Simoncelli pyramid [Sim92]. Next, it is shown how these decompositions can be extended to three dimensions by adding a temporal decomposition step. The 3D decompositions provide good separation of the signal and the noise. Which of the two scale-space-time decompositions is most suited for noise reduction by coring is investigated.

5.3.1 Nondecimated wavelet transform

The *discrete wavelet transform* (DWT) is a popular tool for obtaining scale space representations of data. A popular implementation of the DWT is the decimated DWT in which the transformed data have the same number of coefficients as the input data. A problem with this transform, however, is that shifting of the input image spatially, may lead to entirely different distributions of the signal energy over the transform coefficients [Sim92], [Vet95]. This is caused by the critical subsampling applied in decimated wavelet transforms. Therefore, shifting the input image can lead to significantly different filtered results. This is undesirable because it can lead to temporal artifacts when in the processing of image sequences.

Shift invariance is obtained by nondecimated DWTs. An algorithm that generates nondecimated DWTs is the algorithm à trous ("algorithm with holes") [Hol89]. Because no subsampling is applied in this scheme, the decomposition is significantly overcomplete. For example, a three-level decomposition of an image with N pixels gives 10N transform coefficients.

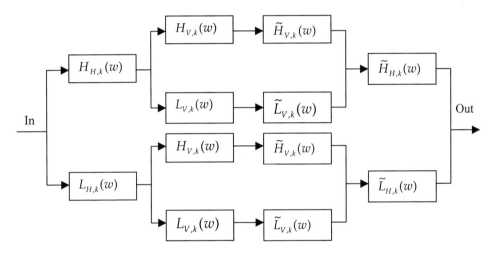

Figure 5.2: *Overview of the algorithm à trous: a 2D wavelet analysis/synthesis scheme. The total decomposition is obtained by inserting the complete filter bank into the white spot near the bottom of the figure recursively. At each recursion level, index is incremented.*

Figure 5.2 gives a schematic overview of the algorithm à trous. First the input image is filtered twice in horizontal direction; once with the high-pass analysis filter and once with the low-pass analysis filter. Next, the filtered data are filtered again with the same high-pass and low-pass analysis filters, but now in a vertical direction. The data that result from low-pass analysis in both the horizontal and vertical directions are decomposed again with the same analysis filter banks. However, this time the analysis filters are dilated by inserting 2^{k-1} zeros between each of the filter coefficients at recursion level k, with $k = 1, 2, \ldots$ for the recursion levels. Initially, the algorithm starts out with $k = 0$, and no zeros are inserted between the filter coefficients. For the synthesis part of the filter bank, again 2^{k-1} zeros are inserted between each of the coefficients of the high-pass and low-pass synthesis filters at each recursion level k.

The algorithm à trous uses bi-orthogonal wavelet pairs. This means that synthesis filters used in the reconstruction phase are not identical to the analysis filters. Table 5.1 gives the filter coefficients for the analysis and synthesis filters. These are symmetric FIR filters, therefore they are linear phase filters. This is a useful property in image processing because nonlinear phase filters degrade edges [Ant94].

Figure 5.3 gives the transform coefficients of the 2D algorithm-a-trous image decomposition of a test image. One half of this image consists of a frequency sweep, the other half shows half a disc that is partially contaminated by additive white gaussian noise. Figure 5.3 shows a number of things. First of all, the signal energy is concentrated in different "frequency" bands, depending on the orientation and the frequency of the local signal components. Furthermore, the spatial location of signal components is preserved; the spatial location of various signal components are clearly visible in Figure 5.3. This is in contrast to the Fourier transform, which indicates the presence of specific frequencies within a signal, but their localization is not known. Finally, the noise energy is spread out over all frequency bands and orientations.

The filter banks used by the algorithm à trous are quite short and they are therefore not ideal in terms of cut-off frequency and signal suppression in the stop bands. The result is *spectral leakage*. Figure 5.3 shows that energy from high-frequency signal components are visible in low-pass subbands and vice versa.

5.3.2 Simoncelli pyramid

The Simoncelli pyramid is a subband decomposition scheme based on directionally sensitive filters [Sim92]. This means that the distribution of signal energy over frequency bands depends on the orientation of structures within the image. Shift invariance is accomplished by avoiding aliasing effects by ensuring that no components with frequencies larger than $\pi/2$ are present before 2:1 subsampling. The Simoncelli decomposition is significantly overcomplete; the number of transform coefficients is much larger than the number of pixels in the original image. For example, a four-level pyramid decomposition with four orientations (four times four sets of high-pass coefficients and one set of low-pass coefficients) of an image with N pixels gives about $9.3N$ coefficients.

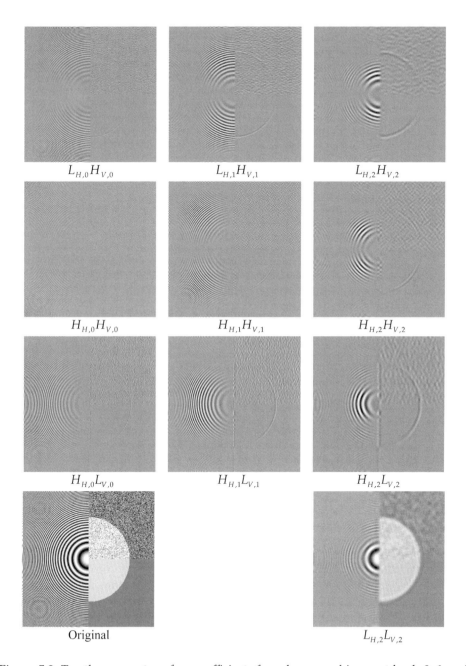

Figure 5.3: *Top three rows: transform coefficients from decomposed image at levels 0, 1, and 2. Bottom right: low-pass residual. Bottom left: original input image. To improve visibility, the contrast has been stretched for all images.*

Low-Pass Analysis	-1/8	2/8	6/8	2/8	-1/8
Low-Pass Synthesis		1/4	1/2	1/4	
High-Pass Analysis		1/8	-2/8	1/8	
High-Pass Synthesis	1/4	2/4	-6/4	2/4	1/4

Table 5.1: *Coefficients for the bi-orthogonal wavelet pairs used by the algorithm à trous.*

Figure 5.4 shows the 2D Simoncelli pyramid (de)composition scheme. The filters $L_k(w)$, $H_k(w)$ and $F_m(w)$ are the 2D low-pass, high-pass, and directional (fan) filters, respectively. The filters $L_0(w)$, $H_0(w)$, $L_1(w)$ and $H_1(w)$ are self-inverting, linear-phase filters. Self inverting-filters have the pleasant property that the analysis and the corresponding synthesis filters are identical.

The following constraints apply to $L_0(w)$, $H_0(w)$, $L_1(w)$ and $H_1(w)$: the aliasing in the low-frequency (subsampled) bands is minimized (5.13), all radial bands have a bandwidth of one octave (5.14), and the overall system has unity response, requiring that low and high-pass filters are power complementary (5.15):

$$L_1(w) \to 0 \quad \text{for} \quad w > \frac{\pi}{2} \tag{5.13}$$

$$L_0(w) = L_1(2w) \tag{5.14}$$

$$|L_i(w)|^2 + |H_i(w)|^2 = 1 \tag{5.15}$$

The 2D filters can be obtained from 1D linear phase FIR filters by means of the McCLellan transform [McC73]. Equation (5.14) can be used to obtain the 2D filter $L_0(w)$ from $L_1(w)$. A conjugate gradient algorithm was used to find the filters $H_0(w)$ and $H_1(w)$ under the constraints set by (5.15) [Pre92].

For practical purposes, the high-pass filters $H_0(w)$ and $H_1(w)$ are directly combined with the fan filters $F_1(w)$, $F_2(w)$, $F_3(w)$ and $F_4(w)$. Taking the 2D Fourier transforms of H_0 and H_1, multiplying the transform coefficients with $f(\theta - \theta_m)$ in (5.16), and taking the inverse Fourier transforms gives the required combination. In (5.16), θ_m is the center of the orientation of the filter.

NOISE REDUCTION BY CORING

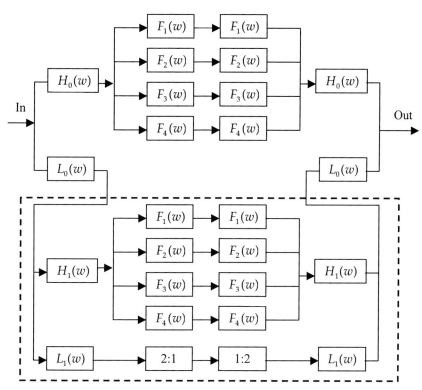

Figure 5.4: *The Simoncelli analysis/synthesis filter bank. The total decomposition is obtained by recursively inserting the contents of the dashed box into the white spot near the bottom of the figure.*

$$f(\theta - \theta_m) = \begin{cases} 1 & \forall |\theta - \theta_m| < \dfrac{\pi}{16} \\ \cos(4|\theta - \theta_m|) & \forall \dfrac{\pi}{16} \le |\theta - \theta_m| \le \dfrac{3\pi}{16} \\ 0 & \text{otherwise} \end{cases} \qquad (5.16)$$

The first filtering stage with filters $L_0(w)$ and $H_0(w)$ is omitted for the experiments in this chapter to reduce the number of computations. This also reduces the number of transform coefficients by $4N$, where N is the number of pixels in a frame. Figure 5.5 shows an example of a decomposition using a pyramid with three levels and the same test image as in the previous section. The 2D filter used banks consisting of 21x21 taps.

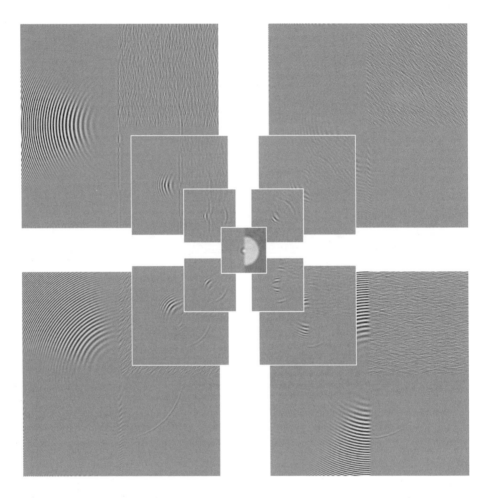

Figure 5.5: *Top: pyramid decomposition of the input image showing the output of the directionally sensitive fan filters and the residual low-pass image. The contrast has been stretched to improve visibility. Note that the local signal energy is concentrated in one or two orientations, whereas the noise energy is spread out over all orientations. Bottom: test image that was also used in Figure 5.3.*

The 2D pyramid decomposition with four orientations has a number of advantages over a 2D nondecimated DWT. First, the local separation between signal and noise is better for the pyramid decomposition than for the DWT. At each level of the pyramid decomposition, the noise energy is distributed over four frequency bands, and the energy of the image structures, such as straight lines, is distributed over one or two frequency bands. In contrast, at each decomposition level of the DWT, the energy of the image structures is distributed over two or three frequency bands, and the noisy energy is also distributed over three frequency bands. Exceptions are horizontal and vertical image structures; their energy is concentrated in one frequency band only. Improved separation between signal and noise means removing more noise and distorting the signal less.

The second advantage of the 2D pyramid decomposition is that the three-level pyramid decomposition gives $5.3N$ coefficients; this is less overcomplete than a shift invariant nondecimated DWT that gives $10N$ coefficients for the same number of levels.

Finally, for the particular implementation of the Simoncelli pyramid in this chapter, there is much less leakage than for the algorithm à trous. This is a result of the constraint set by (5.13) in combination with the relatively large filter banks.

5.3.3 An extension to three dimensions using wavelets

The 2D decorrelating transforms described in the previous sections spatially separate signals from noise. It will be made apparent that this separation can be improved by including motion-compensated temporal information. If the signals are stationary in a temporal direction, the motion-compensated frames from $t-n, ..., t+m$ should all be identical to frame t, except for the noise terms. The (linear) pyramid decompositions of these images should also be identical, again except for the noise terms. This means that a set of transform coefficients at scale-space locations corresponding in a temporal sense should consist of a 1D DC signal plus noise. This signal can be separated into low-pass and a high-pass terms, e.g., by the DWT.

Note that, ideally speaking, one would use long analysis filters to obtain good separation of the signal and noise components in the temporal decomposition step. However, inaccuracies of the motion estimator and the fact that areas become occluded or uncovered form a limiting factor to the length of temporal filter used.

The steps to a 3D spatial-temporal decomposition/reconstruction scheme are summarized by Figure 5.6. Large reductions in computational effort can be obtained for steps 4a and 5a by realizing that, for the purposes of this chapter, it is only necessary to reconstruct the current frame t. Reconstructions of decomposed motion-estimated frames are not of interest here, and therefore they need not be computed.

Analysis:
- 1a. Calculate the motion compensated frames from frames at $t-n, \ldots, t+m$
- 2a. Calculate a 2D decomposition for each (motion compensated) frame.
- 3a. Apply the DWT in the temporal direction to each set of coefficients at t corresponding scale-space locations.

Synthesis:
- 4a. Apply the inverse DWT in the temporal direction to each set of wavelet coefficients to reconstruct the coefficients of the spatially decomposed frame at t.
- 5a. Apply the synthesis stage of the 2D filter bank.

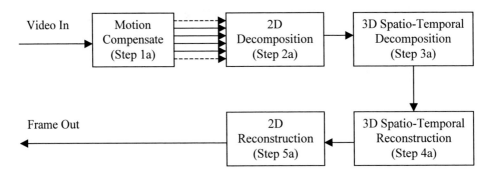

Figure 5.6: *Summary of 3D signal decomposition scheme.*

5.3.4 Noise reduction by coring

The structure of the proposed decomposition/reconstruction algorithm offers several possibilities for coring transform coefficients by inserting the steps summarized in Figure 5.7. This figure represents a framework for 3D scale-space noise reduction by coring.

The generally nonlinear nature of coring makes it difficult to determine the combination of coring characteristics for steps 2b, 3b, and 4b required for optimal noise reduction. Another question is whether optimal coring requires coring in all steps 2b, 3b, and 4b. To exploit the temporal decomposition, coring is certainly necessary in step 3b. However, this alone cannot be optimal as is explained in a moment. The conclusion is that spatial noise reduction is required as well.

Coring the spatio-temporal transform coefficients only (step 3b) is suboptimal because this coring operation can be viewed as a *switching filter* [Kle94] that turns itself on and off automatically, depending on the accuracy of the data. Suppose coring is applied to the spatio-temporally decomposed signal, and suppose there is an error in the motion estimation process that results in large frame differences. In such a case, all the coefficients resulting from the spatio-temporal decomposition have high amplitudes. The spatio-temporal coring function tends to keep high amplitudes intact and will not remove a lot of noise in such circumstances; the filter is effectively switched off locally.

Coring:
- 2b. Core the spatial transform coefficients (except those in the DC band) for all frames.
- 3b. Core the high-pass spatio-temporal transform coefficients.
- 4b. Core the spatial transform coefficients (except those in the DC band) of the current frame.

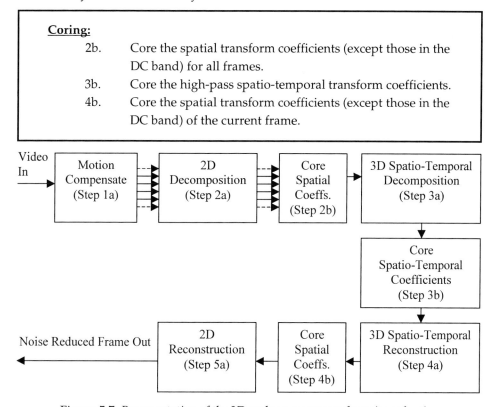

Figure 5.7: *Representation of the 3D scale-space system for noise reduction.*

Step 2b applies optimal coring functions that are computed by (5.8) for each subband of a 2D decomposition. This requires estimating or assuming distributions for the signal and noise coefficients in each subband. In step 3b, hard thresholding is used for coring the spatio-temporal coefficients because it fits in nicely with the switching filter idea. If a spatio-temporally decomposed coefficient is small, it is likely to be noise and it should be removed completely. If the coefficient is large, it is likely that the data were not stationary in a temporal sense and the coefficient should not be altered.

The optimal coring functions in step 4b are much harder to determine because they depend on the spatial coring applied in step 2b and on the spatio-temporal coring applied in step 3b. The effect of the latter is particularly difficult to model due to its dependency on the quality of motion-compensated images. Therefore, rather arbitrarily, soft thresholding is applied in step 4b. Note that soft thresholding is preferred over hard thresholding because the latter tends to give disturbing ringing patterns as discussed earlier.

Threshold selection. A good value for the hard thresholding in the spatio-temporal threshold is $T_{st} = 3\sigma_{hp}$, where σ_{hp}^2 is the estimated variance of the noise in the high-pass coefficients. The motivation for this is the following. If the noise corrupting the image sequence is assumed to be additive, white, and gaussian, and if the motion compensation is perfect, then the high-pass coefficients contain noise energy only. In fact, the high-pass coefficients follow a zero-mean gaussian distribution. Setting all observed coefficients that lie within $\pm 3\sigma_{hp}$ to zero effectively means that noise is removed from 99.7% of the high-pass coefficients. It is assumed that the variance of the high-pass coefficients is much greater than that of the noise, if the motion compensation is not perfect. Therefore, if the temporal intensity differences are large due to imperfect motion compensation, the signal will hardly be affected by the temporal coring.

The threshold T_s for soft thresholding the spatial decomposition coefficients in step 4b is chosen so that the PSNR of the corrected sequence is maximal. In practical situations, T_s cannot be chosen to give the maximum PSNR due to the absence of an unimpaired original to serve as a reference. In this case, the value for T_s that gives the best visual quality of the noise reduced sequence is selected.

Note that no threshold selection is required for the coring of spatial decomposition coefficients in step 2b because the coring functions are completely determined by the signal and noise distributions in each subband.

5.3.5 Perfect reconstruction?

One of the characteristics of wavelets is that they allow for perfect reconstruction. Hence, the algorithm à trous gives perfect reconstruction. Unfortunately, this is not the case for the Simoncelli pyramid. The Simoncelli pyramid is a linear phase function, self-inverting and power complementary in the ideal case. In practice, self-inverting linear-phase FIR filters with more than two taps cannot possess both the power-complementary property and the *perfect reconstruction* (PR) property [Vai92]. If the power-complementary property is retained, the absence of PR is reflected by ringing near sharp edges in reconstructed images. The errors introduced due to the lack of PR is represented by the difference between the original and reconstructed images.

The following investigates how the effects of lack of PR can be minimized. Let Z and $\hat{Y}$ denote a decomposed noisy image before and after coring, respectively. If the (linear) reconstruction operator is denoted by $R[.]$, the noise reduced image $\hat{y}$ is given by:

$$\hat{y} = R[\hat{Y}] = R[Z + \hat{Y} - Z] = R[Z - \hat{N}(Z)] = R[Z] - R[\hat{N}(Z)] \tag{5.17}$$

where $\hat{N}(Z) = Z - \hat{Y}$ can be regarded as an estimate of the noise realization that corrupts the original data. Ideally speaking, $R[Z]$ equals z, therefore:

$$\hat{y} = z - R[\hat{N}(Z)] \tag{5.18}$$

This result shows that reconstructing an image of the noise realization and subtracting it from the noisy input image reduces the effects of lack of PR. This is done instead of directly reconstructing the noise-reduced image from the cored transform coefficients. Hence, the problem of lack of PR for the original image is shifted to lack of PR for the noise realization. This approach, however, introduces no artifacts, such as ringing, that are associated with image structures if the noise is independent of the image contents. Furthermore, because the noise has a lower variance than that of the image contents, the effects of lack of PR for the reconstructed noise signal are much less (or not) visible.

5.3.6 Experiments and results

This section evaluates the noise-reduction capabilities of the wavelet and pyramid noise-reduction schemes described in Section 5.3.4. In both cases, the 2D decompositions are extended to three dimensions by the same bi-orthogonal wavelet used by the algorithm à trous (Table 5.1). To get some indication of the gains achieved by 3D filtering over 2D filtering, the test sequences are processed twice by each filter: once with coring of the spatio-temporal decomposed coefficients (step 3b) and once without. To reduce the computational complexity, no coring is applied to the spatially decomposed motion-compensated frames, i.e., step 2b is omitted.

Sequence	Noisy Sequence (dB)	à trous Spatial Coring (dB)	à trous Temporal+Spatial Coring (dB)	Pyramid Spatial Coring (dB)	Pyramid Temporal+Spatial Coring (dB)
Plane 25	33.0	3.6	4.1	3.8	4.8
Plane 100	27.0	4.8	6.1	5.7	6.7
Plane 225	23.5	6.0	8.3	7.0	8.5
MobCal 25	33.0	1.8	2.6	1.6	2.9
MobCal 100	27.0	3.4	4.7	3.5	4.9
MobCal 225	23.5	4.6	5.9	4.9	6.1

Table 5.2: *PSNR of test sequences and increase in PSNR of noise reduced sequences using the Pyramid and Wavelet decomposition schemes with and without coring of spatio-temporal subband coefficients.*

Two test sequences are evaluated in this section. The first sequence is called *Plane* and shows a plane flying over a landscape. It contains fine detail, sharp edges, uniform regions, and a lot of motion. The sequence was originally recorded with a high-definition camera, and the images are very crisp. There are strong interlacing effects due to motion. The second test sequence is the well-known *MobCal* sequence, which does not display noticeable interlacing effects. Ideally, to avoid the effects of interlacing, one would apply motion-compensated de-interlacing [Del94]. The noise-reduction filters would be applied to the de-interlaced frames. However, motion-compensated de-interlacing adds a lot of complexity to the noise-reduction system. Therefore, the *Plane* sequence is processed on a field-by-field basis instead of on a frame-by-frame basis.

White gaussian noise, with variances 25, 100, and 225, has been added to the test sequences. Figure 5.8 shows an example of a noisy field from the *Plane* sequence and the filtered result obtained by 3D pyramid with both spatio-temporal and spatial coring. Table 5.2 lists the PSNRs of the test sequences and the increase in

PSNR for the filtered results.

Considerable amounts of noise reduction are achieved by the filters. The best results are obtained by coring both the spatio-temporal coefficients and the spatial coefficients (step 3b and step 4b in Section 5.3.4), which gives an improvement ranging from 0.5 to 2.3 decibels over spatial filtering only. The magnitude of the improvements depend on the sequence, the amount of noise, and the spatio-temporal decomposition used. The performance of the pyramid filter is similar or better than that of the shift invariant wavelet filter in terms of PSNR in all cases. Visually speaking, the results given by the pyramid filter are better than those of the wavelet filter; the results are a bit sharper, and artifacts that result from filtering in the form of "low-frequency spatial patterns" are less visible.

Figure 5.8: *Top: noisy field from Plane sequence with noise variance 225 (PSNR = 23.5 dB). Bottom: filtered result from the 3D pyramid filter. (Sequence available by courtesy of the BBC).*

5.4 MPEG2 for noise reduction

Consider a broadcasting environment in which noisy film and video sequences are digitally broadcast with an MPEG2 encoding system, as illustrated by Figure 5.9. It is assumed that no channel errors are introduced. MPEG2 encoding systems try to minimize the coding errors between input $z(i)$ and output $\hat{z}(i)$. However, in the case of noisy image sequences, what they should be doing is minimizing the errors between the original, noise-free image $y(i)$ and the output $\hat{z}(i)$. When doing so, the MPEG2 encoding systems can be considered devices for simultaneous noise reduction and image compression.

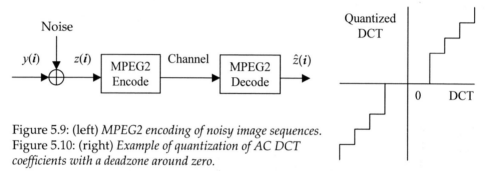

Figure 5.9: (left) *MPEG2 encoding of noisy image sequences.*
Figure 5.10: (right) *Example of quantization of AC DCT coefficients with a deadzone around zero.*

Let $\varepsilon(i)$ denote the error between $y(i)$ and $\hat{z}(i)$. The aim of this section is to adjust an MPEG2 encoding system to minimize the error variance. The error variance can be expressed in terms of DCT coefficients:

$$E[\varepsilon^2(i)] = E[(y(i) - \hat{z}(i))^2] = \sum_{k=1}^{64} E[(Y_k(i') - \hat{Z}_k(i'))^2] \qquad (5.19)$$

Here $Y_k(i')$ and $\hat{Z}_k(i')$, with $k = 1, \ldots, 64$, represent the 64 DCT coefficients of each 8x8 data block within a frame. The column, row, and frame number of a data block is indicated by i'.

Two basic approaches can be followed to minimize (5.19). In theory, these approaches give the same results. The first approach directly minimizes $E[(y(i) - \hat{z}(i))^2]$. As is shown in Appendix C, this approach is equivalent to determining optimal quantizers for the DCT coefficients of a noisy signal. The second approach is based on the fact that the problem of minimizing the overall error variance can be split into two parts for a communication system in which a signal is distorted prior to (lossy) channel encoding [Wol70]. The first part consists of computing the conditional expectation for the true signal given the observed noisy data. The second part consists of designing an encoder that is optimal for the original, noise-free signal.

In the particular case of Figure 5.9, the advantage of the second app roach is that, in principle, the encoder is already optimized for encoding noise-free signals. Therefore, it is not necessary to design new quantization tables as is required for the first approach. All that needs to be done for the second approach is to core the DCT coefficients following (5.8) prior to quantization, i.e., by replacing the observed DCT coefficients with the conditional expectation for the true DCT coefficients. This second approach is investigated further in this section.

In fact, MPEG2 encoders implicitly core noisy DCT coefficients to some extent by incorporating a so-called *dead zone* in the quantizers for the coefficients of the non-intra-coded frames (Figure 5.10) [Mit96]. As a result of the dead zone, DCT coefficients with small magnitudes are mapped to zero. However, note that the use of dead zones is suboptimal for noise reduction because they are not applied to all frames and because they do not address the noise on DCT coefficients with larger amplitudes.

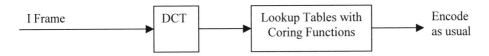

Figure 5.11: *Coring of the DCT coefficients of I frames in an MPEG2 encoder.*

5.4.1 Coring I, P, and B frames

I-frames. The MPEG2 system defines three frame types; namely, I frames and predicted P and B frames. The I frames are encoded by dividing the frames in 8x8 blocks, applying the DCT to the blocks and quantizing the DCT coefficients (Chapter 2). Two basic approaches can be followed towards coring the DCT coefficients of I frames. The first is to estimate the pdf for each DCT coefficient from the observed data for each frame, compute the conditional expectation for each coefficient according to (5.8), and replace the observed coefficients by these values. Computing optimal coring functions for each I frame of an image sequence is expensive in terms of computational complexity, and therefore it is expensive to implement in real-time hardware.

The second approach does not optimize the coring functions for each frame. Instead, fixed sets of coring functions are computed off-line and stored in the encoder as lookup tables (Figure 5.11). The coring functions are computed from a large set of images, so that on average the encoder gives the best results that can possibly be achieved under the condition of static lookup tables. This approach can be implemented in an MPEG2 encoder easily. Section 5.4.2 gives the details of this second approach.

B and P frames. The B and P frames are predicted from frames coded previously. The frame differences between the predicted and current frames are encoded like I frames, i.e., by using DCTs and quantization. Finding the ideal coring coefficients is more difficult now because the signal and noise distributions of the frame differences are not known. These depend on the nonlinear coring and quantization of the frames coded earlier and on the quality of the motion estimation and compensation.

Instead of coring the DCTs of the frame differences, as illustrated in Figure 5.12a, an alternative strategy is preferred in which the DCT and coring operation are performed prior to subtracting the current and predicted frames from each other. Figure 5.12b illustrates this alternative strategy. Note that the coring functions in Figure 5.12a,b are different from each other and that also the results given by the two approaches generally speaking are not identical.

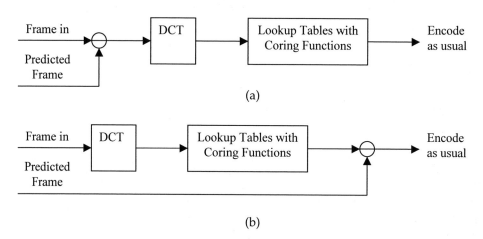

Figure 5.12: *(a) Coring function applied to DCT of frame differences, (b) illustration of how, by sliding the DCT and the coring function in front of the subtraction, B and P frames can be cored as the I frames are. Note that the predicted frame is extracted from a coded frame that has already been noise reduced and need not be cored again.*

NOISE REDUCTION BY CORING

Two points about the scheme in Figure 5.12b are noteworthy. First, the predicted frames have already been coded and hence they have already been noise reduced earlier on. Therefore it is not necessary to core the predicted frames again. Second, the optimal coring characteristics are identical to those computed before for the I frames. This means that only one set of lookup tables is required for the I, P and B frames.

5.4.2 Determining the DCT coring functions

This sections deals with computing the coring functions for the I, P, and B frames. As indicated in the previous section, the coring functions are computed from a large set of images, so that the encoder gives the best results that can be achieved on average with static lookup tables. Computing the coring functions consists of two steps. First, the distributions of the signal and the noise have to be determined. Next, the coring functions can be computed from (5.8) and (5.9).

The noise corrupting the image sequences is assumed to be additive, white, and gaussian with known variance. The distributions of each of the 63 AC DCT coefficients are sometimes modeled by laplacian distributions [Jos95], [Rei83]. In practice, the generalized gaussian is more accurate [Bar98], [Sha95]. The DC coefficients are not cored because their conditional expectation depends too much on the specific sequence. The generalized gaussian distribution is given by:

$$P(x) = a \exp(-b|x|^c) \qquad (5.20)$$

with:

$$a = \frac{bc}{2\Gamma(1/c)} \quad \text{and} \quad b = \frac{1}{\sigma}\sqrt{\frac{\Gamma(3/c)}{\Gamma(1/c)}} \qquad (5.21)$$

where $\Gamma(.)$ is the gamma function and σ is the standard deviation of the distribution. It can be seen from (5.20) and (5.21) that the generalized gaussian is completely determined by the shape parameter c and the noise variance σ^2. The well-known gaussian distribution is obtained by letting $c = 2$; the laplacian distribution is obtained by letting $c = 1$.

An efficient method for estimating the shape parameter c from a set of data based on second-order statistics is given in [Sha95]. Let Y_k denote DCT coefficients with coefficient number $k = 1, \ldots, 64$. The mean and the variance μ_k and σ_k^2 of a set of observed DCT coefficients with coefficient number k can be estimated directly from the observed data. Let ρ_k be:

$$\rho_k = \frac{\sigma_k^2}{E^2[|Y_k - \mu_k|]} \tag{5.22}$$

The shape parameter c_k for the distribution of DCT coefficient k is found by solving:

$$\frac{\Gamma(1/c_k)\Gamma(3/c_k)}{\Gamma^2(2/c_k)} = \rho_k \tag{5.23}$$

Equation (5.23) can be solved efficiently with a lookup table that is generated by letting c_k vary over the range of values that could possibly be expected for this parameter in small steps. Let c_k vary from 0.1 to 2.5 with a step size of, say, 0.01 for these steps. Then the generalized gaussian approximations to the distributions of the observed DCT coefficients are readily obtained from the c_k and the σ_k^2 with (5.21) and (5.20).

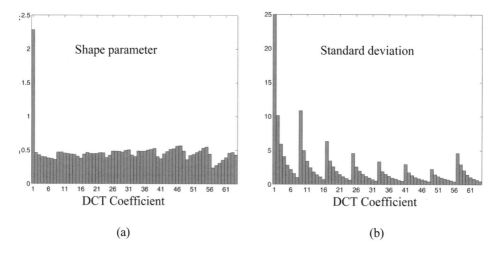

Figure 5.13: (a) Shape parameters and (b) standard deviations estimated for the DCT coefficient.

Figure 5.13 shows the c_k and the σ_k that are estimated from the DCT coefficients obtained from a set of 18 different images. The scanning order in a 2D block of DCTs is taken from left to right (increasing horizontal frequency) and from top to bottom

NOISE REDUCTION BY CORING

(increasing vertical frequency); see Figure 5.14. Except for the first DCT coefficient, the DC component, it can be seen that c_k is a bit smaller than 0.5. The standard deviation of the coefficients decreases with increasing frequency, which is consistent with the well-known fact that natural images contain less energy in high frequencies than in low frequencies.

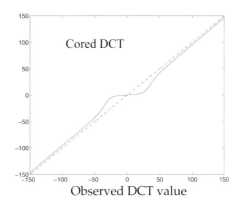

Figure 5.14: (left) *Numbering of DCT coefficients in an 8x8 block.*
Figure 5.15: (right) *Coring function for DCT coefficient 8, computed for noise with variance 100 corrupting the image.*

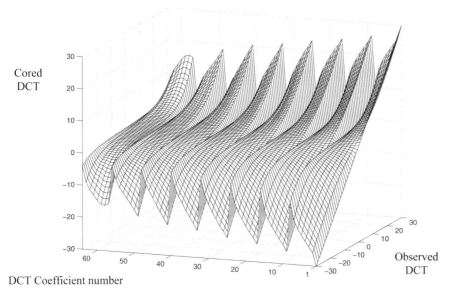

Figure 5.16: *Plot of part of the coring functions for all 64 DCT coefficients, computed for noise with variance 100 corrupting the image.*

Figure 5.15 shows the coring function computed for DCT coefficient number 8 for noise with variance 100 corrupting the image. In this figure, small values are cored towards zero; larger values are altered less. This confirms the intuitive assumption that data with small amplitudes are noisy and unreliable, and they should therefore be discarded. Figure 5.16 plots the coring functions for all 64 DCT coefficients, again with noise with variance 100 corrupting the image. As already mentioned, the DC terms are not cored; hence the 45 degree line for this DCT coefficient. It can be seen that coefficients representing higher spatial frequencies are cored towards zero more strongly than coefficients representing lower spatial frequencies. This, again, matches well with the fact that natural images contain less energy in high frequencies than in low frequencies.

The coring functions depend on the noise variance. A number of lookup tables are computed for different noise variances in a practical situation. The MPEG2 encoder selects the lookup table that corresponds best with the actual noise variance in an image sequence.

5.4.3 Experiments and results

For the experiments, the standard *test model 5* (TM5) MPEG2 encoder [IEC93] was adjusted so that the DCT coefficients are cored using lookup tables, as described in the previous sections. This section describes two sets of experiments. The same test sequences are used in Section 5.3.6.

The first set of experiments evaluates the performance of the adjusted TM5 encoder in terms of the PSNR when applied to test sequences with varying amount of noise. Figure 5.17a shows the scheme used for measuring the PSNR of the corrected sequences. Figure 5.17b,c plots the PSNRs for bitrates ranging from 2 Mbit/s to 15 Mbit/s. The results show that the PSNRs of the filtered and coded sequences are considerably higher at the higher bitrates than those of the noisy input sequences.

The PSNRs of the corrected sequences increase more rapidly with increasing bitrate at low bitrates than at high bitrates. Specifically, the curves for test sequence with noise variance 100 and 225 are quite flat over the range from 4 Mbit/s to 15 Mbit/s. This contrasts to the PSNRs for noise free sequences, which increase steadily with increasing bitrate. This implies that there is an "early" *saturation point* for the bitrate in noisy image sequences. Encoding with bitrates above the saturation point gives only marginal improvements in image quality.

By comparing the results in Figure 5.17, at for instance 8 Mbit/s, to those in Table 5.2, it can be seen that the 3D pyramid and wavelet filters outperform the adjusted MPEG2 encoder in terms of PSNR. However, the adjusted MPEG2 encoder is basically a 2D filter. It can be seen that its performance is similar to that of the 2D pyramid and 2D wavelet filters.

The second set of experiments investigates whether the adjusted TM5 encoder performs better than the standard encoder in combination with prefiltering, e.g., with the 3D pyramid noise-reduction system. It could be imagined that even though the 3D pyramid filter and the 3D wavelet filter on their own outperform the adjusted MPEG2 encoder, their superior quality may be lost due to quantization errors introduced by the standard encoder. Another question is how the performance of the adjusted MPEG2 encoder compares to the standard TM5 MPEG2 encoder with a dead zone when it is applied to a noisy sequence.

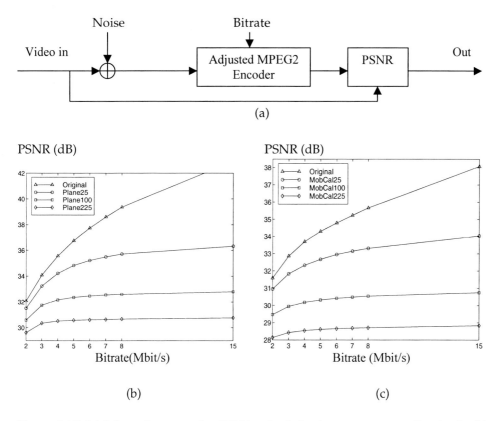

Figure 5.17: *(a) Scheme for measuring PSNRs of coded noisy test sequences. Results for (b) Plane and (c) MobCal sequences with the adjusted MPEG2 encoder and coring. The noise variance in the noisy sequences were 25, 100, and 225, which correspond to PSNRs of 33.0, 27.0, and 23.5 dB, respectively.*

These questions are investigated, using the *Plane* and *MobCal* sequences to which a moderate amount of noise (variance 100) was added. Figure 5.18 plots the PSNRs as a function of the bitrate of the noisy test sequences after encoding by the standard TM5 with and without prefiltering by the 3D pyramid filter. The PSNRs that result from applying the adjusted TM5 coder to the noisy sequences are also shown. Finally, the PSNRs of the coded original, noise-free sequences are plotted as a reference of what can maximally be obtained.

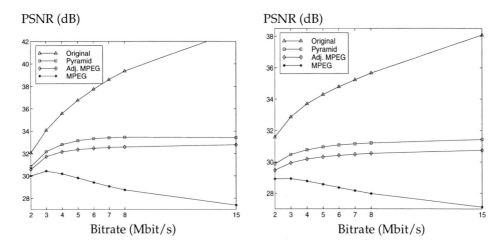

Figure 5.18: *(a) PSNR vs. bitrate for the original Plane sequence, noisy Plane sequence (noise variance 100), and noise-reduced Plane sequence (filtered by the 3D pyramid filter) encoded by the standard TM5 MPEG2 encoder. Also shown is the PSNR of the noisy Plane sequence that was encoded and noise reduced simultaneously by the adjusted MPEG2 encoder with coring. (b) As before, but now for the MobCal sequence.*

Figure 5.18 indicates that prefiltering sequences with a moderate amount of noise prior to encoding with the standard TM5 encoder gives a PSNR that is maximally one decibel higher than when simultaneous filtering and encoding is done by the adjusted TM5 MPEG2 encoder. It can also be seen from Figure 5.18 that the standard TM5 encoder (without prefiltering) also functions as a noise reducer at low bitrates. At 3 Mbit/s, the PSNR of the coded noisy *Plane* sequence is about 3.5 dB higher than that of the noisy original. This number is 1.5 dB for the MobCal sequence. The PSNRs decrease for these sequences at higher bitrates. This behavior is not surprising. The encoder applies a coarse quantization at low bitrates and much noise energy is removed by the dead zone. The encoder is capable of encoding the signal and the noise more accurately at higher bitrates, so that the noise part of the signal is preserved better. In the limiting case, at very high bitrates, the noisy sequence is encoded without errors, and the PSNR equals 23.5 dB.

5.5 Conclusions

This chapter shows that coring is a powerful technique for noise reduction. A 2D shift invariant wavelet filter and the 2D Simoncelli pyramid were introduced. These filters were extended in the temporal dimension so that temporal information, as well as spatial information, in image sequences could taken into account in the noise reduction process. The spatio-temporal decomposition allows temporal filtering of the DC bands of the 2D Simoncelli pyramid and the 2D DWT transforms without introducing severe blur or other artifacts. Two-dimensional scale-space noise reduction filters have no way of filtering the DC bands by means of coring.

The noise reduction capabilities of the Simoncelli pyramid outperforms those of the shift invariant DWT due to the minimal aliasing and its enhanced directional sensitivity. However, the difference in performance in terms of increase in PSNR can be considered marginal if one takes into account the increase in complexity for the pyramid filter compared to the wavelet filter.

Even though the 3D pyramid filter as presented in this chapter is a complex and expensive filter to implement, it is nevertheless a useful one. Visually speaking, the results obtained by the pyramid filter are better than those obtained from the shift invariant wavelet filter. It can be applied when good quality noise reduction is absolutely necessary, i.e., when processing time is less important than image quality. It can also be used as a benchmark for the results obtained by other filters.

This chapter also shows that the MPEG2 scheme can easily be adapted to perform simultaneous noise reduction and compression. The extra costs of the adapted scheme, compared to a standard MPEG2 encoder, consist of implementing lookup tables and an extra DCT operation for the B and P frames. This is a cheaper solution than the pyramid filter or the wavelet filter and it gives reasonable performance. In fact, the experiments indicate that, if a noisy image sequence is to be encoded, the difference between encoding the prefiltered sequence and encoding the noisy sequence with the adapted encoder is less than one decibel over a large range of bitrates. In this case, whether or not prefiltering is a cost-effective solution depends on the required quality of service.

Chapter 6

Evaluation of restored image sequences

6.1 Introduction

Chapter 1.1 explains the motivation for restoration of archived film and video. It is stated there that image restoration improves the perceived (subjective) quality of film and video sequences and that restoration also leads to more efficient compression. This chapter experimentally verifies the validity of these two assertions. Section 6.2 describes the methodology that is used in two sets of experiments for validating the assumptions mentioned. The first set of experiments is aimed at verifying that image restoration indeed improves the perceived quality of impaired image sequences. These experiments are done with test panels. The second set of experiments is aimed at verifying that image restoration indeed improves the coding efficiency. This can be done with test panels, or, as is done in this chapter, by numerical evaluation. Section 6.3 describes and discusses the experimental results.

6.2 Assesment of restored image sequences

6.2.1 Influence of image restoration on the perceived image quality

An important reason for image restoration is that it improves the image quality as perceived by humans. Whether the underlying assumption is indeed true can only be determined by having human observers compare restored sequences to the corresponding impaired sequences. So far, automatic validation (without human beings) is not possible: there are no mathematical models that can adequately model human perception of images in all their aspects.

The *International Telecommunication Union* (ITU) has standardized a number of methods for evaluating image sequences by test panels. For instance, the *double-stimulus continuous quality-scale* (DSCQS) method is well-known [ITU95]. This method measures the relative difference in quality of an impaired sequence given the original, unimpaired image as a reference. The DSCQS method is useful for comparing the performance of various restoration systems.

At this point, the aim is not to compare the performance of different restoration systems. Here, the central question is whether the image restoration algorithms presented in previous chapters improve the perceived image quality. A method simpler than the DSCQS method can be used for finding an answer to this question. The method used here is the *two alternatives forced choice* (2AFC) method [All83]. The 2AFC method is often used in television broadcasting environments to determine at what point a transmission system introduces visible distortions in the transmitted images or image sequences. In the context of image restoration, this method is not used to determine whether there are visible differences between two sequences, but to determine which of the two sequences have the highest perceived quality.

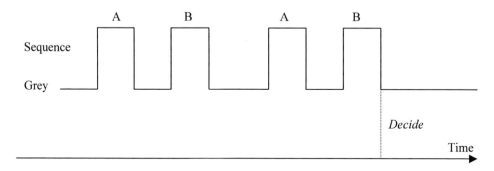

Figure 6.1: *Overview of 2AFC testing.*

In the 2AFC method, the members of a test panel are shown pairs of image sequences A and B twice, as illustrated by Figure 6.1. One of the sequences is the impaired sequence, the other is the restored sequence. Which is which is random. The duration of each sequence is approximately 10 seconds. Between showing sequences A and B, the screen is blanked to a mid-gray value for 2 seconds. After a pair of image sequences has been viewed for the first time, the screen is blanked to a mid-gray value for 5 seconds. Then, sequences A and B are shown again. If was the impaired sequence in the first viewing, then it is also the impaired sequence in the second viewing. The same is true for B. After viewing the sequences the second time, the assessors must indicate which sequence has the better visual quality.

For all experiments in this chapter, differences between the impaired and the corrected sequences are clearly visible. The outcome of 2AFC testing is determined by one of two cases. In the first case, a majority of the votes is given to either A or B. This indicates a general consensus on whether the perceived quality of the corrected sequence is better than that of the impaired sequence. In the second case, about 50% of the votes is given to each of the sequences, and there is no general consensus on which sequence (impaired or restored) is better. The second case can occur, for example, in the case of noise reduction. It is well known that some people prefer a noisy image over a slightly blurred noise-free image. The noise gives an illusion of increased sharpness. Other people prefer a noise-reduced image, even if it is slightly blurred.

6.2.2 Influence of image restoration on the coding efficiency

This section describes experiments that can be carried out to verify that image restoration indeed does lead to more efficient image compression. Before it can be determined how much more efficient one image sequence is compressed with respect to another, a definition for the *increase in coding efficiency* is required.

Let ΔQ denote the increase in coding efficiency between a corrected image sequence and an impaired image sequence. ΔQ can be defined in two ways. The first definition relates ΔQ to the distortion introduced by a codec set to a fixed bitrate. The second definition relates ΔQ to the bandwidth required by a codec to compress a sequence given the allowable distortion.

ΔQ in terms of coding accuracy. Figure 6.2 proposes an experimental setup that can be used for measuring the increase in coding efficiency in terms of how accurately the corrected and the impaired image sequence are coded with respect to each other.

Let $\hat{y}_0(i)$ and $\hat{y}_c(i)$ be restored image sequences before and after coding, respectively. Similarly, let $z_0(i)$ and $z_c(i)$ be impaired image sequences before and after coding, respectively. In Figure 6.2, the restored image sequence is encoded at a constant bitrate. The PSNR computed between the codec input and output is given by $PSNR[\hat{y}_0, \hat{y}_c]$. The impaired image sequence is encoded at the same bitrate. In this case, the PSNR computed between codec input and decoded output is given by $PSNR[z_0, z_c]$. ΔQ is now defined by:

$$\Delta Q = PSNR[\hat{y}_0, \hat{y}_c] - PSNR[z_0, z_c] =$$

$$= 10 \log \left(\frac{224^2}{\frac{1}{N} \sum_i (\hat{y}_0(i) - \hat{y}_c(i))^2} \right) - 10 \log \left(\frac{224^2}{\frac{1}{N} \sum_i (z_0(i) - z_c(i))^2} \right) \quad (6.1)$$

$$= 10 \log \left(\frac{\sum_i (z_0(i) - z_c(i))^2}{\sum_i (\hat{y}_0(i) - \hat{y}_c(i))^2} \right)$$

From (6.1) it can be seen that ΔQ is a function of the ratio between the energy of the coding errors in the impaired image sequence to the energy of the coding errors in the corrected image sequence. If $\Delta Q > 0$, the corrected sequence is coded with fewer errors than the impaired sequence. If $\Delta Q < 0$, the corrected sequence is more difficult to code than the impaired sequence and the compression errors are larger.

It is emphasized here that the coding errors in $\hat{y}_c(i)$ and $z_c(i)$ are computed between the input and output of the codec. The errors are not computed with respect to a ground truth, i.e., an unimpaired original. In practice, no unimpaired references exist for archived film and video material.

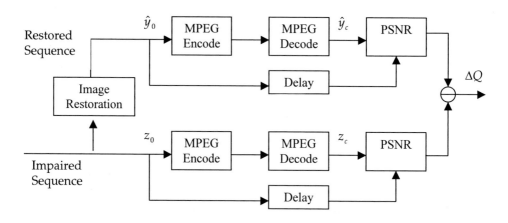

Figure 6.2: *Method for measuring the difference in coding efficiency on the basis of PSNR.*

ΔQ in terms of bandwidth. The definition of ΔQ in terms of bandwidth is given by the difference in bitrate for the coded corrected image sequence and for the coded impaired image sequence:

$$\Delta Q = Bitrate[impaired] - Bitrate[corrected] \qquad (6.2)$$

Here, if $\Delta Q > 0$, the corrected sequence requires fewer bits for coding than the impaired sequence. If $\Delta Q < 0$, the corrected sequence is more difficult to code than the impaired sequence and it requires more bits. Obviously, ΔQ can only be given a meaningful interpretation if it is measured on condition that the bitrates selected for coding the impaired and corrected sequences are related in a meaningful way. The constraint set here for measuring (6.2) requires that the codec introduces the same amount of distortion to the impaired as to the corrected sequence.

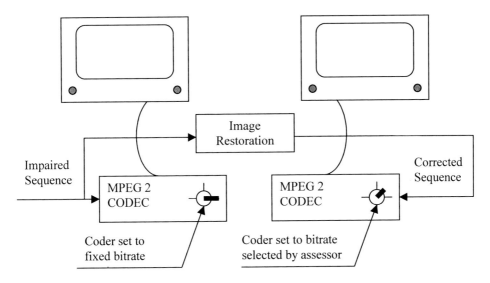

Figure 6.3: *Setup for measuring the increase in coding efficiency using human assessors.*

This raises the question of how the distortion introduced by a codec should be measured. Ideally, the measured distortion is related to the perceived image quality. This requires involving human observers to determine (6.2) with, for instance, the setup proposed in Figure 6.3. In this setup, the impaired image sequence is coded by an MPEG2 codec set to a fixed bitrate. The impaired sequence is restored and coded by an MPEG2 codec of which the bitrate is controlled by an assessor. The codecs are synchronized to compensate for the delay introduced by the restoration system. Their outputs are displayed on two calibrated monitors. During the experiment, the task of the assessor is to set the bitrate of the codec he/she controls to a level such that the perceived quality of the coded corrected sequence is equal to that of the coded impaired sequence. The difference in bitrate of the two codecs gives the increase (or decrease) in coding efficiency given by the image restoration process.

Note that the type of artifacts in the coded impaired and corrected sequences can be different at the bitrates at which the assessor rates the perceived image quality the same. For instance, consider a noisy image sequence coded at a bitrate at which the codec does not introduce visible distortions. The corrected, noise free sequence can be coded at a lower bitrate. At a certain point this bitrate is so low that blocking artifacts start to appear. It is around this point that the assessor will begin to prefer the coded noisy sequence over the coded corrected sequence.

The method for measuring the improvement in coding efficiency with human assessors requires a fair amount of calibrated and synchronized equipment. An alternative method is to measure the distortion with mathematical measures based on the MSE. Obviously, the results will be different from those obtained by human assessors. In this case, a scheme similar to that in Figure 6.2 is used for measuring ΔQ. First, the corrected sequence is coded at a fixed bitrate. Next, the bitrate for coding the impaired sequence is searched so that $PSNR[\hat{y}_0, \hat{y}_c]$ equals $PSNR[z_0, z_c]$, i.e., so the same amount of compression errors have been introduced into the corrected and impaired sequence. Again, as in (6.2), ΔQ is given by the difference in bitrates.

As a final remark, it should be mentioned that ΔQ, measured either in dB or in Mbit/s, can only be meaningful if the restored image sequence consists of sensible data that represent the true image data in a reasonable manner. For example, it is assumed that the restored sequence is not a collection of black frames if the original data is clearly not a collection black frames, but, for instance, a recording of a zoo.

6.3 Experiments and results

This section experimentally verifies that the algorithms proposed in this book indeed improve the perceived image quality by presenting the impaired and restored image sequences to a test panel. The influence of image restoration on the perceived image quality is assessed in two circumstances. In the first circumstance, pairs of impaired and corrected sequences are used. In the second circumstance, pairs of MPEG2 encoded impaired and corrected sequences are used. The latter circumstance verifies the assumption that image restoration improves the perceived image quality also holds in a digital broadcasting environment.

This section also verifies that the algorithms developed in this book improve the coding efficiency. The increase in coding efficiency, ΔQ, is determined by numerical evaluation, both in terms of PSNR and in terms Mbit/s.

Sequence	Amount of Flicker	Number of Blotches	Visibility of Noise
Plane (100 Frames)	High	High	High
Chaplin (112 Frames)	Medium	High	Medium
Charlie (48 Frames)	High	High	Low
Mine (404 Frames)	Medium	Very Low	High
VJ Day (49 Frames)	None	Low/Medium	Low
Soldier (227 Frames)	Medium	Very Low	Low

Table 6.1: *List of impaired sequences used for subjective and objective evaluations with an indication of the severity of the various degradations. Note that the Plane sequence contains artificial degradations.*

Sequence	Flicker Correction	Blotch Correction	Noise Reduction
Plane	X	X	X
Chaplin	X	X	X
Charlie	X	X	
Mine	X		X
VJ Day		X	
Soldier	X		

Table 6.2: *Corrections applied to the test sequences.*

6.3.1 Test sequences

To get an impression of the effects of removing different combinations of artifacts on the perceived image quality and on the increase in coding efficiency, test sequences were selected with various combinations of impairments. The test sequences consist of one artificially degraded sequence and five naturally degraded sequences.

Table 6.1 lists the sequences and gives an indication of the severity of the degradations that impair them. The test sequences are also used in Chapters 3 to 5 and have already been described. An exception is the *Chaplin* sequence, which has not been used before for any experiment. Three frames from this sequence are shown in Chapter 1, Figure 1.1. Table 6.2 lists the artifacts that were corrected in each of the test sequences by the restoration system depicted in Figure 1.2 with the restoration methods developed in this book. The various control parameters of the restoration algorithms were set to values that give good visual results.

6.3.2 Experiments on image restoration and perceived quality

The subjective experiments were done in a dimly lit room. The viewing distance was six times the height of the display used. The test panel consisted of 25 people, all of whom had good vision with a visus of 0.8 or better. Before the actual experiments, the assessors were trained for their task by being shown some examples of sequences with and without flicker, blotches, and noise. Each assessor assessed all the test sequences once. They were asked the following question: "Which sequence do you find more pleasing to view, *A* or *B*?".

As already mentioned, each test sequence should be approximately 10 seconds in duration. Because most of the test sequences are shorter than 10 seconds, they were repeated (looped) a number times so that the duration of the looped sequence was approximately 10 seconds. Only the first 10 seconds of the 16-second *Mine* sequence were shown.

Table 6.3 gives the results for the first set of experiments in which the assessors indicated which sequence they prefer: the impaired image sequence or the restored image sequence. This table shows that, for all test sequences, the majority of the votes is given to restored image sequences. This proves that the image restoration algorithms presented in this book increase the perceived image quality of impaired image sequences.

The restored *Mine* sequence received relatively fewer votes than the other restored sequences. When questioned about this, some of the test panel members indicated they considered the corrected *Mine* sequence to be overly smooth, and, therefore, they preferred the flickering, noisy original. The smoothing was caused by the noise reduction algorithm that was set to achieve a great amount of noise reduction. It is a well-known fact that there is a trade-off between noise reduction and introducing blur. Had the noise reducer been set for less noise reduction, less smoothing would have been introduced, and the assessors in question might well have preferred the corrected sequence.

Table 6.4 gives the results for the second set of experiments in which the assessors indicate which sequence they prefer: the MPEG2 encoded impaired image sequence or the MPEG2 encoded restored image sequence. The standard TM5 MPEG2 encoder was used for all experiments [IEC93]. The coder was set to the main profile and the GOP size was 12. This table shows that for all test sequences, the majority of the votes is given to MPEG2 encoded restored image sequences. This proves that the increase in perceived quality, obtained from the image restoration algorithms presented in this book, are not lost due to coding artifacts introduced by an MPEG2 encoder at 4 Mbit/s. Therefore, image restoration is beneficial in digital broadcasting environments in which films are broadcast in compressed format.

Sequence	Votes for Corrected Sequence (in Percentages)	Votes for Impaired Sequence (in percentages)
Plane	84	16
Chaplin	92	8
Charlie	84	16
Mine	72	28
VJ Day	88	12
Soldier	92	8

Table 6.3: *Results of subjective evaluations for the first set of experiments in which impaired and restored sequences are compared.*

Sequence	Votes for Corrected Sequence (in Percentages)	Votes for Impaired Sequence (in percentages)
Plane	88	12
Chaplin	100	0
Charlie	88	12
Mine	68	32
VJ Day	80	20
Soldier	96	4

Table 6.4: *Results of subjective evaluations for the second set of experiments in which impaired and restored sequences are compared after MPEG compression at 4 Mbit/s*

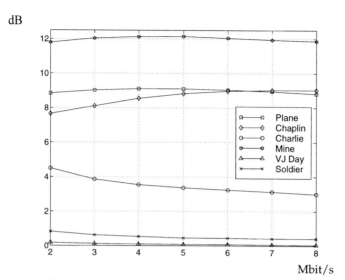

Figure 6.4: ΔQ *measured in dB versus bitrate.*

Sequence	PSNR of $\hat{y}(i)$ at 4Mbit/s (in dB)	Bitrate for $z(i)$ with same PSNR (in Mbit/s)	ΔQ (in Mbit/s)	Savings in bandwith by restoration (in %)
Plane	36.4	15.5	11.5	74.2
Chaplin	39.1	19.9	15.9	79.9
Charlie	40.1	9.0	5.0	55.6
Mine	44.3	38	34.0	89.5
VJ Day	34.0	4.2	0.2	4.8
Soldier	36.8	4.8	0.8	16.7

Table 6.5: *Results of numerical evaluations of ΔQ measured in Mbit/s.*

6.3.3 Experiments on image restoration and coding efficiency

This section presents the results of two sets of numerical evaluations. The first set applies the scheme shown in Figure 6.2 to measure the increase coding efficiency in dB. ΔQ was evaluated for bitrates ranging from 2 Mbit/s to 8 Mbit/s. For all experiments the standard TM5 MPEG2 encoder was used. The coder was set to the main profile and the GOP size was 12.

Figure 6.4 plots the results for this first set of experiments. The curves indicate that image restoration leads to more efficient compression over the range of investigated bitrates; at identical bitrates, the restored image sequences can be compressed with fewer errors than the impaired sequences. This proves that image restoration gives more efficient compression for the artifacts considered.

The gains are smallest for the VJ Day sequence. Only the blotches were restored in this sequence. Because the blotches cover only a small percentage of the total image area in this sequence, removing them has little influence on the overall coding efficiency. The gains for the Soldier sequence, which was corrected for intensity flicker, are somewhat larger. The intensity flicker is a global effect and has a larger influence on the coding efficiency. The Charlie sequence contained much flicker and many blotches. Restoring this sequence gives large gains. The restored Plane, Chaplin, and Mine sequences give the largest increases in coding efficiency. Unlike the other test sequences, these sequences were noise reduced. Noise is difficult to code and removing it simplifies the coder's task (unless, of course, the adjusted coder described in Chapter 5 is used). ΔQ is largest for the Mine sequence. As mentioned in the previous section, the corrected Mine sequence is quite smooth. Hence it can be coded with many fewer errors than the impaired original.

The second set of experiments in this section measures ΔQ in terms of bandwidth, i.e., in Mbit/s. At the time the experiments were carried out, the equipment for measuring ΔQ with human assessors, as described in Section 6.2.2, was not available. The numerical method, also described in Section 6.2.2, was used.

The experiment was set up as follows. First, the PSNR ratios were computed over the encoded/decoded restored image sequences coded at 4 Mbit/s (broadcast quality). Next, the impaired sequences were coded at bitrates so that the PSNRs over the coded/decoded impaired sequences were identical to those of the corrected sequences. The differences in bitrate gives the increase in coding efficiency. The standard TM5 MPEG2 encoder was used for all experiments. The coder was set to the main profile or, for bitrates greater than 15 Mbit/s, to the high profile, and the GOP size was 12.

Table 6.5 lists the results from the second set of experiments. Again, it is concluded that image restoration leads to more efficient compression. Considerable savings in bandwidth can be achieved by restoring impaired image sequences. Again, the largest gains were obtained for the test sequences to which noise reduction was applied. The last column in this table was computed by:

$$percentage = \frac{Bitrate[z(i)] - 4}{Bitrate[z(i)]} x 100\% \qquad (6.3)$$

6.3.4 Discussion of experimental results

The experimental results verify that the image restoration algorithms developed in this book improve the perceived image quality of impaired image sequences. The experiments also verify that image restoration improves the coding efficiency. Therefore, the benefits of restoration of archived film and video is confirmed, and the assumptions underlying the work presented in previous chapters are validated.

A question is how well the numerical experiments for determining the increase in coding efficiency correspond to human perception. ΔQ, as defined in this chapter, reflects an increase in image quality terms of PSNR or in terms of how many bits of irrelevancy have been removed. It is well known that numerical measures do not necessarily correlate well with subjective perception. For instance, ΔQ is a global measure, whereas human observers are very sensitive to local variations in image quality. An example that illustrates this is given by the experimental results for the VJ Day sequence. This sequence was corrected for local artifacts, namely blotches. The results from the test panel evaluation shows that a majority of 88% prefers the corrected sequence over the impaired sequence. The large number of votes implies a clearly visible improvement in the perceived image quality. In contrast, ΔQ the computed for this sequence in terms of PSNR and in terms of bandwidth are small; 0.1 dB and 0.2 Mbit/s, respectively. Therefore, they suggest a marginal improvement only.

Appendix A
Hierarchical motion estimation

Full-search block matching is a well-known method for estimating motion from a source frame to a reference frame. In this method, the source frame is subdivided into image blocks of or pixels. An exhaustive search is performed for each image block to find the optimal match within the reference frame. The *summed squared difference* (SSD) and the *summed absolute difference* (SAD) are often used as matching criteria. The displacement that gives the optimal match yields the motion estimate [Han92], [Tek95].

Full search block matching is very intensive from a computational point of view. Furthermore, the motion vectors obtained from this technique do not necessarily represent (a projection onto two dimensions) of the true motion. They merely represent displacements that give optimal matches.

A method that suffers less from these drawbacks is hierarchical block matching [Bie88], [Haa92], [Tek95]. Figure A.1 shows the principle of this method. First, initial, coarse motion vectors are estimated by applying (full-search) block matching to subsampled images. Next, the initial motion estimates are propagated to the next level with higher resolution and refined. Instead of full-search block matching, the refinements consist of doing a limited search in the region centered around the initial, coarse motion estimate. Again, the refined motion vectors are then propagated to the next level with higher resolution. The refinement process is repeated until the motion vectors have been computed for the source image at full resolution.

As a result of the subsampling and the limited search strategies, hierarchical block matching requires fewer computations than full search block matching. Therefore it is faster. Furthermore, the final hierarchical motion estimates are closer to the true motion and the motion vectors are more consistent locally than the full search block matching motion estimates.

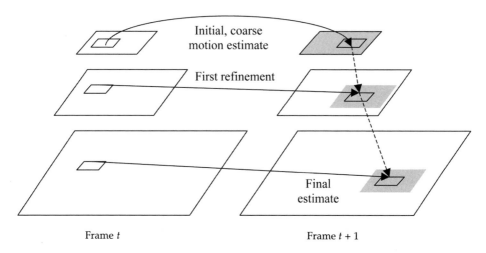

Figure A.1: *General principle of hierarchical block matching. The gray areas indicate the search region for the block matching process.*

The reason for this is that the initial motion estimates are done on coarse images. An 8x8 image region in an image subsampled horizontally and vertically by a factor 4 corresponds to an image region of 32x32 in the high-resolution image. Therefore, the initial motion estimates computed by the hierarchical block matcher take more context into account than a full-search block matcher that uses 8x8 image blocks. Because motion estimates are propagated from coarse resolution levels to finer resolution levels, the refined estimates for adjacent blocks in the higher resolution images are made on the basis of the same initial vectors. Therefore, the final motion estimates are consistent locally.

As is explained in Chapter 2, hierarchical motion estimators are relatively robust to common artifacts in video and film sequences.

Appendix B
Derivation of conditionals

B.1 Introduction

Section 4.5.2 stated that draws have to be taken from the conditionals:

$$a \sim P[a|\sigma_e^2, o(i), z_+, d],$$
$$\sigma_e^2 \sim P[\sigma_e^2|a, o(i), z_+, d], \quad \text{(B.1)}$$
$$o(i) \sim P[o(i)|a, \sigma_e^2, z_+(i), d(i), O]$$

This section shows that in the case of drawing samples from a conditional, it is not necessary that the conditional be known exactly. It suffices that the distribution of the samples follows a function that is proportional to the conditional. The following sections derive such functions for the conditionals in (B.1).

Bayes' rule states:

$$P[A|B] = \frac{P[B|A]\,P[A]}{P[B]} \quad \text{(B.2)}$$

The goal is to draw samples for A, given B, from $P[A|B]$. Because B is given, $P[B]$ can be regarded as a normalizing constant. It is therefore only necessary that the draw for A be proportional to:

$$P[A|B] \propto P[B|A]\,P[A] \quad \text{(B.3)}$$

This means that, when deriving expressions for conditionals from which samples are to be drawn, it is not necessary to compute the normalizing constants. Let B indicate a collection of random variables, b_1, b_2, ..., b_n, and suppose that b_1 is independent of A. Then:

$$P[A|B] \propto P[b_1, b_2, ..., b_n | A] P[A]$$
$$= P[b_1 | b_2, ..., b_n] P[b_2, ..., b_n | A] P[A] \qquad (B.4)$$
$$\propto P[b_2, ..., b_n | A] P[A]$$

It can be seen from (B.4) that it is only necessary to consider terms involving A for drawing random samples for A, given B.

Before deriving the conditionals in (B.1), first a quick word about notation. In this section, bold faced characters describe matrices (capital letters) or vectors (small letters). For instance, z represents a vector into which an observed frame $z(i)$ has been scanned in a lexicographic fashion. Analogous to Chapter 4, z_+ indicates a vector containing the motion-compensated previous, current, and next frame.

B.2 Conditional for AR coefficients

Each image region with missing data is modeled by a 2D AR process that uses a single set of coefficients a. The conditional for a is given by:

$$P[a | \sigma_e^2, o, z_+, d] = \frac{P[a, \sigma_e^2, o | z_+, d]}{\int_a P[a, \sigma_e^2, o | z_+, d] da} \qquad (B.5)$$

At first glance this might seem to be a very complex distribution. Fortunately, as is shown in [Kok98], it turns out that (B.5) is proportional to a multivariate gaussian distribution. The derivation of [Kok98] is repeated here.

First, it is noted that the denominator in (B.5) is independent of a and hence it can be considered as a normalizing constant that can safely be ignored. Therefore:

$$P[a | \sigma_e^2, o, z_+, d] \propto P[a, \sigma_e^2, o | z_+, d] \propto P[z_+ | a, \sigma_e^2, o, d] P[a]$$
$$\propto \frac{1}{(2\pi\sigma_e^2)^{N/2}} \exp\left(\frac{-e^T e}{2\sigma_e^2}\right) P[a] \qquad (B.6)$$

where the identity in (4.33) has been used. Note that (4.33) applies to single pixels whereas (B.6) applies to image blocks with N pixels. Hence the factor N in the last line of (B.6). In this equation, e is a vector with prediction errors and T indicates the transpose operator. This prediction error vector is given by reformulating (4.30) in vector-matrix notation:

$$\hat{y} = A\hat{y} + e = \hat{Y}a + e \tag{B.7}$$

The top line in (B.7) gives the usual vector-matrix representation of an AR image model in which a sparse matrix A that contains the prediction coefficients is multiplied with an image vector. Here, for convenience, the definition in the bottom line in (B.7) is used where the AR coefficients are placed in a, and $\hat{y}(i)$ is scanned into matrix $\hat{Y}$ such that $\hat{Y}a = A\hat{y}$.

The term $e^T e$ in (B.6) is examined more closely now:

$$\begin{aligned} e^T e &= (\hat{y} - \hat{Y}a)^T (\hat{y} - \hat{Y}a) = \hat{y}^T \hat{y} - 2\hat{y}^T \hat{Y}a + a^T \hat{Y}^T \hat{Y}a = \\ &= (a - (\hat{Y}^T \hat{Y})^{-1} \hat{Y}^T \hat{y})(\hat{Y}^T \hat{Y})(a - (\hat{Y}^T \hat{Y})^{-1} \hat{Y}^T \hat{y}) + \hat{y}^T \hat{y} - \hat{y}^T \hat{Y}(\hat{Y}^T \hat{Y})^{-1} \hat{Y}^T \hat{y} \end{aligned} \tag{B.8}$$

Substituting those terms in (B.8) that involve a into (B.6), and also keeping in mind that $P(a)$ has a uniform distribution assigned to it, i.e., that it is a constant, gives:

$$P[a|\sigma_e^2, o, z_+, d] \propto \frac{1}{(2\pi\sigma_e^2)^{N/2}} \exp\left(-\frac{(a - (\hat{Y}^T \hat{Y})^{-1} \hat{Y}^T \hat{y})(\hat{Y}^T \hat{Y})(a - (\hat{Y}^T \hat{Y})^{-1} \hat{Y}^T \hat{y})}{2\sigma_e^2}\right) \tag{B.9}$$

This can be recognized as proportional to a multivariate gaussian and can be denoted compactly as:

$$P[a|\hat{y}, \sigma_e^2, o, z_+, d] \propto N(\hat{a}, \sigma_e^2 (\hat{Y}^T \hat{Y})^{-1}) \tag{B.10}$$

where $\hat{a} = (\hat{Y}^T \hat{Y})^{-1} \hat{Y}^T \hat{y}$ is the least squares estimate for the AR coefficients. $\hat{Y}^T \hat{Y}$ and $\hat{Y}^T \hat{y}$ can be recognized as estimates for the autocorrelation matrix $R_{\hat{y}\hat{y}}$ and the autocorrelation vector $r_{\hat{y}\hat{y}}$. These are necessary for solving the normal equations [Lag94], [The92]. The pdf for a is thus shown to be proportional to a well-known distribution.

B.3 Conditional for the prediction error variance

A single error variance parameter σ_e^2 is associated with each image region with missing data. The conditional for σ_e^2 is given by:

$$P[\sigma_e^2|a,o,z_+,d] = \frac{P[a,\sigma_e^2,o|z_+,d]}{\int_{\sigma_e^2} P[a,\sigma_e^2,o|z_+,d]d\sigma_e^2} \tag{B.11}$$

Again, the denominator can be viewed as a normalizing constant that can safely be ignored:

$$P[\sigma_e^2|a,o,z_+,d] \propto P[a,\sigma_e^2,o|z_+,d] \propto P[\hat{y}|a,\sigma_e^2]P[\sigma_e^2]$$
$$= \frac{1}{(2\pi\sigma_e^2)^{N/2}} \exp\left(\frac{-e^T e}{2\sigma_e^2}\right) P[\sigma_e^2] \tag{B.12}$$

In [Kok98] an equation is derived that is very similar to that in (B.12) and it is noted there that the result is proportional to an inverted gamma distribution $IG(x|\psi,w)$ with parameters ψ and w:

$$IG(x|\psi,w) = \frac{w^\psi}{\Gamma(\psi)x^{\psi+1}} \exp\left(-\frac{w}{x}\right) \tag{B.13}$$

If $x = \sigma_e^2$, $\psi = N/2$, and $w = e^T e/2$, then (B.12) is proportional to (B.13), which means that:

$$P[\sigma_e^2|a,o,z_+,d] \propto IG\left(\sigma_e^2 | \frac{N}{2}, \frac{e^T e}{2}\right) \tag{B.14}$$

B.4 Conditional for the direction of interpolation

Unlike the AR model parameters, the direction of interpolation is computed on a pixel-by-pixel basis instead of on a block-by-block basis. The conditional for $o(i)$ is derived here. At each particular site i the conditional is given by:

$$P[o(i)|a,\sigma_e^2,z_+,d,O] = \frac{P[a,\sigma_e^2,o(i)|z_+,d,O]}{\int_o P[a,\sigma_e^2,o|z_+,d,O]do} \tag{B.15}$$

Here O indicates the direction of interpolation for the pixels in the local region surrounding $o(i)$. Collecting those terms that are proportional to the variables of interest gives:

DERIVATION OF CONDITIONALS

$$P[o(i) \mid a, \sigma_e^2, z_+, d, O] \propto P[z_+(i) \mid a, o(i), \sigma_e^2, d] \, P[o(i) \mid O]$$

$$\propto \exp\left(-\frac{-e^2(i)}{2\sigma_e^2}\right) \exp\left(-\sum_k \beta \mid o(i) - o(i+q_k) \mid\right)$$

$$\propto \exp(-\sum_{i \in S} [(1 - d(i))(z(i) - AR(\hat{y}, i, a))^2 + \tag{B.16}$$

$$d(i)(o(i) z_{mc}(i, t+1) + (1 - o(r)) z_{mc}(i, t-1) - AR(\hat{y}, i, a))^2 +$$

$$\sum_k \beta \mid o(i) - o(i+q_k) \mid])$$

As in Chapter 4, $AR(\hat{y}, a, i)$ denotes the prediction of the corrected image $\hat{y}$ at location i. $AR(\hat{y}, a, i)$ is determined completely by $z_+(i)$, $o(i)$, a, σ_e^2 and d. The eight-connected neighbors of $o(i)$ are indicated by $o(i+q_k)$, with $k = 1, \ldots, 8$.

Drawing samples from (B.16) with the Gibbs sampler is very easy. It involves evaluating (B.16) at a specific site i for $o(i) = 0$ and for $o(i) = 1$, while keeping the other values for the direction mask and the $\hat{y}(i)$ fixed. The results are assigned to c_1 and c_2, respectively. Next, a value for $o(i)$ (and thereby the corresponding $\hat{y}(i)$) is chosen at random, with a probability $c_1 / (c_1 + c_2)$ that $o(i) = 0$ and with a probability $c_2 / (c_1 + c_2)$ that $o(i) = 1$. A single update of an image region consists of applying the Gibbs sampler to each site in that region in turn, following, for instance, a checkerboard scanning pattern.

Appendix C

Optimal Quantizers for Encoding Noisy Image Sequences

This appendix shows that minimizing the error variance $E[\varepsilon^2(i)]$ between input $y(i)$ and output $\hat{z}(i)$ for a communication system as depicted in Figure 5.9 is equivalent to designing optimal quantizers for the MPEG2 encoder. It is assumed that the channel is error free. Work related to this topic is given in [Eph88], [Fin65]. The equation for the optimal quantizers is derived. For ease of notation, spatial indices i are omitted in this appendix.

In the absence of channel errors, the scheme in Figure 5.9 can be simplified to that in Figure C.1 in which the noisy signal is transformed by the DCT, quantized, inverse quantized and inverse transformed. Figure C.2 gives an example of a quantizer with L_k representation levels. The error variance is related to the quantization error in the coded DCT coefficients:

$$E[\varepsilon^2] = \sum_{k=1}^{64} E[(Y_k - \hat{Z}_k)^2] = \sum_{k=1}^{64} E[(Y_k - Q_k[Z_k])^2]$$
$$= \sum_{k=1}^{64} \int_{-\infty}^{\infty}\int_{-\infty}^{\infty} (Y_k - Q_k[Z_k])^2 P_{Y_k,Z_k}[Y_k,Z_k] dz_k dy_k$$

(C.1)

Here Y_k, $\hat{Z}_k$ and Z_k, with $k = 1, ..., 64$, indicate DCT coefficients with coefficient number k obtained from 8x8 image blocks. The quantizer for DCT coefficient k is indicated by $Q[.]$. The joint probability distribution $P_{Y_k,Z_k}[Y_k,Z_k]$ is given by:

$$P_{Y_k,Z_k}[Y_k,Z_k] = P_{Z_k}[Z_k|Y_k]P_{Y_k}[Y_k] = P_N[Y_k - Z_k]P_{Y_k}[Y_k] \qquad (C.2)$$

where $P_N[.]$ is the distribution of the additive noise.

Because the MPEG2 coding standard defines the representation levels of the inverse quantizer in the decoder, the only free parameters in the chain from input to output are the decision levels of the quantizers in the encoder. Minimizing (C.1) is therefore equivalent to selecting optimal decision levels $d_{k,m}$ in the quantizer, given the representation levels $r_{k,l}$, with $l = 1, \ldots, L_k$ and $m = 1, \ldots, L_k + 1$.

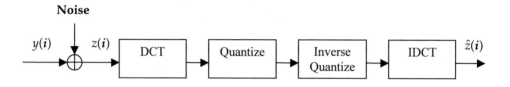

Figure C.1: *Simplification MPEG2 encoding/decoding over a noise free channel.*

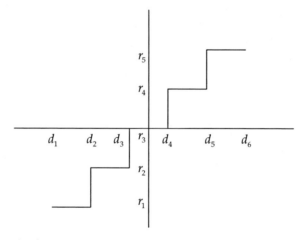

Figure C.2: *Example of quantizer with representation levels r_1 to r_5 and decisions levels d_1 to d_6. Note that d_1 and d_6 lie at plus and minus infinity.*

Without loss of generality, let $d_{k,1} = -\infty$ and $d_{k,L+1} = \infty$. Equation (C.1) can then be broken down into L partial integrals over the L decision intervals:

$$E[\varepsilon^2] = \sum_{k=1}^{64} \int_{-\infty}^{\infty} \sum_{l=1}^{L_k} \int_{d_l}^{d_{l+1}} (Y_k - Q_k[Z_k])^2 P_N[Y_k - Z_k] P_{Y_k}[Y_k] dZ_k dY_k$$

$$= \sum_{k=1}^{64} \sum_{l=1}^{L_k} \int_{-\infty}^{\infty} (Y_k - r_{k,l})^2 P_{Y_k}[Y_k] \int_{d_l}^{d_{l+1}} P_N[Y_k - Z_k] dZ_k dY_k$$

(C.3)

Equation (C.3) is always positive. Hence, it is minimized by minimizing each of its 64 terms, i.e., by selecting the optimal decision levels given the representation levels for all the individual quantizers Q_k. The optimal decision levels are obtained by setting the derivatives with respect to $r_{k,l}$ to zero. This yields the following decision levels $d_{k,m}$, with $2 \leq m \leq L_k$:

$$\int_{-\infty}^{\infty} (((r_{k,m-1} - Y_k))^2 - (r_{k,m} - Y_k)^2) P_{Y_k}[Y_k] P_N[Y_k - d_{k,m}] dY_k = 0 \qquad (C.4)$$

The optimal quantizers are now defined. Some concluding remarks can now be made. First, note that in an MPEG2 encoder, the input signal $y(i)$ in Figure C.1 can be either a true image or an image representing prediction errors, depending on whether an I, P, or B frame is coded. The statistics for these images vary, and therefore different quantizers need to be computed for each situation. Second, depending on the amount of bits that are available, an MPEG2 encoder selects a quantizer with a certain number of quantization levels. To get minimum error variance, multiple optimal quantizers have to be computed to accommodate this freedom of the encoder.

Bibliography - Part I

[Abr96] E. Abreu, M. Lightstone, S.K. Mitra, and K. Arakawa, "A New Efficient Approach for the Removal of Impulse Noise from Highly Corrupted Images", IEEE Trans. on Image Processing, Vol. 5, No. 6, pp. 1012-1025, 1996.

[Ade84] E.H. Adelson, C.H. Anderson, J.R. Bergen, P.J. Burt, and J. M. Ogden, "Pyramid Methods in Image Processing", RCA Engineer, Vol. 29, No. 6, pp. 33-41, 1984.

[All83] J. Allnat. "Transmitted-picture Assessment", John Wiley & Sons, 1983.

[Ant94] M.Antonini, T. Gaidon, P. Mathieu, and M. Barlaud, "Wavelet Transform and Image Coding", in "Wavelets in Image Communication", M. Barlaud. ed, Elsevier, pp. 65-119, 1994.

[Arc91] G. R. Arce, "Multistage Order Statistic Filters for Image Sequence Processing", IEEE Trans. on Signal Processing, Vol. 39, No. 5, pp. 1146-1163, 1991.

[Arm98] S. Armstrong, A.C. Kokaram, and P.J.W. Rayner, "Reconstructing Missing Regions in Colour Images using Multichannel Median Models", Proc. of EUSIPCO 98, Vol. II, pp. 1029-1034, Rhodes, Greece, 1998.

[Ast90] J. Astola, P. Haavisto, and Y. Neuvo, "Vector Median Filters", Proc. of the IEEE, Vol. 78, No. 4, pp. 678-689, 1990.

[Ban97] M.R. Banham and A.K. Katsaggelos, "Digital Image Restoration", IEEE Signal Processing Magazine, Vol. 14, No. 2, pp. 24-41, 1997.

[Bar98] M. Barni, F. Bartolini, A. Piva, and F. Rigacci, "Statistical Modelleing of Full Frame DCT coefficients", Signal Processing IX, Vol. 3, pp. 1513-1516, Rhodes, Greece, 1998.

[Bie87]   J. Biemond, L. Looijenga, D.E. Boekee, and R.H.J.N. Plompen. "A pel-recursive Wiener Based displacement estimation Algorithm", Signal Processing, Vol. 13, No.4, pp. 399-412, 1987

[Bie88]   M. Bierling, "Displacement Estimation by Hierarchical Block Matching", SPIE VCIP, pp. 942-951, Cambridge, U.K., 1988.

[Bil75]   F.C. Billingsley, "Noise Considerations in Digital Image Processing Hardware" in "Topics in Applied Physics", Vol. 6, e.d. T.S. Huang, Springer-Verlag, Berlin, 1975.

[Bra95]   J.C. Brailean, R.P. Kleihorst, S.N. Efstratiadis, A.K. Katsaggelos, and R.L. Lagendijk, "Noise Reduction Filters for Dynamic Image Sequences: A review", Proc. of the IEEE, Vol 83, no. 9, pp. 1272-1292, 1995.

[Bur83]   P.J. Burt and E.H. Adelson, "The Laplacian Pyramid as a Compact Image Code", IEEE Trans. on Comm., Vol. 31, pp. 532-540, 1983.

[Can86]   J. Canny, "A Computational Approach to Edge Detection", IEEE PAMI, Vol. 8, No. 6, 1986.

[Che97]   M.J. Chen, L.G. Chen, and R. Weng, "Error Concealment of Lost Motion Vectors with Overlapped Motion Compensation", IEEE Trans. on Circuits and Systems for Video Technology, Vol. 7, No. 3, June 1997.

[Dav91]   J. Davidse, "Analoge Singaalbewerkingstechniek", DUM, Delft, The Netherlands, 1991.

[Dav92]   J. Davidse, "Televisie Techniek en Beeldversterking", DUM, Delft, The Netherlands, 1992.

[Del94]   P. Delogne, L. Cuvelier, B. Maison, B. Van Caillie, and L. Vandendorpe, "Improved Interpolation, Motion Estimation, and Compensation for Interlaced Pictures", IEEE Trans. on Image Processing, Vol. 3, No. 5, 1994.

[Don95]   D.L. Donoho, "De-noising by Soft-Thresholding", IEEE Trans. on Information Theory, Vol 41, No. 3, pp. 613-627, 1995.

[Don94a]   D.L. Donoho and I.M. Johnstone, "Ideal Denoising in an Orthonormal Basis hosen from a Library of Bases", Technical Report 461, 1994.

[Don94b]   D.L. Donoho and I.M. Johnstone, "Ideal Spatial Adaptation via Wavelet Shrinkage", Biometrika, Vol. 81, pp. 425-455, 1994.

[Dub94]   E. Dubois and S. Sabri, "Noise Reduction in Image Sequences using Motion Compensated Temporal Filtering", IEEE Trans. on Communications, no. 32, pp. 826-831, 1984.

[Eph88]   Y. Ephraim and R.M. Gray, "A Unified Approach for Encoding Clean and Noisy Sources by Means of Waveform and Autoregressive Model Vector Quantization", IEEE Trans. on Information Theory, Vol. 34, No. 4, 1988.

BIBLIOGRAPHY - PART I

[Erd98]   A. T. Erdem, and C. Eroglu, "The Effect of Image Stabilization on the Performance of the MPEG-2 Video Coding Algorithm", Proc. of VCIP-98, Vol. 1, pp. 272-277, San Jose, California, USA, 1998.

[Fer96]   D. Ferrandiere, "Motion Picture Restoration using Morphological Tools", International Symposium on Mathematical Morphology (ISMM), pp. 361-368, Kluwer Academic Press, 1996.

[Fin65]   T. Fine, "Optimum Mean-Square Quantization of a Noisy Input", IEEE Trans. on Information Theory, pp. 293-294, 1965.

[Gem84]   S. Geman and D. Geman, "Stochastic Relaxation, Gibbs Distribution, and Bayesian Restoration of Images", IEEE Trans. on Pattern Recognition and Machine Intelligence, Vol. 6, pp. 721-741, 1984.

[Goh96]   W.B. Goh, M.N. Chong, S. Kalra, and D. Krishnan, "Bi-Directional 3D Auto-Regressive Model Approach to Motion Picture Restoration", Proc. of ICASSP 96, pp. 2277-2280, 1996.

[Gol51]   P.C. Goldmark and J.M. Hollywood, "A New Technique for Improving the Sharpness of Television Pictures", JSMPTE, Vol. 57, pp. 382-396, 1951.

[Haa92]   G. de Haan, "Motion Estimation and Compensation", Ph.D. Thesis, TU Delft, Delft, The Netherlands, 1992.

[Has92]   P. Haskell and D. MesserSchmitt, "Resynchronization of Motion Compensated Video Affected by ATM Cell Loss", Proc. of ICASSP 1992, Vol. 3, pp. 545-548, San Francisco, USA, 1992.

[Ha97]   B.G. Haskell, A. Puri, and A. N. Netravali, "Digital Video: An Introduction to MPEG-2", Chapman & Hall, 1997.

[Hat67]   J.G. Hayes, "Numerical Approximation to Functions and Data", Canterbury, England, 1967.

[Hir89]   H.P. Hiriyannaiah, G.L. Bilbro, and W.E. Snyder, "Restoration of Piecewise-Constant Images by Mean-Field Annealing", J. Opt. Soc. Amer., pp. 1901-1912, 1989.

[Hol89]   M. Holschneider, R. Kronland-Martinet, and J. Morlet, "A real-time algorithm for signal analysis with the help of the wavelet transform", in Wavelets, Time-Frequency Methods and Phase Space, Berlin: springer, IPTI, pp. 286-297, 1989.

[IEC1]   IEC/ISO 13818-1 IS "General Coding of Moving Pictures and Associated Audio- Part 1: System".

[IEC2]   IEC/ISO 13818-2 IS "General Coding of Moving Pictures and Associated Audio- Part 2: Video".

[IEC3]   IEC/ISO 13818-3 IS "General Coding of Moving Pictures and Associated Audio- Part 3: Audio".

[IEC93]   IEC/ISO JTC1/SC29/WG11/93-400, "Test Model 5", Test Model Editing Committee, 1993.

[ITU95]   ITU-R Rec. BT.500-7, "Methodology for the Subjective Assessment of Quality of Television Pictures", ITU, 1995.

[Jai89]   A.K. Jain, "Fundamentals of Digital Image Processing", Prentice Hall, 1989.

[Jos95]   R.L. Joshi and T.R. Fisher, 'Comparison of Generalized Gaussian and Laplacian Modeling in DCT Image Coding", IEEE Signal Processing Letters, Vol 2, No. 5, pp. 81-82, 1995.

[Jus81]   B. I. Justusson, "Median Filtering: Statistical Properties", in Topics in Applied Physics 43, T.S. Huang ed., pp.161-196, 1981.

[Kal97]   S. Kalra, M.N. Chong, and D. Krishnan, "A new Auto Regressive (AR) model-based algorithm for Motion Picture Restoration", Proc. of ICASSP 97, Vol. 4, pp. 2557-2560, Munich, Germany, 1997.

[Kle94]   R.P. Kleihorst, "Noise Filtering of Image Sequences", Ph. D. Thesis, TU Delft, Delft, The Netherlands, 1994.

[Kok98]   A.C. Kokaram, "Motion Picture Restoration", Springer Verlag, 1998.

[Kok95a]   A.C. Kokaram, R.D. Morris, W.J. Fitzgerald, and P.J.W. Rayner, "Detection of Missing Data in Image Sequences", IEEE Trans. on Image Processing, pp. 1496-1508, Vol. 4, No. 11, 1995.

[Kok95b]   A.C. Kokaram, R.D. Morris, W.J. Fitzgerald, and P.J.W. Rayner, "Interpolation of Missing Data in Image Sequences", IEEE Trans. on Image Processing, pp. 1509-1519, Vol. 4, No. 11, 1995.

[Kok97]   A.C. Kokaram, P.M.B. van Roosmalen, P.J.W. Rayner, and J. Biemond, "Line Registration of Jittered Video", ICASSP-97, pp. 2553-2556, Munich, Germany, 1997.

[Kon92]   J. Konrad and E. Dubois, "Bayesian Estimation of Motion Vector Fields", IEEE Trans. on PAMI, Vol. 14, No. 9, 1992.

[Lag91]   R.L. Lagendijk and J. Biemond, "Iterative Identification and Restoration of Images", Kluwer Academic Publishers, 1991.

[Lag94]   R.L. Lagendijk and J. Biemond, "Statistische Signaalverwerking", DUM, The Netherlands, 1994.

[Lam93]   W. Lam, A.R. Reibman, and B. Liu, "Recovery of Lost or Erroneously Reveived Motion Vectors", Proc. of ICASSP 1993, vol. 5, pp.417-420, Minneapolis, USA, 1993.

[Leo94]   A. Leon-Garcia, "Probability and Random Processes for Electrical Engineering", 2nd. Ed., Addison-Wesley, 1994.

[Mal89]   S. Mallat, "A Theory for Multiresolution Signal Decomposition: The Wavelet Representation", IEEE Trans. on Pattern Anal. and Mach. Intell., Vol. 11, No 7, 1989.

[Mar98]   P. Maragos, "Morphological Signal and Image Processing", in The Digital Signal Processing Handbook, V. Madisettii and D.B. Williams eds., CRC Press, 1998.

[Mar95]	J.B. Martens, "Adaptive Contrast Enhancement through Residue-Image Processing", Signal Processing, Vol. 44, pp. 1-18, 1995.

[McC73]	J.H. McClellan, "The design of two-dimensional filters by transformations", Proc. 7th Annual Princeton conference of Sciences and Systems, pp. 247-251, 1973.

[McM68]	R.H. McMann and A.A. Goldberg, "Improved Signal Processing Techniques for Color Television Broadcasting", JSMPTE, Vol. 77, pp. 221-228, 1968.

[Mit96]	J. L. Mitchell, W.B. Pennebaker, C. E. Fog, and D. J. LeGall, "MPEG Video Compression Standard", Chapman & Hall, USA, 1996.

[Mor96]	R.D. Morris, W.J. Fitzgerald, and A.C. Kokaram, "A Sampling Based Approach to Line Scratch Removal from Motion Picture Frames", Proc. of ICIP-96, vol I, pp. 801-804, Lausanne, Switzerland, IEEE, 1996.

[Mul96]	H. Muller-Seelich, W. Plaschzug, P. Schallauer, S. Potzman, and W. Haas, "Digital Restoration of 35mm Film", Proc. of ECMAST 96, Vol. 1, pp. 255-265, Louvain-la-Neuve, Belgium, 1996.

[Nad97]	M.J. Nadenau and S.K. Mitra, "Blotch and Scratch Detection in Image Sequences based on Rank Ordered Differences", in Time-Varying Image Processing and Moving Object Recognition, V. Cappelini ed., Elsevier, 1997.

[Nar93]	A. Narula and J.S. Lim, "Error Concealment Techniques for an All-Digital High-Definition Television System", Proc. of SPIE, Vol. 2094, pp. 304-315, 1993.

[Nat95]	B.K. Natarajan, "Filtering Random Noise from Deterministic Signals via Data Compression", IEEE Trans. on Signal Processing, Vol. 43, No. 11, pp. 2595-2605, 1995.

[Odo95]	J.M. Odobez and P. Bouthemy, "Robust Multiresolution Estimation of Parametric Motion Models", Journal of Visual Communications and Image Representation, pp. 348-365, 1995.

[Rua96]	J.J.K. Ó Ruanaidh and W.J. Fitzgerald, "Numerical Bayesian Methods Applied to Signal Processing", Springer, USA, 1996.

[Özk92]	M.K. özkan, A.T. Erdem, M.I. Sezan, and A.M. Tekalp, "Efficient Multiframe Wiener Restoration of Blurred and Noisy Image Sequences, IEEE Trans. on Image Processing, Vol 1, no 4, pp 453-476, 1992.

[Özk93]	M.K. özkan, M.I. Sezan, and M. Tekalp, "Adaptive Motion Compensated Filtering of Noisy Image Sequences", Trans. on Circuits and Systems for Video Technology, Vol 3, No. 4, IEEE, 1993.

[Pea77]	J.J. Pearson, D.C. Hines, S. Goldsman, and C.D. Kuglin, "Video Rate Image Correlation Processor", SPIE, Vol. 119, Application of Digital Image processing, 1977.

[Pow82]	P.G. Powell and B.E. Bayer, "A method for the Digital Enhancement of Unsharp, Grainy Photographic Images", IEE Int. Conf. on Electronic Image Processing, No. 214, pp. 179-183, 1982.

[Pra91]	W.K. Pratt, "Digital Image Processing", John Whiley & Sons, 2nd Ed., 1991.

[Pra75] W.K. Pratt, "Vector Space Formulation of Two-Dimensional Signal Processing Operations", Computer Graphics and Image Processing 4, pp. 1-24, 1975.

[Pre92] W.H. Press, S.A. Teukolsky, W.T. Vetterling, and B.P. Flannery, "Numerical Recipes in C", 2nd Ed., Cambridge University Press, U.K., 1992.

[Rei83] R.C. Reiniger and J.D. Gibson, "Distributions of the Two-Dimensional DCT coefficients for Images", IEEE Trans. on Communications, Vol. 31, No. 6, pp. 835-839, 1983.

[Ric95] P. Richardson and D. Suter, "Restoration of Historic FIlm for Digital Compression: A Case Study", Proc. of ICIP-95, Vol. II, pp. 49-52, Washington D.C. USA, IEEE, 1995.

[Rob87] R.A. Roberts, C.T. Mullis, "Digital Signal Processing", Addison-Wesley, USA, 1987.

[Roo] P.M.B. van Roosmalen, J. Biemond, and R.L. Lagendijk, "Restoration and Storage of Film and Video Archive Material", to appear in NATO-ASI Signal Processing for Multimedia.

[Roo99a] P.M.B. van Roosmalen, A.C. Kokaram, and J. Biemond, "Fast High Quality Interpolation of Missing Data in Image Sequences using a Controlled Pasting Scheme", Proc. of ICASSP-99, USA, 1999.

[Roo98a] P.M.B. van Roosmalen, A.C. Kokaram, and J. Biemond, "Noise Reduction of Image Sequences as PreProcessing for MPEG2 Encoding", Signal Processing IX, pp. 2253-2256, Rhodes, Greece, 1998.

[Roo98b] P.M.B. van Roosmalen, R.L. Lagendijk, and J. Biemond, "Improved Blotch Detection by Postprocessing", Proc. of IEEE Signal Processing Symposium SPS'98, pp. 223-226, Leuven, Belgium, 1998.

[Roo99b] P.M.B. van Roosmalen, R.L. Lagendijk, and J. Biemond, "Correction of Intensity Flicker in Old Film Sequences", IEEE Trans. on Circuits and Systems, to appear Dec. 1999.

[Roo97] P.M.B. van Roosmalen, R.L. Lagendijk, and J. Biemond, "Flicker Reduction in Old Film Sequences", in: Time-Varying Image Processing and Moving Object Recognition, 4, V. Cappellini, ed., Elsevier Science, pp. 11-18, 1997.

[Roo96] P.M.B. van Roosmalen, S.J.P. Westen, R.L. Lagendijk, and J. Biemond, "Noise Reduction for Image Sequences using an Oriented Pyramid Thresholding technique", Proc. of ICIP-96, Vol. I, pp. 375-378, Lausanne, Switzerland, IEEE, 1996.

[Ros82] A. Rosenfeld and A. Kak, "Digital Picture Processing", 2nd. Ed., Vol. 1, Academic Press, 1982

[Ros78] J. P. Rossi, "Digital Techniques for Reducing Television Noise", JSMPTE, Vol. 87, pp. 134-140, 1978.

[Sch98] A. van der Schaaf, "Natural Image Statistics and Visual Processing", Ph. D. Thesis, Rijksuniversiteit Groningen, Groningen, The Netherlands, 1998.

[Sha95]	K. Sharifi and A. Leon-Garcia, "Estimation of Shape Parameter for Generalized Gaussian Distributions in Subband Decomposition of Video", IEEE Trans. on Circuits and Systems, Vol. 5, No. 1, pp. 52-56, 1995.
[Sim96]	E.P. Simoncelli and E. H. Adelson, "Noise Removal Via Bayesian Wavelet Coring", Proc. of ICIP-96, Vol. I, pp. 379-382, Lausanne, Switzerland, IEEE, 1996.
[Sim92]	E. P. Simoncelli, W. T. Freeman, E. H. Adelson, and D. J. Heeger, "Shiftable Multiscale Transforms", IEEE Trans. on Information Theory, Vol. 38, No. 2, pp. 587-909, 1992.
[Str88]	G. Strang, "Linear Algebra and its Applications", 3rd Ed., Harcourt Brace Jovanovich, USA, 1988.
[Tek95]	A. M. Tekalp, "Digital Video Processing", Prentice Hall, USA, 1995.
[The92]	C.W. Therrien, "Discrete Random Signals and Statistical Signal Processing", Prentice Hall, USA, 1992.
[Vai92]	P.P. Vaidyanathan, "Multirate systems and filter banks", Prentice Hall, pp. 337-338, 1992.
[Vel88]	R.N.J. Veldhuis, "Adaptive Restoration of Unknown Samples in Discrete-Time Signals and Digital Images", Ph. D. Thesis, Katholieke Universiteit Nijmegen, Nijmegen, The Netherlands, 1988.
[Vet95]	M. Vetterli and J. Kovacevic, "Wavelets and Subband Coding", Prentice Hall, USA, 1995.
[Vla96]	T. Vlachos and G. Thomas, "Motion Estimation for the Correction of Twin-Lens Telecine Flicker", Proc. of ICIP-96, Vol. I, pp. 109-112, Lausanne, Switzerland, IEEE, 1996.
[Vor93]	S. Voran and S. Wolf, "An Objective Technique for Assessing Video Impairments", IEEE Pacific Rim Conference on Communications, Computers and Signal Processing, 1993.
[Wan95]	B.A. Wandell, "Foundations of Vision", Sinauer Associates, USA, 1995.
[Wan]	J.Y.A. Wang and E. Adelson, "Representing Moving Images with Layers", IEEE Trans. on Image Processing, Vol. 3, pp. 625-638.
[Wes95]	S.J.P. Westen, R.L. Lagendijk, and J. Biemond, "Perceptual Image Quality based on a Multiple Channel HVS Model", Proc. of ICASSP-95, pp. 2351-2354, USA, 1995.
[Wol70]	J.K. Wolf and J. Ziv, "Transmission of Noisy Information to a Noisy Receiver With Minimum Distortion", IEEE Trans. on Information Theory, Vol. 16, No. 4, pp. 406-411, 1970.
[Won68]	E. Wong, "Two-Dimensional random fields and representation of Images", SIAM, Journal of Applied Mathematics, Vol. 16, No. 4, pp. 756-770, 1968.
[Woo72]	J. Woods, "Two-Dimensional discrete Markov Random Fields", IEEE Trans. on Information Theory, Vol. 18, No 2, pp. 232-240, 1972.

Image and Video Databases

Part II: Watermarking

Chapter 7
Introduction to Watermarking

7.1 The need for watermarking

In the past few years there has been an explosion in the use and distribution of digital multimedia data. Personal computers with internet connections have taken the homes by storm, and have made the distribution of multimedia data and applications much easier and faster. Electronic commerce applications and on-line services are rapidly being developed. Even the analog audio and video equipment in the home are in the process of being replaced by their digital successors. As a result, we can see the digital mass recording devices for multimedia data enter the consumer market of today.

Although digital data have many advantages over analog data, service providers are reluctant to offer services in digital form because they fear unrestricted duplication and dissemination of copyrighted material. Because of possible copyright issues, the intellectual property of digitally recorded material must be protected [Sam91]. The lack of such adequate protection systems for copyrighted content was the reason for the delayed introduction of the Digital Versatile Disk (DVD) [Tay97]. Several media companies initially refused to provide DVD material until the copy protection problem had been addressed [Rup96] and [Ren96]. Representatives of the consumer electronics industry and the motion picture industry have agreed to seek legislation concerning digital video recording devices. Recommendations describing ways that would protect both intellectual property and consumers' rights have been submitted to the US Congress [Ren96] and resulted in the Digital Millennium Copyright Act [DCM98], which was signed by President Clinton October 28, 1998.

To provide copy protection and copyright protection for digital audio and video data, two complementary techniques are being developed: encryption and watermarking [Cox97]. Encryption techniques can be used to protect digital data during the transmission from the sender to the receiver [Lan99a]. However, after the receiver has received and decrypted the data, the data is in the clear and no longer protected. Watermarking techniques can complement encryption by embedding a secret imperceptible signal, a watermark, directly into the clear data in such a way that it always remains present. Such a watermark can for instance be used for the following purposes:

- **Copyright protection:** For the protection of intellectual property, the data owner can embed a watermark representing copyright information in his data. This watermark can prove his ownership in court when someone has infringed on his copyrights.

- **Fingerprinting:** To trace the source of illegal copies, the owner can use a fingerprinting technique. In this case, the owner can embed different watermarks in the copies of the data that are supplied to different customers. Fingerprinting can be compared to embedding a serial number that is related to the customer's identity in the data. It enables the intellectual property owner to identify customers who have broken their license agreement by supplying the data to third parties. In Section 7.4 a fingerprinting application is explained in more detail.

- **Copy protection:** The information stored in a watermark can directly control digital recording devices for copy protection purposes [Lan98a]. In this case, the watermark represents a copy-prohibit bit and watermark detectors in the recorder determine whether the data offered to the recorder may be stored or not. A complete copy protection system is discussed in Section 7.4.

- **Broadcast monitoring:** By embedding watermarks in commercial advertisements an automated monitoring system can verify whether advertisements are broadcasted as contracted [And98]. Not only commercials but also valuable TV products can be protected by broadcast monitoring [Kal99]. News items can have a value of over 100.000 USD per hour, which make them very vulnerable to intellectual property rights violation. A broadcast surveillance system can check all broadcast channels and charge the TV stations according to their findings.

- **Data authentication:** *Fragile* watermarks [Wol99a] can be used to check the authenticity of the data. A fragile watermark indicates whether the data has been altered and supplies localization information as to where the data was altered.

Watermarking techniques are not only used for protection purposes. Other applications include:

- **Indexing:** Indexing of video mail, where comments can be embedded in the video content; indexing of movies and news items, where markers and comments can be inserted that can be used by search engines.

- **Medical safety:** Embedding the date and the patient's name in medical images could be a useful safety measure [And98].

- **Data hiding:** Watermarking techniques can be used for the transmission of secret private messages. Since various governments restrict the use of encryption services, people may hide their messages in other data.

7.2 Watermarking requirements

Each watermarking application has its own specific requirements. Therefore, there is no set of requirements to be met by all watermarking techniques. Nevertheless, some general directions can be given for most of the applications mentioned above:

- **Perceptual transparency:** In most applications the watermarking algorithm must embed the watermark such that this does not affect the quality of the underlying host data. A watermark-embedding procedure is truly imperceptible if humans cannot distinguish the original data from the data with the inserted watermark [Swa98]. However, even the smallest modification in the host data may become apparent when the original data is compared directly with the watermarked data. Since users of watermarked data normally do not have access to the original data, they cannot perform this comparison. Therefore, it may be sufficient that the modifications in the watermarked data go unnoticed as long as the data are not compared with the original data [Voy98].

- **Payload of the watermark:** The amount of information that can be stored in a watermark depends on the application. For copy protection purposes, a payload of one bit is usually sufficient.

  According to a recent proposal for audio watermarking technology from the International Federation for the Phonographic Industry, (IFPI), the minimum payload for an audio watermark should be 20 bits per second, independently of the signal level and music type [Int97]. However, according to [Pet98a] this minimum is very ambitious and should be lowered to only a few bits per second. For the protection of intellectual property rights, it seems reasonable to assume that one wants to embed an amount of information similar to that used for ISBN, International Standard Book Numbering, (roughly 10 digits) or better ISRC, International Standard Recording Code, (roughly 12 alphanumeric letters). On top of this, one should also add the year of copyright, the permissions granted on the work and rating for it [Kut99]. This means that about 60 bits [Fri99a] or 70 bits [Kut99] of information should be embedded in the host data, the image, video-frame or audio fragment.

- **Robustness:** A fragile watermark that has to prove the authenticity of the host data does not have to be robust against processing techniques or intentional alterations of the host data, since failure to detect the watermark proves that the host data has been modified and is no longer authentic. However, if a watermark is used for another application, it is desirable that the watermark always remains in the host data, even if the quality of the host data is degraded, intentionally or unintentionally. Examples of unintentional degradations are applications involving storage or transmission of data, where lossy compression techniques are applied to the data to reduce bit rates and increase efficiency. Other unintentional quality-degrading processing techniques include filtering, re-sampling, digital-analog (D/A) and analog-digital (A/D) conversion. On the other hand, a watermark can also be subjected to processing solely intended to remove the watermark [Cox97]. In addition, when many copies of the same content exist with different watermarks, as would be the case for fingerprinting, watermark removal is possible because of collusion between several owners of copies. In general, there should be no way in which the watermark can be removed or altered without sufficient degradation of the perceptual quality of the host data so as to render it unusable.

- **Security:** The security of watermarking techniques can be interpreted in the same way as the security of encryption techniques. According to Kerckhoffs [And98], one should assume that the method used to encrypt the data is known to an unauthorized party, and that the security must lie in the choice of a key. Hence a watermarking technique is truly secure if knowing the exact algorithms for embedding and extracting the watermark does not help an unauthorized party to detect the presence of the watermark [Swa98].

- **Oblivious vs. non-oblivious watermarking:** In some applications, like copyright protection and data monitoring, watermark extraction algorithms can use the original unwatermarked data to find the watermark. This is called *non-oblivious* watermarking [Kut99]. In most other applications, e.g. copy protection and indexing, the watermark-extraction algorithms do not have access to the original unwatermarked data. This renders the watermark extraction more difficult. Watermarking algorithms of this kind are referred to as *public*, *blind* or *oblivious* watermarking algorithms.

The requirements listed above are all related to each other. For instance, a very robust watermark can be obtained by making many large modifications to the host data for each bit of the watermark. However, large modifications in de host data will be noticeable and many modifications per watermark bit will limit the maximum amount of watermark bits that can be stored in a data object. Hence, a trade-off should be found between the different requirements so that an optimal watermark for each application can be developed. The mutual dependencies between the basic requirements are shown in Figure 7.1.

The relation between the basic requirements for a well-designed secure watermark is represented in Figure 7.2. The security of a watermark influences the robustness enormously. If a watermark is not secure, it cannot be very robust.

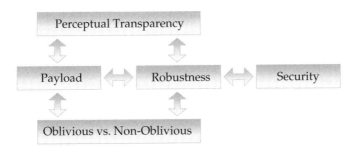

Figure 7.1: *Mutual dependencies between the basic requirements.*

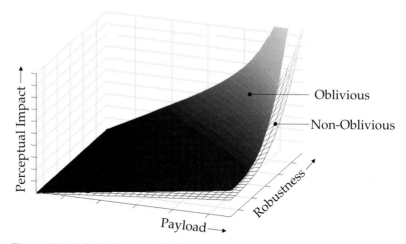

Figure 7.2: *Relation between the basic requirements for a secure watermark.*

7.3 Brief history of watermarking

Watermarking techniques are not new. Watermarking forms a particular group in the steganography field. *Steganography* stems from the Greek words στεγανος for "covered" and γραφω for "to write", and means covered or secret writing. While classical cryptography is about rendering messages unintelligible to unauthorized persons, steganography is about concealing the existence of the messages. Kahn has traced the roots of steganography to Egypt 4000 years back, where hieroglyphic symbol substitutions were used to inscribe information in the tomb of a nobleman, Khnumhoteb II [Kah67] and [Swa98].

Herodotus wrote about how the Greeks received a warning of Xerxes' hostile intentions through a message underneath the wax of a writing tablet [Her72]. Another secret writing method he described was to shave the head of a messenger and tattoo a message or image on the messenger's head. After the hair had grown back, the message would be undetectable until the head was shaved again [Joh98] and [Kob97].

A method suggested by Aenas the Tactician was to mark successive letters in a cover text with secret ink, barely visible pin pricks or small dots and dashes [Kah67]. The marked letters formed the secret message.

Johannes Trithemius (1462-1526), a German monk, was the first who used the term steganography. He encoded letters as religious words in such a way as to turn covert messages into apparently meaningful prayers. As a reward for this artifice the first printing of his manuscript Steganographia in 1606 was placed on the Vatican's prohibited Index and was characterized as "full of peril and superstition" [Kah67] and [Lea96].

Figure 7.3: *Title page of Porta's book: De occultis notis.*

In 1593, Giovanni Baptista Porta published a book about cryptography under the title: *De occultis literarum notis seu artis animemi occulte alijs significadi, aut ab alijs significata expiscandi enodandique. Libri III* (Figure 7.3). In his book, he describes amongst others a method for concealing a secret text message in a cover message by means of a mask. In the following example the secret message can be extracted by ignoring the masked (gray) text [Por93]:

> Honor Militiae tuus suit Carolus pater, nam cum infini to victus est, cum mini 1a exercitu inuitus parte hostis fugit, ac prope ultimum diem iniurius peribit, necabi nt Bere illum; atque extemplo puer Arato peribit, res omnes deprehensae bonae si su it, ante Sillam, & optimo capite non poenitentias amplius decidere sperabit. Vale.

In the 17th century it was not unusual to publish manuscripts anonymously, especially if it concerned the writing of histories. The risk of offending powerful political parties, which could have severe consequences to the author, was far too great. Therefore, Bishop Francis Godwin coded his name as the initial capital letters of each chapter of his manuscript [Lea96]. This is an early example of copyright protection.

An example of embedding copyright or authorship information in musical scores was practiced by Bach, who embedded his name in many of his pieces. For instance, in his organ chorale "Vor deinem Thron", he used null cipher coding by spelling out *B-A-C-H* in notes, where B-flat represents *B*, and B represents *H* or by counting the number of occurrences of a note, one occurrence for *A*, two for *B*, three for *C* and eight for *H* [Swa98].

In World War II steganogaphic techniques were widely used [Kah67] and [Joh98]. In the USA the post banned a large class of objects that could conceal messages, like chess games, crosswords and newspaper clippings. Other objects were changed before these were delivered, lovers' Xs were deleted, watch hands were shifted, loose stamps and blank paper were replaced. Censors even rephrased telegrams to prevent that people hid secret messages in normal text messages. In one case, a censor changed "father is dead" to "father is deceased", which resulted in the reply "is father dead or deceased?". Thousands of people were involved in reading mail, looking for language which appeared to be forced. For example, the following message was actually sent by a German spy [Kah67]:

> Apparently neutral's protest is thoroughly discounted and ignored. Isman hard hit. Blockade issue affects pretext for embargo on by-products, ejecting suets and vegetable oils.

Extracting the second letter in each word reveals the following message:

> Pershing sails from NY June 1.

During the 1980s steganographic techniques were used for fingerprinting. Prime Minister Margaret Thatcher became so irritated at press leaks of cabinet documents that she had the word processors reprogrammed to encode the user's identity in the word spacing, so that disloyal ministers could be traced [And98].

From this brief history overview we can conclude that most applications mentioned in Section 7.1 are nothing else than variations on the historical ones.

7.4 Scope of Part II

There are many types of watermarking techniques. This book concentrates on techniques for *real-time embedding of watermarks in and the extraction of watermarks from*

INTRODUCTION TO WATERMARKING 165

compressed image and video data. These watermarking techniques can for instance be used in fingerprinting and copy protection systems for home-recording devices.

- **Fingerprinting:** A consumer can receive digital services, like pay TV or video on demand, by cable or satellite dish using a set-top box and a smart card, which he has to buy and can therefore be related to his identity. To prevent other non-paying consumers to make use of the same services, the service provider encrypts the data, for which he uses one or more keys. This protects the services during transmission. The set-top box in the home of the consumer decrypts the data if a valid smart card is used, and adds a watermark, representing the identity of the user, to the compressed clear data. The fingerprinted data can now be fed to the internal video decoder to view the data or the data can be stored in compressed form.

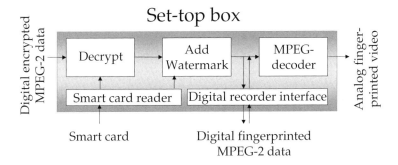

Figure 7.4: *Set-top box with fingerprinting capabilities.*

The service provider can now identify consumers who supply data to third parties breaking their license agreement. The complete scheme of a set-top box with fingerprinting facilities is depicted in Figure 7.4.

- **Copy protection:** Service providers are reluctant to accept digital recording devices, because of they fear unrestricted copying of services like Pay TV, Pay-Per-View and Video-On-Demand. However, digital video recorders enable consumers to use services on another time than the time the services are actually broadcasted (*time-shifting*), or to insert longer breaks in a movie. A compromise between the conflicting desires of the service providers and the consumers would be the embedding of an SCMS-like [IEC958] copy protection system in each digital recorder [Han96].

Using the Serial Copy Management System, consumers can make copies of any digital source, but they cannot make copies of copies. An example of an SCMS-like copy protection scheme using watermarking techniques is shown in Figure 7.5.

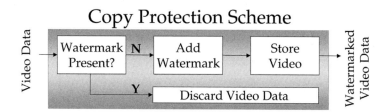

Figure 7.5: *A copy protection scheme for digital recorders.*

This copy protection system checks all incoming video streams for a predefined copy-prohibit watermark. If such a watermark is found, the incoming video must already have been copied before and is therefore refused by the recorder. If the copy-prohibit watermark is not found, the watermark is embedded and the watermarked video is stored. This means that video data stored on this recorder always contains a watermark and cannot be duplicated if a recorder is used equipped with such a copy protection system.

Besides the basic requirements mentioned in Section 7.2, a watermarking technique should meet the following extra requirements to qualify as a real-time technique for compressed image and video data applicable to recording devices:

- **Oblivious:** It should be possible to extract watermark information without using the original unwatermarked data, since a recorder and a set-top box do not have the original data at their disposal.

- **Low complexity:** There are two reasons why the watermarking techniques cannot be too complex: they are to be processed in real time, and as they are to be used in consumer products, they must be inexpensive. This means that fully decompressing the data, adding a watermark and finally compressing the data is not an option for embedding a watermark.

- **Preserve host data size:** The watermark should not increase the size of the compressed host data. For instance, if the size of a compressed MPEG-video stream increases, transmission over a fixed bit rate channel can cause problems, the buffers in hardware decoders can run out of space, or the synchronization of audio and video can be disturbed.

Protection systems that make use of watermarking techniques consist in general of a chain of cryptographic techniques. The watermark information can be encrypted first. Subsequently, the processed watermark information is added to the host data by means of embedding techniques. The encryption and embedding techniques use keys; these keys may vary in time. Cryptography protocols have to take care of the key-management problem. In Figure 7.6 the involved fields of cryptography are represented graphically. The subjects of encryption and protocol development are outside the scope of this book. The focus is on developing, analyzing and testing the embedding techniques for watermarks.

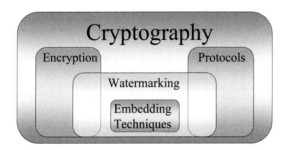

Figure 7.6: *Fields of cryptography involved in watermarking applications.*

7.5 Overview of Part II

Part II of this book is structured as follows. In Chapter 8 the state of the art in watermarking techniques for digital image and video data is presented. Since the most commonly used watermarking techniques use additive noise for watermark embedding and correlation techniques for watermark detection, the correlation-based techniques are discussed in full detail here. Various correlation-based techniques are explained for embedding video content dependent or independent watermarks representing one bit, multiple bits or logos in the spatial, Fourier,

Discrete Cosine or Discrete Wavelet Transform domain which do or do not use Human Visual System models to maximize the watermark energy. In addition extra measures are discussed that make these watermarks resistant to lossy compression techniques and geometrical transformations. Other non-correlation-based techniques, like least significant bit modification, DCT coefficient ordering, salient point modification and fractal-based techniques are briefly explained at the end of this chapter.

In Chapter 9 the state of the art in real-time watermarking algorithms for compressed video data is discussed. Furthermore, two new algorithms are proposed and evaluated that are computationally highly efficient and very suitable for consumer applications requiring moderate robustness. These real-time watermarking algorithms are based on the basic Least Significant Bit (LSB) modification principle, which is here directly applied to MPEG compressed video streams. Since the watermarking methods discussed in this chapter rely heavily on the MPEG video compression standard, this chapter starts with a brief description of the relevant parts of the MPEG standard.

In Chapter 10 the slightly more complex Differential Energy Watermarking (DEW) concept is proposed which is applicable for real-time consumer applications requiring more robustness. The DEW concept is suitable for directly embedding watermarks in and extracting watermarks from MPEG/JPEG or embedded zero tree wavelet encoded video and image data. The DEW algorithm embeds the label bits of the watermark by selectively discarding high frequency coefficients in certain video frame regions. The label bits of the watermark are encoded in the pattern of energy differences between DCT blocks or hierarchical wavelet trees.

Chapter 11 describes how a statistical model is derived and experimentally validated to find optimal parameter settings for the DEW algorithm. The performance of the DEW algorithm has been defined as its robustness against re-encoding attacks, its label size, and its visual impact. We show analytically how the performance is controlled by three embedding parameters. The derived statistical model gives us an expression for the label bit error probability as a function of these three parameters. Using this expression, we show how we can optimize a watermark for robustness, label size or visibility and how we can add adequate error correcting codes to the label bits.

In Chapter 12 the DEW algorithm is evaluated. For this purpose, benchmarking approaches for watermarking algorithms and watermark removal attacks described in literature are discussed. Next, the performance of the DEW algorithm for MPEG compressed video data is compared to a real-time spread spectrum technique for MPEG compressed video data. Finally, the DEW algorithm for JPEG compressed and uncompressed still images is compared to a basic spread spectrum method, which is not specially designed for real-time operation on compressed data. The real-time aspect is neglected in this comparison and for the evaluation the guidelines of the benchmarking methods from literature are followed and the removal attacks are taken into account.

Chapter 8

State-of-the-Art in Image and Video Watermarking

8.1 Introduction

In order to embed watermark information in host data, watermark embedding techniques apply minor modifications to the host data in a perceptually invisible manner, where the modifications are related to the watermark information. The watermark information can be retrieved afterwards from the watermarked data by detecting the presence of these modifications.

A wide range of modifications in any domain can be used for watermarking techniques. Prior to embedding or extracting a watermark, the host data can be converted to, for instance, the spatial, the Fourier, the Wavelet, the Discrete Cosine Transform or even the Fractal domain, where the properties of the specific transform domains can be exploited. In these domains modifications can be made like: Least Significant Bit modification, noise addition, coefficient re-ordering, coefficient removal, warping or morphing data parts and block similarities enforcing. Further, the impact of the modifications can be minimized with the aid of Human Visual Models, whereas modifications can be adapted to the anticipated post-processing techniques or to the compression format of the host data.

Since the most commonly used techniques use additive noise for watermark embedding and correlation techniques for watermark detection, we discuss the oblivious correlation-based techniques extensively in this chapter, together with all its possible variations. Other oblivious techniques are briefly explained at the end of this chapter. The cryptographic security of the methods described here lies in the key that is used to generate a pseudorandom watermark pattern or to pseudorandomly select image regions or coefficients to embed the watermark. In general, the robustness of the watermark against processing techniques depends on the embedding depth and the amount of information bits of the watermark.

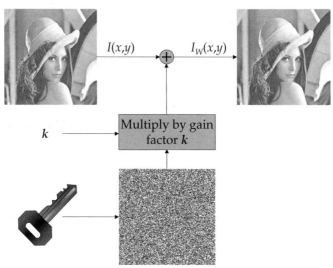

$W(x,y)$: Pseudo Random Pattern {-1,0,1}

Figure 8.2.1: *Watermark embedding procedure.*

8.2 Correlation-based watermark techniques

8.2.1 Basic technique in the spatial domain

The most straightforward way to add a watermark to an image in the spatial domain is to add a pseudorandom noise pattern to the luminance values of its pixels. Many methods are based on this principle [Sch94], [Ben95], [Pit95], [Car95], [Har96], [Lan96a], [Pit96a], [Smi96], [Wol96], [Lan97a], [Wol97], [Zen97], [Fri99b], [Wol98], [Wol99a] and [Kal99]. In general, the pseudorandom noise pattern consists of the integers {-1,0,1}, however also floating-point numbers can be used. The pattern is generated based on a key using, for instance, seeds, linear shift registers or randomly shuffled binary images. The only constraints are that the energy in the pattern is more or less uniformly distributed and that the pattern is not correlated with the host image content. To create the watermarked image $I_w(x,y)$ the pseudorandom pattern $W(x,y)$ is multiplied by a small gain factor k and added to the host image $I(x,y)$, as illustrated in Figure 8.2.1.

$$I_w(x,y) = I(x,y) + k \cdot W(x,y) \quad (8.2.1)$$

To detect a watermark in a possibly watermarked image $I'_w(x,y)$ we calculate the correlation between the image $I'_w(x,y)$ and the pseudorandom noise pattern $W(x,y)$. In general, $W(x,y)$ is normalized to a zero mean before correlation. If the correlation R_{XY} exceeds a certain threshold T the watermark detector determines that image $I'_w(x,y)$ contains watermark $W(x,y)$:

$$R_{I'_w(x,y)W(x,y)} \begin{array}{l} > T \\ < T \end{array} \begin{array}{l} \rightarrow \quad W(x,y) \text{ detected} \\ \rightarrow \quad \text{No } W(x,y) \text{ detected} \end{array} \quad (8.2.2)$$

If $W(x,y)$ only consists of the integers $\{-1,1\}$ and if the number of -1s equals the number of 1s, we can estimate the correlation as:

$$R_{I'_w(x,y)W(x,y)} = \frac{1}{Z}\sum_{i=1}^{Z} I'_{W_i}(x,y)W_i(x,y) = \frac{1}{Z}\sum_{i=1}^{Z/2} I'_{W_i} W_i^+ + \frac{1}{Z}\sum_{i=1}^{Z/2} I'_{W_i} W_i^-$$
$$= \frac{1}{2}\{\mu[I_W^+(x,y)] - \mu[I_W^-(x,y)]\} \quad (8.2.3)$$

Where Z is the number of pixels in the image I'_w, and $^{+,-}$ indicates the set of pixels where the corresponding noise pattern is positive or negative, and $\mu[I'_w{}^+(x,y)]$ represents the average value of set pixels in $I'_w{}^+(x,y)$. From Equation 8.2.3 it follows that the watermark detection problem corresponds to testing the hypothesis whether two randomly selected sets of pixels in a watermarked image have the same mean.

Figure 8.2.2 shows that the watermark detector can make two types of errors. In the first place, it can detect the existence of a watermark, although there is none. This is called a false positive. In the second place, the detector can reject the existence of the watermark, even though there is one. This is called a false negative. In [Kal98a] the probabilities of these two types of errors are derived based on a first-order autoregressive image model:

$$P_{fp} = \frac{1}{2}\mathrm{erfc}(\frac{T\sqrt{Z}}{\sigma_W \sigma_I \sqrt{2}}) \quad \text{and} \quad P_{fn} = \frac{1}{2}\mathrm{erfc}(\frac{(\sigma_W^2 - T)\sqrt{Z}}{\sigma_W \sigma_I \sqrt{2}}) \quad (8.2.4)$$

where $\mathrm{erfc}(x) = \dfrac{1}{\sqrt{2\pi}} \int\limits_{x}^{\infty} e^{-t^2/2} dt$

Here, σ_w^2 represents the variance of the watermark pixels and σ_I^2 denotes the variance of the image pixels. If the watermark pattern $W(x,y)$ only consists of the integers $\{-1,1\}$ and the number of -1s equals the number of 1s, the variance of the watermark σ_w^2 equals k^2. The errors P_{fp} and P_{fn} can be minimized by increasing the gain factor k. However, using larger values for the gain factor decreases the visual quality of the watermarked image.

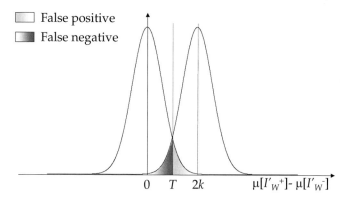

Figure 8.2.2: *Watermark detection procedure.*

Since the image content can interfere with the watermark, especially in the low frequency components, the reliability of the detector can be improved by applying matched filtering before correlation [Dep98], [Sch94], [Lan96a]. This decreases the contribution of the original image to the correlation. For instance, a simple edge-enhance FIR filter F_{edge} can be used, where F_{edge} is given by the following convolution kernel:

$$F_{edge} = \begin{pmatrix} -1 & -1 & -1 \\ -1 & 10 & -1 \\ -1 & -1 & -1 \end{pmatrix} / 2 \qquad (8.2.5)$$

The experimental results presented in the next section show that applying this filter before correlation reduces the error probability significantly, even when the visual quality of the watermarked image was affected seriously before correlation [Lan96a], [Lan97a].

8.2.2 Extensions to embed multiple bits or logos in one image

From the watermark detector's point of view, an image I can be regarded as Gaussian noise, which distorts the watermark information W. Further, the watermarked image I_w can be seen as the output of a communication channel subject to Gaussian noise over which the watermark information is transmitted. In this case, reliable transmission of the watermark is theoretically possible if its information rate does not exceed the channel capacity, which is given by [Sha49]:

$$C_{ch} = W_b \log_2\left(1 + \frac{\sigma_W^2}{\sigma_I^2}\right) \text{ bit/pixel} \tag{8.2.6}$$

Here, C_{ch} is given in units of watermark information bits per image pixel and the available bandwidth W_b is equal to 1 cycle per pixel. However, for practical systems a tighter empirically lower bound can be determined [Smi96]:

$$C_{ch} = W_b \log_2\left(1 + \frac{\sigma_W^2}{\alpha \cdot \sigma_I^2}\right) \text{ bit/pixel} \tag{8.2.7}$$

Here, α is a small headroom factor, which is larger than 1 and typically around 3. Since the signal-to-noise ratio σ_W^2/σ_I^2 is significantly smaller than 1, Equation 8.2.7 can be approximated by:

$$C_{ch} \approx \frac{1}{\ln 2}\left(\frac{\sigma_W^2}{\alpha \cdot \sigma_I^2}\right) \text{ bit/pixel} \tag{8.2.8}$$

According to this equation, it should be possible to store much more information in an image than just 1 bit using the basic technique described in the previous section. For instance, a watermark consisting of the integers {-k, k} added to the 512x512 "Lena image" (Figure 8.2.1) can carry approximately 50, 200 or 500 bits of information for k=1, 2 or 3 respectively and for α=3.

There are several ways to increase the payload of the basic watermarking technique. The simplest way to embed a string of l watermark bits $b_0 b_1 \ldots b_{l-1}$ in an image is to divide the image I into l sub-images $I_0 I_1 \ldots I_{l-1}$ and to add a watermark to each sub-image, where each watermark represents one bit of the string [Smi96], [Lan96a] and [Lan97a]. This procedure is depicted in Figure 8.2.3.

Using Equation 8.2.8 we can calculate the number of pixels P required per sub-image for reliable detection of a single bit in a sub-image:

$$P \approx \frac{\alpha \sigma_I^2 \ln 2}{\sigma_W^2} \text{ pixels} \tag{8.2.9}$$

The watermark bits can be represented in several ways. A pseudorandom pattern can be added if the watermark bit equals one, and the sub-image can be left unaffected if the watermark bit equals zero. In this case, the detector calculates the correlation between the sub-image and the pseudorandom pattern and assigns the value 1 to the watermark bit if the correlation exceeds a certain threshold T; otherwise the watermark bit is assumed to be 0.

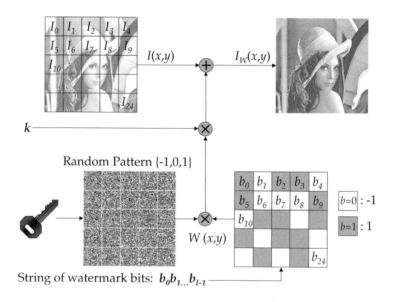

Figure 8.2.3: *Watermark bit string embedding procedure.*

The use of a threshold can be circumvented by adding two different pseudorandom patterns RP_0 and RP_1 for watermark bit 0 and 1. The detector now calculates the correlation between the sub-image and the two patterns. The bit value corresponding with the pattern that gives the highest correlation is assigned to the watermark bit. In [Smi96] the two patterns are chosen in such a way that they only differ in sign, $RP_0 = -RP_1$. In this case, the detector only has to calculate the correlation between the sub-image and one of the patterns; the sign of the correlation determines the watermark bit value.

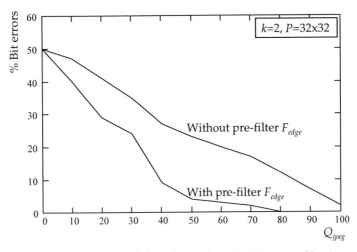

Figure 8.2.4 *Watermark detection with and without pre-filtering.*

To investigate the effect on the robustness of the watermark of the pre-filter in the detector, the gain factor k, and the number of pixels P per watermark bit, we perform the following experiments. We first add a watermark to an image with the method of [Smi96]. Next, we compress the watermarked image with the JPEG algorithm [Pen93], where the quality factor Q_{jpeg} of the compression algorithm is made variable. Finally, the watermark is extracted from the decompressed image and compared bit by bit with the originally embedded watermark bits. From this experiment, we find the percentages of watermark bit errors due to JPEG compression as a function of the JPEG quality factor.

The first experiment shows the effect of applying the pre-filter given by Equation 8.2.5 before detection of a watermark embedded with a gain factor $k=2$, and $P=32\times32$ pixels per watermark bit. In Figure 8.2.4 the percentages bit errors caused by JPEG compression are plotted for a detector that uses this pre-filter and for a plain detector. It can clearly be seen that pre-filtering significantly increases the robustness of the watermark.

The second experiment shows the effect of increasing the gain factor k for a watermark embedded with $P=32\times32$ pixels per watermark bit and detected using a pre-filter. From Figure 8.2.5 it follows that the robustness of a watermark can be improved significantly by increasing the gain factor.

The third experiment shows the influence of the number of pixels P per watermark bit on the robustness of a watermark embedded with a gain factor $k=2$

and detected using a pre-filter. From Figure 8.2.6 it follows that decreasing the payload of the watermark by increasing P improves the robustness significantly.

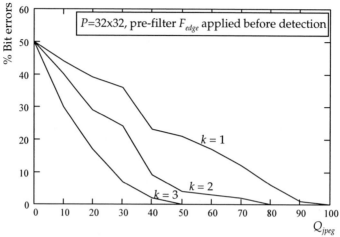

Figure 8.2.5: *Influence of the gain factor k on the robustness of a watermark.*

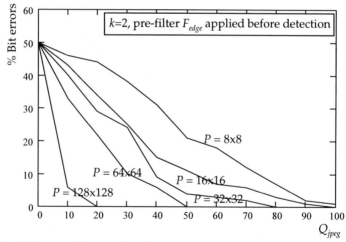

Figure 8.2.6. *Influence of the number of pixels per watermark bit P on the robustness of a watermark.*

Another way to increase the payload of the basic watermarking technique is the use of Direct Sequence Code Division Multiple Access (DS-CDMA) spread spectrum communications [Rua98a] and [Rua98b]. Here, for each bit b_j out of the watermark

STATE-OF-THE-ART IN IMAGE AND VIDEO WATERMARKING 179

bit string $b_0 b_1 \ldots b_{l-1}$ a different stochastically independent pseudorandom pattern RP_j is generated that has the same size as the image. This pattern is dependent on the bit value b_j. Here we use the pattern $+RP_j$ if b_j represents a 0 and $-RP_j$ if b_j represents a 1. The summation of all l random patterns $\pm RP_j$ forms the watermark. Prior to adding the watermark to an image, we can scale the watermark by a gain factor or limit it to a certain small range. An example of the 1-dimensional watermark generation is presented in Figure 8.2.7. This example uses 7 different pseudorandom patterns to embed the 7 watermark bits 0011010.

```
RP₀:-1  1  1-1-1  1-1-1  1  1-1    b₀:0  →   +RP₀:-1  1  1-1-1  1-1-1  1  1-1
RP₁:  1  1-1-1  1-1-1  1  1-1  1    b₁:0  →   +RP₁:  1  1-1-1  1-1-1  1  1-1  1
RP₂:  1-1-1  1-1-1  1  1-1  1-1    b₂:1  →   -RP₂:-1  1  1-1  1  1-1-1  1-1  1
RP₃:-1-1  1-1-1  1  1-1  1-1-1    b₃:1  →   -RP₃:  1  1-1  1  1-1-1  1-1  1  1
RP₄:-1  1-1-1  1  1-1  1-1-1  1    b₄:0  →   +RP₄:-1  1-1-1  1  1-1  1-1-1  1
RP₅:  1-1-1  1  1-1  1-1-1  1  1    b₅:1  →   -RP₅:-1  1  1-1-1  1-1  1  1-1-1
RP₆:-1-1  1  1-1  1-1-1  1  1  1    b₆:0  →   +RP₆:-1-1  1  1-1  1-1-1  1  1  1  +
                                               W    :-3  5  1-3  1  3-7  1  3-1  3
```

Figure 8.2.7: *Example of a CDMA watermark generation for 7 bits* $b_0 b_1 \ldots b_7$.

Each bit b_j out of the watermark bit string $b_0 b_1 \ldots b_{l-1}$ can be extracted by calculating the correlation between the normalized image I'_W and the corresponding pseudorandom pattern RP_j. If the correlation is positive, the value 0 is assigned to the watermark bit, otherwise the watermark bit is assumed to be 1. Figure 8.2.8 shows as an example the extraction of the embedded watermark bits in Figure 8.2.7.

```
W   : -3    5   1 -3   1   3 -7   1   3 -1   3
I   : 98   98  97  98  97  96  97  96  95  94  94  +
I_w : 95  103  98  95  98  99  90  97  98  93  97
```

$$E[(RP_0 - E[RP_0]) \cdot (I_W - E[I_W])] = +15.6 \rightarrow b_0 = 0$$
$$E[(RP_1 - E[RP_1]) \cdot (I_W - E[I_W])] = +16.4 \rightarrow b_1 = 0$$
$$E[(RP_2 - E[RP_2]) \cdot (I_W - E[I_W])] = -26.4 \rightarrow b_2 = 1$$
$$E[(RP_3 - E[RP_3]) \cdot (I_W - E[I_W])] = -3.1 \rightarrow b_3 = 1$$
$$E[(RP_4 - E[RP_4]) \cdot (I_W - E[I_W])] = +21.6 \rightarrow b_4 = 0$$
$$E[(RP_5 - E[RP_5]) \cdot (I_W - E[I_W])] = -23.6 \rightarrow b_5 = 1$$
$$E[(RP_6 - E[RP_6]) \cdot (I_W - E[I_W])] = +0.4 \rightarrow b_6 = 0$$

Figure 8.2.8: *Example of CDMA watermark extraction, compare to Figure 8.2.7.*

Both ways of extending the watermark payload have their advantages and disadvantages. If each watermark bit has its own image tile, there is no interference between the bits and only a small number of multiplications are required to calculate the correlations. However, if the image is cropped, the watermark bits located at the border are lost. If CDMA techniques are used, the probability that all bits can be recovered after cropping the image is high. However, the watermark bits may interfere with each other and many multiplications are required to calculate the correlations, since each bit is completely spread over the image.

The watermark bits embedded using the methods mentioned above can represent anything: copyright messages, serial numbers, plain text, control signals etc. The content represented by these bits can be compressed, encrypted and protected by error correcting codes. In some cases it may be more useful to embed a small logo instead of a bit string as a watermark. If the watermarked image is distorted, the watermark logo will also be affected. But now the sophisticated pattern-recognition capabilities of the human visual system can be exploited to detect the logo [Bra97], [Hsu96] and [Voy96]. For instance, we can embed a binary watermark logo with 128x32 pixels in an image with 512x512 pixels using the techniques described in this section. Each logo pixel is embedded in an image tile of 8x8 pixels by adding the pseudorandom pattern +RP or –RP to the image tile for a black or white logo pixel respectively. As an example in Figure 8.2.9 the results are shown of the logos extracted after the watermarked image has been degraded with the lossy JPEG [Pen93] compression algorithm using several quality factors. From Figure 8.2.9 it can be seen that, although it is heavily corrupted, the logo can still be recognized.

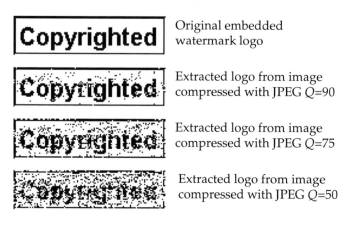

Figure 8.2.9: *Extracted watermark logos from a JPEG distorted image.*

8.2.3 Techniques for other than spatial domains

The techniques described in the previous section can also be applied in other non-spatial domains. Each transform domain has it own advantages and disadvantages. In [Rua96c] the phase of the Discrete Fourier Transform (DFT) is used to embed a watermark, because the phase is more important than the amplitude of the DFT values for the intelligibility of an image. Putting a watermark in the most important components of an image improves the robustness of the watermark, since tampering with these important image components to remove the watermark will severely degrade the quality of the image. The second reason to use the phase of the DFT values is that it is well known from communications theory that often phase modulation possesses superior noise immunity in comparison with amplitude modulation [Rua96c].

Many watermarking techniques use DFT amplitude modulation because of its translation or shift invariant property [Her98a], [Her98b], [Per99], [Rua96a], [Rua97], [Rua98a] and [Rua98b]. Because cyclic translations of the image in the spatial domain do not affect the DFT amplitude, the watermark embedded in this domain will be translation invariant and, in case a CDMA watermark is used, it is even slightly resistant to cropping. Furthermore, the watermark can directly be embedded in the most important middle band frequencies, since modulation of the lowest frequency coefficients results in visible artifacts while the highest frequency coefficients are very vulnerable to noise, filtering and lossy compression algorithms. Finally the watermark can easily be made image content dependent by modulating the DFT amplitude coefficients $|I(u,v)|$ in the following way [Cox95]:

$$|I_W(u,v)| = |I(u,v)| \cdot (1 + k \cdot W(u,v)) \qquad (8.2.10)$$

Here, $W(u,v)$ represents a CDMA watermark, a 2-dimensional pseudorandom pattern, and k denotes the gain factor. Now, the modification of a DFT coefficient is not fixed but proportional to the amplitude of the DFT coefficient. Small DFT coefficients are hardly affected, whereas larger DFT coefficients are affected more severely. This complies with Weber's law [Jai81]. The human visual system does not perceive equal changes in images equally, but visual sensitivity is nearly constant with respect to relative changes in an image. If ΔI is a just noticeable difference, then $\Delta I / I =$ constant. Rewriting Equation 8.2.10 gives:

$$\frac{|I_W(u,v)| - |I(u,v)|}{|I(u,v)|} = \frac{\Delta I(u,v)}{|I(u,v)|} = k \cdot W(u,v) \cong \text{constant} \qquad (8.2.11)$$

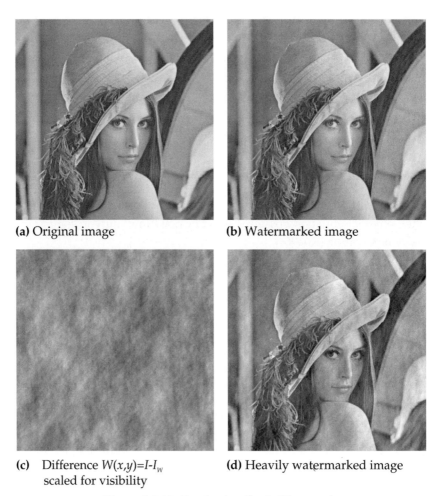

(a) Original image

(b) Watermarked image

(c) Difference $W(x,y)=I-I_w$ scaled for visibility

(d) Heavily watermarked image

Figure 8.2.10: *Fourier Amplitude Watermark.*

Since the watermark is here mainly embedded in the larger DFT coefficients, the perceptually most significant components of the image, the robustness of the watermark improves.

Note that the symmetry of the Fourier coefficients must be preserved to ensure that the image data is still real valued after the inverse transform to the spatial domain. If the coefficient $|I(u,v)|$ in an image with NxM pixels is modified according to Equation 8.2.10, its counterpart $|I(N-u,M-v)|$ must be modified in the same way. In Figure 8.2.10b an example is given of an image in which a watermark is embedded using all DFT amplitude coefficients according to Equation 8.2.10 and using a relatively small gain factor k. Figure 8.2.10c presents the strongly amplified difference between the original image and the watermarked image. Figure 8.2.10d shows an image watermarked using a large value for the gain factor k.

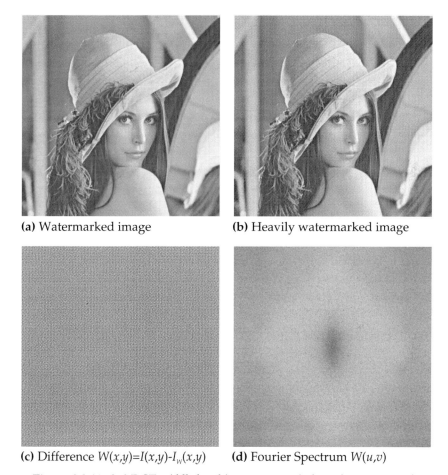

(a) Watermarked image (b) Heavily watermarked image

(c) Difference $W(x,y)=I(x,y)-I_w(x,y)$ (d) Fourier Spectrum $W(u,v)$

Figure 8.2.11: *8x8 DCT middle band image content independent watermark.*

Another commonly used domain for embedding a watermark is the Discrete Cosine Transform (DCT) domain [Bol95], [Cox95], [Cox96a], [Cox96b], [Hsu96], [Piv97], [Pod97], [Tao97], [Rua96b] and [Wol99c]. Using the DCT an image can easily be split up in pseudo frequency bands, so that the watermark can conveniently be embedded in the most important middle band frequencies. Furthermore, the sensitivity of the human visual system (HVS) to the DCT basis images has been extensively studied, which resulted in a default JPEG quantization table [Pen93]. These results can be used for predicting and minimizing the visual impact of the distortions caused by the watermark. Finally, the block-based DCT is widely used for image and video compression. By embedding a watermark in the same domain we can anticipate lossy compression and exploit the DCT decomposition to make real-time watermark applications.

In Figure 8.2.11a an example is given of an image in which a 2-dimensional CDMA watermark W is embedded in the 8x8 block DCT middle band frequencies. The 8x8 DCT coefficients $F(u,v)$ are modulated according to the following Equation:

$$I_{W_{x,y}}(u,v) = \begin{cases} I_{x,y}(u,v) + k \cdot W_{x,y}(u,v) & u,v \in F_M \\ I_{x,y}(u,v) & u,v \notin F_M \end{cases} \quad x,y=0,8,16\ldots \quad (8.2.12)$$

Here F_M denotes the middle band frequencies, k the gain factor, (x,y) the spatial location of an 8x8 pixel block in image I and (u,v) the DCT coefficient in the corresponding 8x8 DCT block (Figure 8.2.12).

In Figure 8.2.11c the strongly amplified difference between the original image and the watermarked image is presented. Figure 8.2.11d shows the Fourier Spectrum of the watermark. Here, it can clearly be seen that watermark only affects the middle band frequencies.

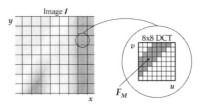

Figure 8.2.12. *Definition of the middle band frequencies in a DCT block.*

The watermark can be made image dependent by changing the modulation function to:

$$I_{W_{x,y}}(u,v) = \begin{cases} I_{x,y}(u,v) \cdot (1 + k \cdot W_{x,y}(u,v)) & u,v \in F_M \\ I_{x,y}(u,v) & u,v \notin F_M \end{cases} \quad x,y=0,8,16,\ldots \quad (8.2.13)$$

If this modulation function is applied, the results from Figure 8.2.11 change into the results shown in Figure 8.2.13. From Figure 8.2.13b and c it appears that most distortions introduced by the watermark are located around the edges and in the textured areas.

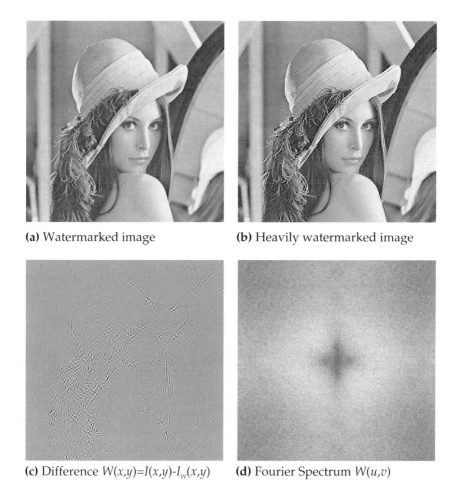

(a) Watermarked image

(b) Heavily watermarked image

(c) Difference $W(x,y)=I(x,y)-I_w(x,y)$

(d) Fourier Spectrum $W(u,v)$

Figure 8.2.13: *8x8 block DCT middle band image content dependent watermark.*

If watermarking techniques can exploit the characteristics of the Human Visual System (HVS), it is possible to hide watermarks with more energy in an image, which makes watermarks more robust. From this point of view the Digital Wavelet

Transform (DWT) is a very attractive tool, because it can be used as a computationally efficient version of the frequency models for the HVS [Bar99]. For instance, it appears that the human eye is less sensitive to noise in high resolution DWT bands and in the DWT bands having an orientation of 45° (i.e. *HH* bands). Furthermore, DWT image and video coding, such as embedded zero-tree wavelet (EZW) coding, will be included in the up-coming image and video compression standards, such as JPEG2000 [Xia97]. By embedding a watermark in the same domain we can anticipate lossy EZW compression and exploit the DWT decomposition to make real-time watermark applications. Many approaches apply the basic techniques described at the beginning of this section to the high resolution DWT bands, LH_1, HH_1 and HL_1 (Figure 8.2.14) [Bar99], [Bol95], [Kun97], [Rua96b], [Xia97].

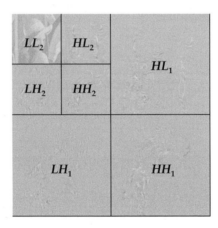

Figure 8.2.14: *DWT 2-level decomposition of an image.*

In Figure 8.2.15a an example is given of an image in which a 2-dimensional CDMA watermark *W* is embedded in the LH_1, HH_1 and HL_1 DWT bands using a large gain factor *k*. The DWT coefficients in each of the three DWT bands are modulated as follows:

$$I_W(u,v) = I(u,v) + k \cdot W(u,v) \quad (8.2.14)$$

Figure 8.2.15b shows the strongly amplified difference between the original image and the watermarked image.

(a) Heavily watermarked image (b) Difference $W(x,y)=I(x,y)-I_w(x,y)$

Figure 8.2.15: *DWT image content independent watermark.*

The DWT watermark can be made image dependent by modulating the DWT coefficients in each of the three DWT bands as follows:

$$I_W(u,v) = I(u,v) \cdot (1 + k \cdot W(u,v)) \tag{8.2.15}$$

In Figure 8.2.16a an example is given of an image in which the same CDMA watermark W is embedded in the LH_1, HH_1 and HL_1 DWT bands using Equation 8.2.15 with a large gain factor k. Figure 8.2.16b shows the strongly amplified difference between the original image and the watermarked image.

8.2.4 Watermark energy adaptation based on HVS

The robustness of a watermark can be improved by increasing the energy of the watermark. However, increasing the energy degrades the image quality. By exploiting the properties of the Human Visual System (HVS), the energy can be increased locally in places where the human eye will not notice it. As a result, by exploiting the HVS, one can embed perceptually invisible watermarks that have higher energy than if this energy were to be distributed evenly over the image.

If a visual signal is to be perceived, it must have a minimum amount of contrast, which depends on its mean luminance and frequency. Furthermore, a signal of a given frequency can mask a disturbing signal of a similar frequency

[Wan95] and [Bar98]. This masking effect is already used in the image-dependent DCT watermarking method described in the previous section, where the DCT coefficients are modulated by means of Equation 8.2.13. Here, to each sinusoid present in the image (masking signal), another sinusoid (watermark) is added, having an amplitude proportional to the masking signal. If the gain factor k is properly set, frequency masking occurs.

The HVS is less sensitive to changes in regions of high luminance. This fact can be exploited by making the watermark gain factor luminance dependent [Kut97]. Furthermore, since the human eye is least sensitive to the blue channel, a perceptually invisible watermark embedded in the blue channel can contain more energy than a perceptually invisible watermark embedded in the luminance channel of a color image [Kut97].

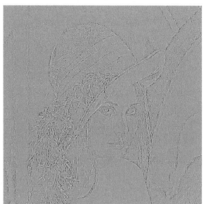

(a) Heavily watermarked image **(b)** Difference $W(x,y)=I(x,y)-I_w(x,y)$

Figure 8.2.16: *DWT image content dependent watermark.*

Around edges and in textured areas of an image, the HVS is less sensitive to distortions than in smooth areas. This effect is called spatial masking and can also be exploited for watermarking by increasing the watermark energy locally in these masked image areas [Mac95]. The basic spatial watermarking techniques described in Sections 8.2.1 and 8.2.2 can be extended with spatial masking compensation by, for instance, using the following modulation function.

$$I_W(x,y) = I(x,y) + Msk(x,y) \cdot k \cdot W(x,y) \qquad (8.2.16)$$

Here $W(x,y)$ represents the 2-dimensional pseudorandom pattern of the watermark, k denotes the fixed gain factor and $Msk(x,y)$ represents a masking image. The values of the masking image range from 0 to k'_{max} and give a measure of insensitivity to distortions for each corresponding point in the original image $I(x,y)$. In [Kal99] the masking image Msk is generated by filtering the original image with a Laplacian high-pass filter and by taking the absolute values of the resulting filtered image.

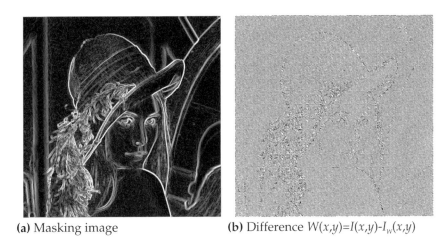

(a) Masking image   (b) Difference $W(x,y)=I(x,y)-I_w(x,y)$

Figure 8.2.17: *Watermarking using masking image based on Prewitt operator.*

In Figure 8.2.17a a mask is shown for the "Lena image" (Figure 8.2.10a) which is generated by a simple Prewitt edge detector. Figure 8.2.17b shows the strongly amplified watermark modulated with this mask.

Experiments have shown that a perceptually invisible watermark modulated with a gain factor locally adapted to such a mask can contain twice as much energy as a perceptually invisible watermark modulated with a fixed gain factor.

To investigate the effect of this energy doubling on the robustness of the watermark we perform the following experiment. We add a watermark $W_{fixed}(x,y)$ to the "Lena image" with the method of [Smi96] using a fixed gain factor $k=2$. Increasing this fixed gain factor causes visible artefacts in the resulting watermarked image. Next, we add a watermark $W_{var}(x,y)$ to another "Lena image" with the same method, but now we use a variable gain factor locally adapted to the masking image presented in Figure 8.2.17a. Although the watermark $W_{var}(x,y)$ contains about twice as much energy as $W_{fixed}(x,y)$ the watermark is not noticeable in the resulting watermarked image. Then we compress both watermarked images with the JPEG algorithm [Pen93], where the quality factor Q_{jpeg} of the compression algorithm is made variable. Finally, the watermarks are extracted from the decompressed image and compared bit by bit with the originally embedded watermark bits. From this experiment, we find the percentages of watermark bit errors due to JPEG compression as a function of the JPEG quality factor. In Figure 8.2.18 the error curves are plotted for both watermarks $W_{fixed}(x,y)$ and $W_{var}(x,y)$. It can be seen that the robustness can be slightly improved by applying a variable gain factor adapted to the HVS.

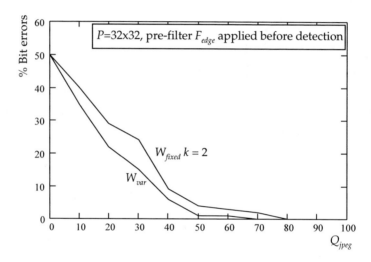

Figure 8.2.18: *Influence of a variable gain factor adapted to the HVS on the robustness of a watermark.*

In [Ng99] the squared sum of the 8x8 DCT AC-coefficients is used to generate a masking image. Figure 8.2.19a shows a mask generated using this DCT-AC energy for the "Lena image". Figure 8.2.19b presents the strongly amplified watermark modulated with this mask.

Spatial masking can also be applied if the watermark is embedded in another domain e.g. DFT, DCT or DWT. In this case, the non-spatial watermark is first embedded in an image I, resulting in the temporary image I_{Wt}. The watermarked image I_W is now constructed by mixing the original image I and this temporary image I_{Wt} by means of a masking image Msk as described above [Bar98] and [Piv97]:

$$I_W(x,y) = (1 - Msk(x,y))I(x,y) + Msk(x,y) \cdot I_{Wt}(x,y) \qquad (8.2.17)$$

Here the masking image must be scaled to values in the range from 0 to 1. Watermarking methods based on more sophisticated models for the HVS can be found in [Bar98], [Bar99], [Fle97], [Gof97], [Kun97], [Piv97], [Pod97], [Swa96a], [Swa96b], [Wol99b] and [Wol99c].

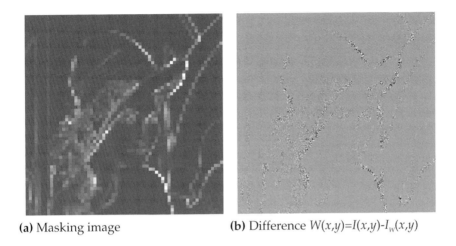

(a) Masking image   (b) Difference $W(x,y)=I(x,y)-I_w(x,y)$

Figure 8.2.19: *Watermarking where a masking image is used based on DCT-AC energy.*

8.3 Extended correlation-based watermark techniques

8.3.1 Anticipating lossy compression and filtering

Watermarks that have been embedded in an image by means of the spatial watermarking techniques described in Sections 8.2.1 and 8.2.2 cannot be detected reliably after the watermarked image has been highly compressed with the lossy JPEG compression algorithm. This is due to the fact that such watermarks consist essentially of low-power, high frequency noise. Since JPEG allocates fewer bits to the higher frequency components, such watermarks can easily be distorted. Furthermore, these watermarks can also be affected severely by low-pass operations like linear or median filters.

The robustness to JPEG compression can be improved in several ways. In [Smi96] the pseudorandom pattern W is first compressed and then decompressed using the JPEG algorithm. The energy of the resulting pattern W is increased to compensate for the energy lost through the compression. Finally, this pattern is added to the image to generate the watermarked image. The idea here is to use the compression algorithm to filter out in advance all the energy that would otherwise be lost later in the course of the compression. It is assumed that a watermark formed in this way is invariant to further JPEG compression that uses the same quality factor, except for small numerical artifacts. Analogous pre-distortion of the watermark pattern, such as filtering, can be applied to prevent other anticipated degradations of the watermarked image.

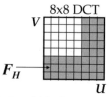

Figure 8.3.1: DCT bands F_H in which the watermark energy Φ is minimized.

In [Nik96] the energy of the watermark pattern is shifted to the lower frequencies by calculating an individual gain factor $k_{x,y}$ for each pixel of the watermark pattern instead of using the same gain factor k for all pixels. First a pseudorandom pattern $W(x,y)$ is generated consisting of the integers 0 and k. Next, the pattern is divided into 8x8 blocks and the DCT transform $W(u,v)$ is calculated for each 8x8 block. The non-zero elements in the 8x8 blocks are now regarded as gain factors $k_{x,y}$ and are adapted in such a way that the energy Φ in the vulnerable high frequency DCT bands F_H is minimized (Figure 8.3.1):

$$\Phi = \sum_{u,v \in F_H} \sum W(u,v)^2 \qquad F_H=\{u,v \mid 5 < u \leq 8, 5 < v \leq 8\} \qquad (8.3.1)$$

The energy Φ is minimized under the following constraints:

$$\sum_{x=1}^{8}\sum_{y=1}^{8} W(x,y) \cdot k = \sum_{x=1}^{8}\sum_{y=1}^{8} W(x,y) \cdot k_{x,y} \qquad k_{min} \leq k_{x,y} \leq k_{max} \qquad (8.3.2)$$

The effect of this high-energy minimization on the watermark pattern is illustrated in Figure 8.3.2. Figure 8.3.2a shows the watermark pattern within an 8x8 block, where a constant gain factor of $k=3$ is used. After the high-energy minimization with $k_{min}=0$ and $k_{max}=6$ the watermark pattern fades smoothly to zero (Figure 8.3.2.b) although the sum of the non-zero pixels still equals to the sum of the non-zero pixels in the original pattern.

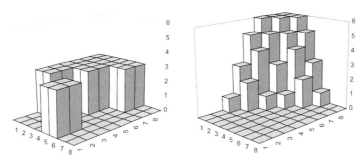

Figure 8.3.2: **(a)** *Original watermark block* **(b)** *Low frequency watermark block.*

In [Lan96a] and [Lan97a] JPEG compression immunity is obtained by deriving a different gain factor k for each 32x32 pixel block based on a lower quality JPEG compressed image. A 32x32 pseudorandom pattern representing a watermark bit is added to an 32x32 image tile. A copy of this watermarked image tile is degraded according to the JPEG standard for which end a relatively low quality factor is used. If the watermark bit cannot be extracted correctly from this degraded copy, the watermark pattern is added to the image by means of a higher gain factor and a new degraded copy is formed to check the bit. This procedure is repeated iteratively for each bit until all bits can be extracted reliably from the degraded copies. A watermark formed in this way is resistant to JPEG compression using a quality factor equal to or greater than the quality factor used to degrade the copies. In Figure 8.3.3 an example of such a watermark is shown, amplified for visibility purposes.

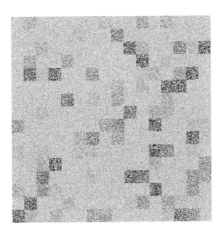

Figure 8.3.3: *Watermark where the local gain factor per block is based on a lower quality image.*

8.3.2 Anticipating geometrical transforms

A watermark should not only be robust to lossy compression techniques, but also to geometrical transformations such as shifting, scaling, cropping, rotation etc. Geometrical transforms hardly affect the image quality, but they do make most of the watermarks that have been embedded by means of the techniques described in the previous sections undetectable for the watermark detectors. Since geometrical transforms affect the synchronization between the pseudorandom pattern of the watermark and the watermarked image, the synchronization must be retrieved before the detector performs the correlation calculations.

The most obvious way to achieve shift invariance is using the DFT amplitude modulation technique described in Section 8.2.3. However if for some reason another watermarking embedding domain is preferred and shift invariance is required, a marker can be added in the spatial domain to determine the translation. This marker can be a pseudorandom pattern like the watermark itself. The detector first determines the spatial position of this marker by shifting the marker over all possible locations in the image and calculating the correlation between the marker and the corresponding image part. The translation with the highest correlation defines the spatial position of the marker. Finally, the image is shifted back to its original position and the normal watermarking detection procedure is applied.

An exhaustive search for a marker is computationally quite demanding. Therefore, in [Kal99] a different approach is proposed: adding a pseudorandom pattern twice, but at different locations in the image. The content of the watermark, i.e. the watermark bits, is here embedded in the relative positions of the two watermark patterns. To detect the watermark, the detector computes the phase correlation between the image and the watermark pattern using the fast Fourier transform and it detects the two correlation peaks of the two patterns. The content of the watermark is derived from relative position of the peaks. If the whole image is shifted before detection, the absolute positions of the correlation peaks will change, but the relative positions will remain unchanged, leaving the watermark bits readable for the detector.

In [Fle97] a method is proposed to add a grid to an image that can be used to scale, rotate and shift an image back to its original size and orientation. The grid is represented by a sum of sinusoidal signals, which appear as peaks in the FFT frequency domain. These peaks are used to determine the geometrical distortions.

In [Kut98] a method is proposed which embeds a pseudorandom pattern multiple times at different locations in the spatial domain of an image. The detector estimates the watermark W' by applying a high-pass filter F_{HP} to the watermarked image:

$$W' = I_W \otimes F_{HP} \, , \qquad F_{HP} = \begin{pmatrix} 0 & 0 & 0 & -1 & 0 & 0 & 0 \\ 0 & 0 & 0 & -1 & 0 & 0 & 0 \\ 0 & 0 & 0 & -1 & 0 & 0 & 0 \\ -1 & -1 & -1 & 12 & -1 & -1 & -1 \\ 0 & 0 & 0 & -1 & 0 & 0 & 0 \\ 0 & 0 & 0 & -1 & 0 & 0 & 0 \\ 0 & 0 & 0 & -1 & 0 & 0 & 0 \end{pmatrix} / 12 \qquad (8.3.3)$$

Next, the autocorrelation function of the estimated watermark W' is calculated. This function will have peak values at the center and the positions of the multiple embedded watermarks. If the image has undergone a geometrical transformation, the peaks in the autocorrelation function will reflect the same transformation, and hence provide a grid that can be used to transform the image back to its original size and orientation.

In [Her98a], [Her98b], [Rua97], [Per99], [Rua98a] and [Rua98b] a method is proposed that embeds the watermark in a rotation, scale and translation invariant domain using a combination of Fourier Transforms (DFT) and a Log Polar Map (LPM). Figure 8.3.4 presents a scheme of this watermarking method.

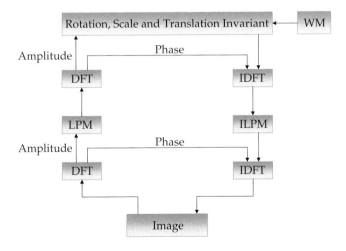

Figure 8.3.4: *Rotation, scale and translation invariant watermarking scheme.*

First the amplitude of the DFT is calculated to get a translation invariant domain. Next, for every point (u,v) of the DFT amplitude a corresponding point in the Log Polar Map (μ,θ) is determined:

$$u = e^{\mu} \cos(\theta) \qquad v = e^{\mu} \sin(\theta) \qquad (8.3.4)$$

This coordinate system of the Log Polar Map converts rotation and scaling into translations along the horizontal and vertical axis. By taking the amplitude of the DFT of this Log Polar map, we obtain a rotation, scale and translation invariant domain. In this domain a CDMA watermark can be added, for instance by modulating the coefficients using Equation 8.2.10.

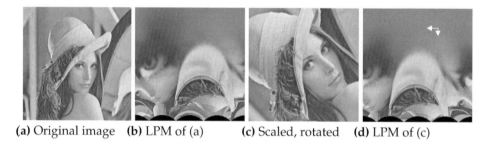

(a) Original image (b) LPM of (a) (c) Scaled, rotated (d) LPM of (c)

Figure 8.3.5: *Example of the properties of the Log Polar Map.*

Figure 8.3.5 demonstrates an example of the properties of the Log Polar Map. Figure (b) shows the Log Polar Map of the "Lena image" (a). Figure (c) depicts a rotated and scaled version of the "Lena image" and Figure (d) shows its corresponding Log Polar Map. It can clearly be seen that the rotation and scaling are converted into translations.

In practice it has proven to be difficult to implement a watermarking scheme as illustrated in Figure 8.3.4. The authors therefore propose a different approach, where a CDMA watermark is embedded in the translation invariant amplitude DFT domain as described in Section 8.2.3. To make the watermark scale and rotation invariant, they embed a second watermark, a template, in this domain. To extract the watermark, they first determine the scale and orientation of the watermarked image by using the template in the following way:

- The DFT of the watermarked image is calculated.
- The Log Polar Map of the DFT amplitudes and the template pattern is calculated.

- The horizontal and vertical offsets between the two log polar maps are calculated using exhaustive search and cross-correlation techniques, resulting in a scale and rotation factor.

Next, the image is transformed back to its original size and orientation, and the information-carrying watermark is extracted.

8.3.3 Correlation-based techniques in the compressed domain

Not only robustness, but also computational demands play an important role in real-time watermarking applications. In general image data is transmitted in compressed form. To embed a watermark in real time the compressed format must be taken into account, because first decompressing the data, adding a watermark and then re-compressing the data is computationally too demanding. In [Har96], [Har97a], [Har97b], [Har97c] and [Wu97] a method is proposed that adds a DCT transformed pseudorandom pattern directly to selected DCT coefficients of an MPEG compressed video signal. To extract the watermark they decompress the video data and apply the correlation techniques described in Section 8.2. Since the scope here is real-time watermarking algorithms, the above-mentioned method and novel alternatives are described in full in Chapters 9 and 10.

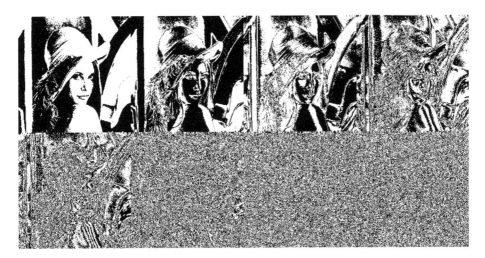

Figure 8.4.1: *Bit planes for the "Lena image"*.

8.4 Non-correlation-based watermarking techniques

8.4.1 Least significant bit modification

The simplest example of a spatial domain watermarking technique that is not based on correlation is the least significant bit modification method. If each pixel in a gray level image is represented by an 8-bit value, the image can be sliced up in eight bit planes. In Figure 8.4.1 these eight bit planes are represented for the "Lena image", where the upper left image represents the most significant bit plane and the lower right image represents the least significant bit plane.

Since the least significant bit plane does not contain visually significant information, it can easily be replaced by an enormous amount of watermark bits. More sophisticated watermarking algorithms that make use of LSB modifications can be found in [Sch94], [Aur95], [Aur96], [Hir96] and [Fri99c]. These watermarking techniques are not very secure and not very robust to processing techniques because the least significant bit plane can easily be replaced by random bits, effectively removing the watermark bits.

8.4.2 DCT coefficient ordering

In [Koc95], [Zha95], [Koc94] and [Bur98] a watermarking method is proposed that adds a watermark bit string in the 8x8 block DCT domain. To watermark an image, the image is divided into 8x8 blocks. From these 8x8 blocks the DCT transform is calculated and two or three DCT coefficients are selected in each block in the middle band frequencies F_M (Figure 8.4.2). The selected coefficients are quantized using the default JPEG quantization table [Pen93] and a relatively low JPEG quality factor. The selected coefficients are then adapted in such a way that their magnitudes form a certain relationship. The relationships among the selected coefficients compose 8 patterns (combinations), which are divided into 3 groups. Two groups are used to represent the watermark bits '1' or '0', and the third group represents invalid patterns. If the modifications which are needed to hold a desired pattern become too large, the block is marked as invalid. For example, if a watermark bit with value '1' must be embedded in a block, the third coefficient should have a lower value than the two other coefficients. The embedding process and the list of patterns are represented in Figure 8.4.2.

In Figure 8.4.3 the heavily amplified difference between the original "Lena image" and the watermarked version is shown. In [Bor96a] and [Bor96b] a similar watermarking method is proposed, but here the DCT coefficients are modified in such a way that they fulfill a linear or circular constraint imposed by the watermark code.

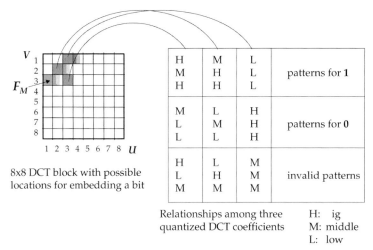

Figure 8.4.2: *Watermarking based on adapting relationship between 3 coefficients.*

Figure 8.4.3: *Watermark $W(x,y)=I(x,y)-I_w(x,y)$ created by adapting relationships between DCT coefficients.*

In the methods described here, the relationships between a few middle band coefficients within an 8x8 DCT block define the watermark bits. In [Lan97a], [Lan97b], [Lan98a] and [Lan99b] a method is proposed that uses the relationship between a large amount of high frequency band DCT coefficients in different DCT blocks to define the watermark bits. This new algorithm, its performance and its statistical modeling are described in full in Chapters 10 and 11.

8.4.3 Salient-point modification

In [Ron99] a watermarking method is proposed that is based on modification of salient points in an image. Salient points are defined as isolated points in an image for which a given saliency function is maximal. These points could be corners in an image or locations of high energy for example.

Figure 8.4.4: *Examples of watermark patterns for salient-point modification.*

To embed a watermark we extract the set of pixels with highest saliency S from the image. Next, a binary pseudorandom pattern $W(x,y)$ with the same dimensions as the image is generated. This can be a line or block pattern as represented in Figure 8.4.4. If this pattern is sufficiently random and covers 50% of all the image pixels, 50% of all salient points in set S will be located on the pattern and 50% off the pattern $W(x,y)$. Finally, the salient points in set S are adapted in such a way that a statistically significant high percentage of them lies on the watermark pattern (i.e. the black pixels in the pattern). There are two ways to adapt the salient points:

- The location of the salient points can be changed by warping the points towards the watermark pattern. In this case small, local geometrical changes are introduced in the image.
- The saliency of the points can be decreased or increased by adding well-chosen pixel patterns to the neighborhood of a salient point.

To detect the watermark we extract the set of pixels with highest saliency S from the image and compare the percentages of the salient points on the watermark pattern and off the pattern. If both percentages are about 50%, no watermark is detected. If there is a statistically significant high percentage of salient points on the pattern, the watermark is detected. The payload of this watermark is 1 bit.

8.4.4 Fractal-based watermarking

Some watermark embedding algorithms are proposed that are based on Fractal compression techniques [Dav96], [Pua96], [Bas98] and [Bas99]. They mainly use block-based local iterated function system coding [Jac92]. We first briefly describe the basic principles of this fractal compression algorithm here. An image is partitioned at two different resolution levels. On the first level, the image is partitioned in range blocks of size $n \times n$. On the second level the image is partitioned in domain blocks of size $2n \times 2n$. For each range block, a transformed domain block is searched for which the mean square error between the two blocks is minimal. Before the range blocks are matched on the domain blocks, the following transformations are performed on the domain blocks. First, the domain blocks are sub-sampled by a factor two to get the same dimensions as the range blocks. Subsequently, the eight isometries of the domain blocks are determined (the original block and its mirrored version rotated over 0, 90, 180 and 270 degrees). Finally, the scale factor and the offset for the luminance values is adapted. The image is now completely described by a set of relations for each range block, by the index number of the best fitting domain block, its orientation, the luminance scaling and the luminance offset. Using this set of relations, an image decoder can reconstruct the image by taking any initial random image and calculating the content of each range block from its associated domain block using the appropriate geometric and luminance transformations. Taking the resulting image as initial image one repeats this process iteratively until the original image content is approximated closely enough.

In [Pua96] a watermarking technique is proposed which embeds a watermark of 32 bits $b_0 b_1 ... b_{31}$ in an image. The embedding procedure consists of the full fractal encoding and decoding process as described above, where the watermark embedding takes place in the fractal encoding process. First, the image $I(x,y)$ is split in two regions $A(x,y)$ and $B(x,y)$. For each watermark bit b_j U range blocks are pseudorandomly chosen from $I(x,y)$. If b_j equals one, the domain blocks to code the U range blocks are searched in region $A(x,y)$. If b_j equals zero, the domain blocks to code the U range blocks are searched in region $B(x,y)$. For range blocks which are not involved in the embedding process, domain blocks are searched in regions $A(x,y)$ and $B(x,y)$. To extract the watermark information, we must select and re-encode the U range blocks for each bit b_j. If most of the best fitting domain blocks are found in region $A(x,y)$, the value 1 is assigned to bit b_j, otherwise the bit is assumed to be zero.

In [Bas98] and [Bas99] a watermark is embedded by forcing range blocks to map exactly on specific domain blocks. The watermark pattern here consists of this specific mapping. This mapping is enforced by adding artificial local similarities to the image. The size of the range blocks may be chosen equal to the size of the domain blocks. In Figure 8.4.5 an example is given of this process.

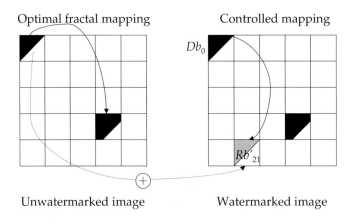

Figure 8.4.5: *Modifying the mapping between range and domain blocks.*

The left image illustrates how a fractal encoder would map the range block Rb_{18} on domain block Db_0 in an unwatermarked image. To embed the watermark, this mapping $Db_0 \rightarrow Rb_{18}$ must for instance be changed to $Db_0 \rightarrow Rb_{21}$. To force the mapping to this form, a block Rb'_{21} is generated from block Db_0 by changing its luminance values. By adding block Rb' to the image, we change the optimal fractal mapping to its desired form $Db_0 \rightarrow Rb_{21}$, because the quadratic error between Db_0, corrected for luminance scale and offset and Rb_{21} is now smaller than the error between Db_0 and Rb_{18}.

To detect the watermark we calculate the optimal fractal mapping between the range blocks and the domain blocks. If a statistically significant high percentage of the mappings between range blocks and domain blocks match the predefined mappings of the watermark pattern, the watermark is detected.

8.5 Conclusions

Not all existing watermarking techniques are discussed in this chapter, because some techniques are specifically designed for e.g. printing purposes, and others are not so extensively represented in literature as the methods described in this chapter. We will therefore only enumerate the most important principles of some of these other methods here:

- For printed images dithering patterns can be adapted to hide watermark information [Tan90] and [Che99].

- Instead of the pixel values, the histogram of an image can be modified to embed a watermark [Col99].

- Quantization can be exploited to hide a watermark. In [Rua96c] a method is proposed in which the pixel values of an image are first coarsely quantized, before some small adaptations are made to the image. To detect these adaptations the watermarked image is subtracted from its coarsely quantized version. In [Kun98] selected wavelet coefficients are quantized using different quantizers for watermark bits 0 and 1.

In this chapter we discussed the two most important classes of watermarking techniques. The first class comprises the correlation-based methods. Here a watermark is embedded by adding pseudorandom noise to image components and detected by correlating the pseudorandom noise with these image components. The second class comprises the non-correlation-based techniques. This class of watermarking methods can roughly be divided into two groups: the group based on least significant bit (LSB) modification and group based on geometrical relations.

Chapter 9

Low Complexity Watermarks for MPEG Compressed Video

9.1 Introduction

The scope of Chapters 9, 10 and 11 is real-time watermarking algorithms for MPEG compressed video. In this chapter the state of the art in real-time watermarking algorithms is discussed and two new computationally highly efficient algorithms are proposed which are very suitable for consumer applications requiring moderate robustness. In Chapter 10 the slightly more complex DEW watermarking algorithm is proposed, which is applicable for applications requiring more robustness. In Chapter 11 a statistical model is derived to find optimal parameter settings for the DEW method.

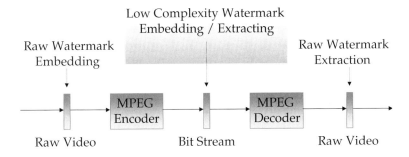

Figure 9.1.1: *Watermark embedding / extraction in raw vs. compressed video.*

A real-time watermarking algorithm should meet several requirements. In the first place it should be an oblivious low complexity algorithm. This means that fully decompressing the video data, adding a watermark to the raw video data and finally compressing the data again is not an option for real-time watermark embedding. The watermark should be embedded and detected directly in the compressed stream to avoid computationally demanding operations as shown in Figure 9.1.1.

Furthermore, the watermark embedding operation should not increase the size of the compressed video stream. If the size of the stream increases, transmission over a fixed bit rate channel can cause problems: the buffers in hardware decoders can run out of space, or the synchronization of audio and video can be disturbed.

Since the watermarking methods discussed in the following chapters heavily rely on the MPEG video compression standard [ISO96] the relevant parts of the MPEG-standard and the different domains in which a low complexity watermark can be added are described in Section 9.2. In Section 9.3 an overview is given of two real-time correlation-based watermarking algorithms from literature. In Sections 9.4 and 9.5 two new computationally highly efficient algorithms are proposed which are very suitable for consumer applications requiring moderate robustness [Lan96b], [Lan97b] and [Lan98a].

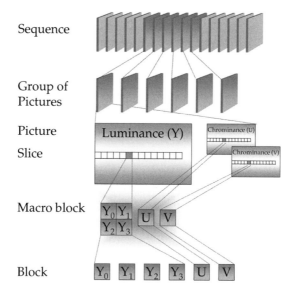

Figure 9.2.1: *The layered MPEG syntax.*

9.2 Watermarking MPEG video bit streams

Before discussing the low complexity watermarking techniques, we first briefly describe the MPEG video compression standard [ISO96] itself. The MPEG video bit stream has a layered syntax. Each layer contains one or more subordinate layers, as illustrated in Figure 9.2.1. A video *Sequence* is divided into multiple *Group of Pictures* (GOPs), representing sets of video frames which are contiguous in display order. Next, the frames are split in slices and macro blocks. The lowest layer, the block layer, is formed by the luminance and chrominance blocks of a macro block.

The MPEG video compression algorithm is based on the basic hybrid coding scheme [Gir87]. As can be seen in Figure 9.2.2 this scheme combines interframe coding (DPCM) and intraframe coding to compress the video data.

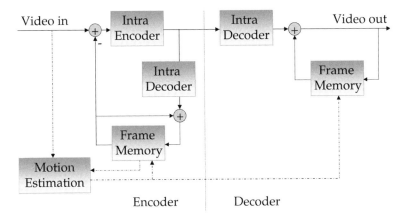

Figure 9.2.2: *Motion compensated hybrid coding scheme.*

Within a GOP the temporal redundancy among the video frames is reduced by the application of temporal DPCM. This means that the frames are temporally predicted by other motion compensated frames. Subsequently, the resulting prediction error, which is called the displaced frame difference, is encoded. Three types of frames are used in the MPEG standard: (I) Intraframes, which are coded without any reference to other frames, (P) Predicted frames, which are coded with reference to past I- or P-frames, and (B) Bi-directionally interpolated frames, which are coded with references to both past and future frames. An encoded GOP always starts with an I-frame, to provide access points for random access of the video stream. In Figure 9.2.3 an example of a GOP with 3 frame types and their references is shown.

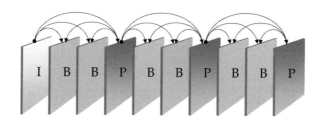

Figure 9.2.3: *GOP with 3 frame types and the references between the frames.*

The spatial redundancy in the prediction error of the predicted frames and the I-frames, represented by the luminance component Y and the chrominance components U and V, is reduced using the following operations: First the chrominance components U and V are subsampled. Next, the DCT transform is performed on the 8x8 pixel blocks of the Y, U and V components, and the resulting DCT coefficients are quantized. Since the de-correlating DCT transform concentrates the energy in the lower frequencies and the human eye is less sensitive to the higher frequencies, the high frequency components can be quantized more coarsely. The DCT coefficient with index (0,0) is called the DC-coefficient, since it represents the average value of the 8x8 pixel block. The other DCT coefficients are called AC-coefficients.

8x8 block	Tuples (*run,level*)	VLC codewords
5 -3 4 0 0 -1 0 -2	(0,5), (0,3), (0,2), (2,4),	001001100
2 0 7 0 0 0 0 0	(1,7), (3,2), (3,1), (2,4),	001010
0 0 0 0 0 0 0 0	(4,1), (4,2)	01000
0 2 4 0 0 0 0 0		0000000101000
0 0 1 0 0 0 0 0		00000010100
0 0 0 0 0 0 0 0		001001000
0 0 0 0 0 0 0 0		001110
0 0 0 0 0 0 0 0		0000000101000
		001100
		00000011110
		10
coefficient domain (*cd*)	run-level domain	bit domain (*bd*)

Figure 9.2.4: *DCT-block representation domains.*

In the lowest MPEG layer, the block layer, the spatial 8x8 pixel blocks are represented by 64 quantized DCT coefficients. Figure 9.2.4 shows the three domains in which the block layer can be divided. The first domain is the *coefficient domain (cd)*, where a block contains 8x8 integer entries that correspond with the quantized DCT coefficients. Many of the entries are usually zero, especially those entries that correspond with the spatial high frequencies. In the *run-level domain*, the non-zero AC coefficients are re-ordered in a zigzag scan fashion and are subsequently represented by a *(run,level)* tuple, where the run is equal to the number of zeros preceding a certain coefficient and the level is equal to the value of the coefficient. In lowest level domain, the *bit domain (bd)*, the *(run,level)* tuples are entropy coded and represented by variable length coded (VLC) codewords. The codewords for a single DCT-block are terminated by an end of block (EOB) marker.

A real-time watermarking algorithm for MPEG compressed video should closely follow the MPEG compression standard to avoid computationally demanding operations, like DCT and inverse DCT transforms or motion vector calculation. Therefore, the algorithm should work on the block layer, the lowest layer of the MPEG stream. A watermarking algorithm that operates on the *coefficient domain* level only needs to perform VLC coding, tuple coding and quantization steps. This process is illustrated in Figure 9.2.5.

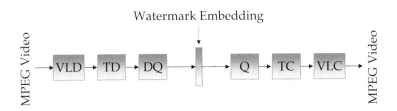

Figure 9.2.5: *Coefficient domain watermarking concept.*

A watermarking algorithm that operates on the *bit domain* level only needs the VLC coding processing step. Here, a complete watermark embedding procedure can consist of VLC-decoding, VLC-modification and VLC-encoding. This process is illustrated in Figure 9.2.6.

In Section 9.3 an overview is given of two real-time correlation-based watermarking algorithms from literature. The first method described in this section is applied in the *coefficient domain*. The second method is more advanced and operates on a slightly higher level than the *coefficient domain*, since it needs a full

MPEG decoding operation for drift compensation and watermark detection, and an additional DCT operation. The new watermarking methods proposed in Sections 9.4 and 9.5 operate on the lowest level domain, the *bit domain*, and are therefore computationally the most efficient methods. The DEW algorithm proposed in Chapters 10 and 11 is applied completely in the *coefficient domain*.

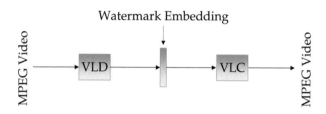

Figure 9.2.6: *Bit domain watermarking concept.*

9.3 Correlation-based techniques in the coefficient domain

9.3.1 DC-coefficient modification

In [Wu97] a method is proposed that adds a DCT transformed pseudorandom pattern directly to the DC-DCT coefficients of an MPEG compressed video stream. The watermarking process only takes the luminance values of the I-frames into account. To embed a watermark the following procedure is performed: First a pseudorandom pattern consisting of the integers {–1,1} is generated based on a secret a key. This pattern has the same dimensions as the I-frames. Next, the pattern is modulated by a watermark bit string and multiplied by a gain factor, as described in Section 8.2.2. Finally, the 8x8 block DCT transform is applied to the modulated pattern and the resulting DC-coefficients are added to the corresponding DC-values of each I-frame. The watermark can be detected using correlating techniques in the DCT domain or in the spatial domain, as described in Section 8.2.2.

The authors report that the algorithm decreases the visual quality of the video stream drastically. Therefore, the gain factor of the watermark has to be chosen very low (<1) and the number of pixels per watermark bit has to be chosen extremely high (>> 100,000) to maintain a reasonable visual quality for the resulting video stream. This is mainly due to the fact that the watermark pattern is embedded in just one of the 64 DCT coefficients, the DC-component. Furthermore, the pattern consists only of low frequency components to which the human eye is quite sensitive. For comparison, the algorithm described in Section 8.2.2 uses a gain factor of 2 and about 1000 pixels per watermark bit.

9.3.2 DC- and AC-coefficient modification with drift compensation

9.3.2.1 Basic watermarking concept

In [Har96], [Har97a], [Har97b], [Har97c] and [Har98] a more sophisticated watermarking algorithm is proposed that embeds a watermark not only in the DC-coefficients, but also in the AC-coefficients of each I-, P- and B-frame. Here the watermark is also a pseudorandom pattern consisting of the integers {–1,1} generated by means of a secret key. This pattern has the same dimensions as the video frames. The pattern is modulated by a watermark bit string and multiplied by a gain factor k as described in Section 8.2.2.

To embed the watermark, the watermark pattern $W(x,y)$ is divided into 8x8 blocks. These blocks are transformed to the DCT domain and denoted by $W_{x,y}(u,v)$, where $x,y=0,8,16...$ and $u,v=0...7$. Next, the two-dimensional blocks $W_{x,y}(u,v)$ are re-ordered in a zigzag scan fashion and become arrays $W_{x,y}(i)$, where $i=0...63$. $W_{x,y}(0)$ represents the DC-coefficient and $W_{x,y}(63)$ denotes the highest frequency AC-coefficient of a 8x8 watermark block. Since the corresponding MPEG encoded 8x8 video content blocks are encoded in the same way as $I_{x,y}(i)$, these arrays can be used directly to add the watermark. For each video block $I_{x,y}(i)$ out of an I-, P-, or B-frame the following steps are performed:

1. The DC-coefficient is modulated as follows:

$$I_{W_{x,y}}(0) = I_{x,y}(0) + W_{x,y}(0) \qquad (9.3.1)$$

Which means that the average value of the watermark block is added to the average value of the video block.

2. To modulate the AC-coefficients, the bit stream of the encoded video block is searched VLC-by-VLC for the next VLC code word, representing the next non-zero DCT coefficient. The run and level of this code word are decoded to determine its position i along the zigzag scan and its amplitude $I_{x,y}(i)$.

A candidate DCT coefficient for the watermarked video block is generated, which is defined as:

$$I_{W_{x,y}}(i) = I_{x,y}(i) + W_{x,y}(i) \quad i \neq 0 \tag{9.3.2}$$

Now the constraint that the video bit rate may not increase comes into play. The size Sz_I of the VLC needed to encode $I_{x,y}(i)$ and the size Sz_{I_W} of the VLC needed to encode $I_{W_{x,y}}(i)$ are determined using the VLC-Tables B.14 and B.15 of the MPEG-2 standard [ISO96]. If the size of VLC encoding the candidate DCT coefficient is equal to or smaller than the size of the existing VLC, the existing VLC is replaced. Otherwise the VLC is left unaffected. This means that the DCT coefficient $I_{x,y}(i)$ is modulated in the following way:

$$\text{If} \quad Sz_{I_W} \leq Sz_I \quad \text{then} \quad I_{W_{x,y}}(i) = I_{x,y}(i) + W_{x,y}(i) \tag{9.3.3}$$
$$\text{else} \quad I_{W_{x,y}}(i) = I_{x,y}(i)$$

This procedure is repeated until all AC-coefficients of the encoded video block have been processed.

To extract the watermark information, the MPEG encoded video stream is first fully decoded and the watermark bits are retrieved by correlating the decoded frames with the watermark pattern $W(x,y)$ in the spatial domain using the standard techniques as described in Section 8.2.2.

9.3.2.2 Drift compensation

A major problem of directly modifying DCT coefficients in an MPEG encoded video stream is drift or error accumulation. In an MPEG encoded video stream predictions from previous frames are used to reconstruct the actual frame, which itself may serve as a reference for future predictions. The degradations caused by the watermarking process may propagate in time, and may even spread spatially. Since all video frames are watermarked, watermarks from previous frames and from the current frame may accumulate and result in visual artefacts. Therefore, a drift compensation signal Dr must be added. This signal must be equal to the difference of the (motion compensated) predictions from the unwatermarked bit stream and the watermarked bit stream. As a drift-compensated watermarking scheme, Equation 9.3.2 becomes:

$$I_{W_{x,y}}(i) = I_{x,y}(i) + W_{x,y}(i) + Dr_{x,y}(i) \qquad (9.3.4)$$

A disadvantage of this drift signal is that the complexity of the watermark embedding algorithm increases substantially, since an additional DCT operation and a complete MPEG decoding step are required to calculate the drift compensation signal. The increase in complexity compared to the complexity of the coefficient domain methods is illustrated in Figure 9.3.1.

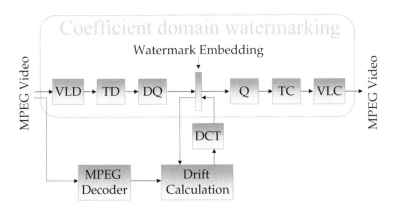

Figure 9.3.1: *Increase of complexity due to drift compensation.*

9.3.2.3 Evaluation of the correlation-based technique

Due to the bit rate constraint, only around 10-20% of the DCT coefficients are altered by the watermark embedding process, depending on the video content and the coarseness of the MPEG quantizer. In some cases, especially for very low bit rate video, only the DC-coefficients are modified. This means that only a fraction of the watermark pattern $W(x,y)$ can be embedded, typically around 0.5...3% [Har98]. Since only existing (non-zero) DCT coefficients of the video stream are watermarked, the embedded watermark is video content dependent. In areas with only low frequency content, the watermark automatically consists of only low frequency components. This complies with the Human Visual System. The watermark energy is mainly embedded in areas containing a lot of video content energy.

The authors [Har98] report that the complexity of the watermark embedding process is much lower than the complexity of a decoding process followed by watermarking in the spatial domain and re-encoding. The complexity is somewhat higher than the complexity of a full MPEG decoding operation. Typical parameter settings for the embedding are $k=1\ldots5$ for the gain factor of the watermark and $P=500,000\ldots1,000,000$ for the number of pixels per watermark bit, yielding watermark label bit rates of only a few bytes per second. The authors claim that the watermark is not visible, except in direct comparison to the unwatermarked video, and that the watermark can withstand linear and non-linear operations like filtering, noise addition and quantization in the spatial or frequency domain.

9.4 Parity bit modification in the bit domain

9.4.1 Bit domain watermarking concept

In Section 8.4.1 we saw that watermarking algorithms based on LSB (least significant bit) modification have an enormous payload and are computationally not demanding. In this section, this LSB modification principle is directly applied in the bit domain of MPEG compressed video, resulting in a computationally highly efficient watermarking algorithm with an extremely high payload [Lan96b], [Lan97b] and [Lan98a].

We embed a watermark consisting of l label bits b_j ($j = 0, 1, 2,\ldots, l-1$) in the MPEG-stream by selecting suitable VLCs and forcing the least significant bit of their *quantized level* to the value of b_j. To ensure that after decoding the change in the VLC is perceptually invisible and the MPEG-bit stream has kept its original size, we select only those VLCs for which another VLC exists with:

- the same run length
- a level difference of 1
- the same code word length

A VLC that meets this requirement is called a *label bit carrying VLC (lc-VLC)*. According to Table B.14 and B.15 of the MPEG-2 standard [ISO96], an abundance of such *lc-VLCs* exists. Furthermore, all fixed length coded DCT coefficients following an Escape code meet the requirement. Some examples of *lc-VLCs* are listed in Table 9.4.1, where the symbol *s* represents the sign bit. This sign bit represents the sign of the DCT coefficient level.

The VLCs in the intracoded and intercoded macro blocks can be used in the watermarking process. The DC coefficients are not used, because they are predicted from other DC coefficients and coded with a different set of VLCs and Escape codes. Furthermore, replacing each DC coefficient in intracoded and intercoded frames can result in visible artefacts due to drift. If only the AC coefficients are taken into account, the watermark is adapted more to the video content and the drift is limited.

Variable length code	VLC size	Run	Level	LSB of Level
0010 0110 s	8 + 1	0	5	1
0010 0001 s	8 + 1	0	6	0
0000 0001 1101 s	12 + 1	0	8	0
0000 0001 1000 s	12 + 1	0	9	1
0000 0000 1101 0 s	13 + 1	0	12	0
0000 0000 1100 1 s	13 + 1	0	13	1
0000 0000 0111 11 s	14 + 1	0	16	0
0000 0000 0111 10 s	14 + 1	0	17	1
0000 0000 0011 101 s	15 + 1	1	10	0
0000 0000 0011 100 s	15 + 1	1	11	1
0000 0000 0001 0011 s	16 + 1	1	15	1
0000 0000 0001 0010 s	16 + 1	1	16	0

Table 9.4.1: *Example of lc-VLCs in Table B.14 of the MPEG-2 Standard.*

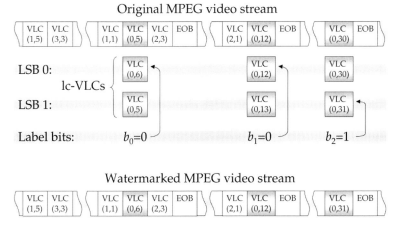

Figure 9.4.1: *Example of the LSB watermarking process.*

To add the label bit stream L to an MPEG-video bit stream, we test the VLCs in each macro block. If an *lc-VLC* is found and the least significant bit of its level is unequal to the label bit b_j ($j=0,1,2,\ldots,l-1$), this VLC is replaced by another, whose LSB-level represents the label bit. If the LSB of its level equals the label bit b_j the VLC is not changed. The procedure is repeated until all label bits are embedded. In Figure 9.4.1 an example is given of the watermarking process, where 3 label bits are embedded in the MPEG video stream.

To extract the label bit stream L we test the VLCs in each macro block. If an lc-VLC is found, the value represented by its LSB is assigned to the label bit b_j. The procedure is repeated for $j=0,1,2,\ldots,l-1$ until lc-VLCs are no longer found.

9.4.2 Evaluation of the bit domain watermarking algorithm

9.4.2.1 Test sequence

The maximum label bit rate is the maximum number of label bits that can be added to the video stream per second. This label bit rate is determined by the number of lc-VLCs in the video stream and is not known in advance. Therefore, we first experimentally evaluate the maximum label bit rate by applying the watermarking technique to an MPEG-2 video sequence. The sequence lasts 10 seconds, has a size of 720 by 576 pixels, is coded with 25 frames per second, has a GOP-length of 12 and contains P-, B- and I-frames. The sequence contains smooth areas, textured areas and sharp edges. During the 10 seconds of the video there is a gradual frame-to-frame transition, but at the end the camera turns fast to another view. A few frames of the sequence are shown in Figure 9.4.2. This sequence is coded at different bit rates (1.4, 2, 4, 6 and 8 Mbit/s) and used for all experiments in Part II of this book. It will be referred to as the "sheep sequence".

Figure 9.4.2: *A few frames of the "sheep sequence".*

9.4.2.2 Payload of the watermark

In Table 9.4.2 the results of the watermark embedding procedure are listed. Only the *lc-VLCs* in the intracoded macro blocks, excluding the DC coefficients, are used to embed watermark label bits. In this table the "number of VLCs" equals the number of all coded DCT coefficients in the intracoded macro blocks, including the fixed length coded coefficients and the DC-values. It appears that it is possible to store up to 7 kbit of watermark information per second in the MPEG streams if only intracoded macro blocks are used.

Video bit rate	Number of VLCs	Number of *lc-VLCs*	Max. label bit rate
1.4 Mbit/s	334,433	1,152 (0.3%)	0.1 kbit/s
2.0 Mbit/s	670,381	11,809 (1.8%)	1.2 kbit/s
4.0 Mbit/s	1,401,768	34,650 (2.5%)	3.5 kbit/s
6.0 Mbit/s	1,932,917	52,337 (2.7%)	5.2 kbit/s
8.0 Mbit/s	2,389,675	69,925 (2.9%)	7.0 kbit/s

Table 9.4.2: *Total number of VLCs and number of lc-VLCs in the intracoded macro blocks of **10 seconds** MPEG-2 video coded using different bit rates and the maximum label bit rate.*

If also the *lc-VLCs* in the intercoded blocks are used, the maximum label bit rate increases to 29 kbit/s. The results of this experiment are listed in Table 9.4.3. In this case the "number of VLCs" equals the number of all coded DCT coefficients in the intracoded and intercoded macro blocks, including the fixed length coded coefficients and the DC-values.

Video bit rate	Number of VLCs	Number of *lc-VLCs*	Max. label bit rate
1.4 Mbit/s	350,656	1,685 (0.5%)	0.2 kbit/s
2.0 Mbit/s	1,185,866	30,610 (2.6%)	3.1 kbit/s
4.0 Mbit/s	4,057,786	135,005 (3.3%)	13.5 kbit/s
6.0 Mbit/s	7,131,539	222,647 (3.1%)	22.3 kbit/s
8.0 Mbit/s	10,471,557	289,891 (2.8%)	29.0 kbit/s

Table 9.4.3: *Total number of VLCs and number of lc-VLCs in the intracoded and intercoded macro blocks of **10 seconds** MPEG-2 video, coded using different bit rates and the maximum label bit rate.*

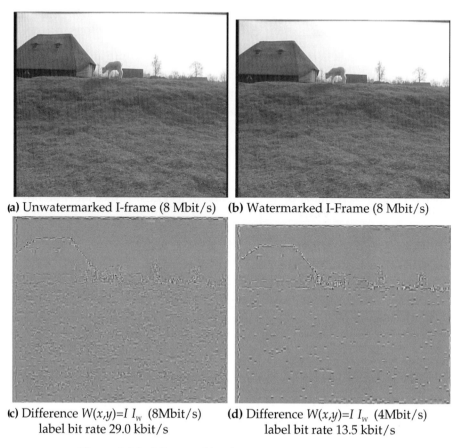

(a) Unwatermarked I-frame (8 Mbit/s) (b) Watermarked I-Frame (8 Mbit/s)

(c) Difference $W(x,y)=I\, I_w$ (8Mbit/s) label bit rate 29.0 kbit/s

(d) Difference $W(x,y)=I\, I_w$ (4Mbit/s) label bit rate 13.5 kbit/s

Figure 9.4.3: *Watermarking by VLC parity bit modification.*

9.4.2.3 Visual impact of the watermark

Informal subjective tests show that the watermarking process does not result in any visible artefacts in the streams coded at 4, 6 and 8 Mbit/s. It was not possible to reliably evaluate the quality degradation due to watermark embedding at less than 2 Mbit/s, because the unwatermarked MPEG-streams are already of poor quality, as these contain many compression artefacts. Although the visual degradation of the video due to the watermarking is not noticeable, the degradations are numerically measurable. In particular the maximum local degradations and the drift due to accumulation are relevant. In Figure 9.4.3a an original I-frame of the "sheep sequence" is represented. The sequence is MPEG-2 encoded at 8 Mbit/s. Figure

9.4.3b shows the corresponding watermarked frame. In Figure 9.4.3c the strongly amplified difference between the original I-frame and the watermarked frame is presented. Figure 9.4.3d shows the difference between the original I-frame coded at 4 Mbit/s and the corresponding watermarked frame. Since more bits are stored in an I-frame of a video stream coded at 8 Mbit/s, more degradations are introduced (Figure 9.4.3c) than in an I-frame of a video stream coded at 4 Mbit/s (Figure 9.4.3d).

According to Figure 9.4.3 most differences are located around the edges and in the textured areas. The smooth areas are left unaffected. In order to explain this effect the location of the *lc-VLCs* is investigated. In Figure 9.4.4 a histogram is shown of the "sheep sequence" coded at 8 Mbit/s. The number of all VLCs (including the fixed length codes) that code non-zero DCT coefficients and the number of *lc-VLCs* are plotted along the logarithmic vertical axis, represented by respectively white and gray bars. The DCT coefficient index scanned in the zigzag order ranging from 0 to 63 is shown on the horizontal axis.

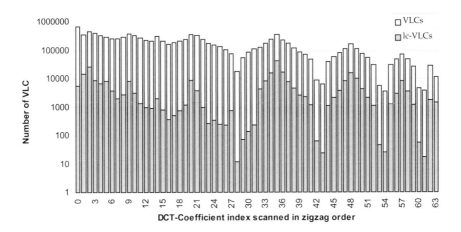

Figure 9.4.4: *Number of VLCs and lc-VLCs in 10s MPEG-2 video coded at 8Mb/s.*

Figure 9.4.4 shows that the *lc-VLCs* are fairly uniformly distributed over the DCT-spectrum. Therefore, we can expect each non-zero DCT coefficient represented by a VLC to have an equal probability of being modified. If we take into account that according to Table 9.4.3 at most 3.3% of all VLCs are *lc-VLCs*, the probability of a VLC being modified can roughly be estimated as follows:

$$P[VLC\ modified] = P[VLC = lc\text{-}VLC] \cdot P[label\ bit \neq LSB\ level\ VLC] \qquad (9.4.1)$$
$$P[VLC\ modified] < 0.033 \cdot \tfrac{1}{2} = 0.016$$

Smooth blocks are coded with only one or a few DCT coefficients. Because only 1.6% of them is replaced, most of the smooth areas are left unaffected. The textured blocks and the blocks containing sharp edges are coded with far more VLCs. These blocks will therefore contain the greater part of the *lc-VLCs*.

The maximum *local* degradation (the number of *lc-VLCs* per block) must be as low as possible. The visual impact of the watermarking process will be much smaller if the degradations introduced by modifying an *lc-VLC* are distributed more or less uniformly over the frame, instead of concentrated and accumulated in a relatively small area of the frame, or even worse, accumulated in a single DCT-block.

In Figure 9.4.5 a histogram is shown of 10 seconds of the watermarked "sheep sequence" coded at 8 Mbit/s. On the vertical axis the number of *lc-VLCs* per 8x8 block is shown. The number of 8x8 blocks that contain this amount of *lc-VLCs* is plotted along the logarithmic horizontal axis.

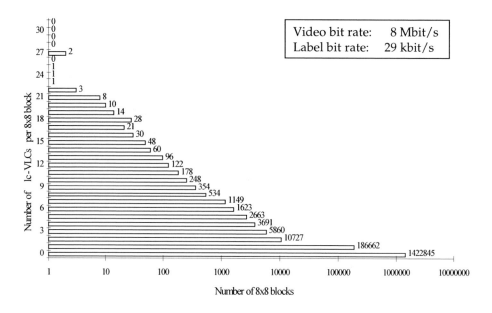

Figure 9.4.5: *Log histogram of the number of lc-VLCs per 8x8 block.*

This figure shows that 87% of all coded 8x8 blocks do not contain any *lc-VLC*. The rest of the coded 8x8 blocks contain one or more *lc-VLCs*. Most blocks (186.662) contain only one *lc-VLC*, which is about 64% of all *lc-VLCs* in the sequence. These numbers can be explained by Table B.14 and B.15 of the MPEG-2 standard [ISO96].

The most frequently occurring run-level pairs are coded with short VLCs. Almost all short VLCs do not qualify as an *lc-VLC*. This means that the chance of a large number of *lc-VLCs* in one 8x8 block is relatively low.

To limit the maximum number of *lc-VLC* replacements per DCT-block to T_m, we can use a threshold mechanism. If the number of *lc-VLCs* exceeds T_m, only the first T_m *lc-VLCs* are used for the watermark embedding; the other *lc-VLCs* are left unchanged. In Table 9.4.4 the label bit rates for the "sheep sequence" coded at 8 Mbit/s are listed for several values of T_m. If at most two *lc-VLC* replacements per block are allowed ($T_m = 2$), the label bit rate is only decreased to 83% of the maximum label bit rate for which $T_m = unlimited$. So by limiting the number of *lc-VLC* replacements per block we can avoid unexpected large local degradations without drastically affecting the maximum label bit rate.

T_m =max. *lc-VLC* replacements per block	Max. label bit rate
2	24.2 Kbit/s
4	26.9 Kbit/s
6	28.1 Kbit/s
8	28.6 Kbit/s
10	28.8 Kbit/s
Unlimited	29.0 Kbit/s

Table 9.4.4: *Label bit rates using a threshold for at most T_m lc-VLC replacements per 8x8 DCT-block (Video bit rate 8 Mbit/s).*

9.4.2.4 Drift

In an MPEG-video stream P-frames are predicted from the previous I- or P-frame. The B-frames are predicted from the two nearest I- or P-frames. Since intracoded and intercoded macro blocks are used for the watermark embedding, errors are introduced in all frames. However, error accumulation (drift) from the frames used for the prediction occurs in the predicted P- and B-frames. The drift can clearly be seen in Figure 9.4.6, where the difference $\Delta MSE = MSE_l - MSE_u$ is plotted. The MSE_u is the Mean-Square-Error (MSE) per frame between the original uncoded "sheep sequence" and the "sheep sequence" coded at 8 Mbit/s. The MSE_l is the MSE per frame between the uncompressed sequence and the watermarked sequence coded at 8 Mbit/s.

In Figure 9.4.6 it can be seen that the I-frames (numbered 1,13,25,37...) have the smallest ΔMSE; in the worst case the ΔMSE of a predicted B-frame is 2 to 3 times larger than the error in the I-frames. The average Peak Signal-to-Noise Ratio (PSNR) between the MPEG-compressed original and the uncompressed original is 37dB. If the watermarked compressed video stream coded at 8 Mbit/s is compared with the original compressed stream, the ΔMSE causes an average $\Delta PSNR$ of 0.1dB and a maximum $\Delta PSNR$ of 0.2dB. From these $\Delta PSNR$ values we conclude that the drift can be neglected and no drift compensation signal is required.

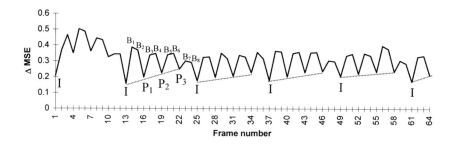

Figure 9.4.6: *ΔMSE of the watermarked "sheep sequence" coded at 8 Mbit/s with a label bit rate of 29.0 kbit/s.*

9.4.3 Robustness

A large label bit stream can be added and extracted in a very fast and simple way, but it can also be removed without significantly affecting the quality of the video. However, it still takes a lot of effort to completely remove a label from a large MPEG video stream. For example, decoding the watermarked MPEG-stream and encoding it again using another bit rate will destroy the label bit string. But re-encoding is an operation that is computationally demanding and requires a high capacity disk.

The easiest way to remove the label is by watermarking the stream again using another label bit stream. In this case the quality is slightly affected. During the re-labeling phase the adapted *lc-VLCs* in the watermarked video stream can either return to their original values or change to VLCs that represent DCTs that differ two quantization levels from the original ones in the unwatermarked video stream. Non-adapted *lc-VLCs* in the watermarked video stream can change to a value that differs one quantization level from the one in the original video stream. This means that there is some extra distortion, although the quality is only slightly affected. Since re-labeling of a large MPEG video stream still requires special hardware or a very powerful computer, the bit domain watermarking method is more suitable for consumer applications requiring moderate robustness.

9.5 Re-labeling resistant bit domain watermarking method

By reducing the payload of the watermark drastically we can easily change the bit domain watermarking algorithm described in Section 9.4.1 to a re-labeling resistant algorithm. The watermark label bits b_j are now not stored directly in the least significant bits of the VLCs, but a 1-dimensional pseudorandom watermark pattern $W(x)$ is generated consisting of the integers {-1,1}, based on a secret key, which is modulated with the label bits b_j as described in Section 8.2.2. The procedure to add this modulated pattern to the video stream is similar to the procedure described in Section 9.4.1.

However, we now select only those VLCs for which two other VLCs exist having the same run length and the same codeword length. One VLC must have a level difference of $+\delta$ and the other VLC must have a level difference of $-\delta$. Most lc-VLCs meet these requirements for a relatively small δ (e.g. δ = 1,2,3). For notational simplicity we call these VLCs pattern-carrying VLCs (*pc-VLCs*).

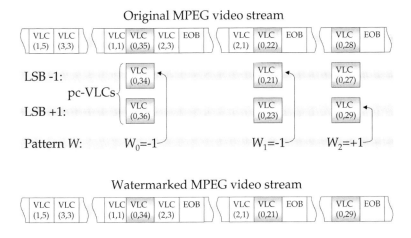

Figure 9.5.1: *Example of the re-labeling resistant watermarking method.*

To embed a watermark in a video stream, we simply add the modulated watermark pattern to the levels of the *pc-VLCs*. To extract the watermark, we collect the *pc-VLCs* in an array. The watermark label bits can now be retrieved by calculating the correlation between this array of *pc-VLCs* and the secret watermark pattern $W(x)$. In

Figure 9.5.1 an example is given of the watermark embedding process. About 1,000...10,000 *pc-VLCs* are now required to encode one watermark label bit b_j, but several watermark label bit strings can be added without interfering with each other, if independent pseudorandom patterns are used to form the basic pattern $W(x)$.

9.6 Conclusions

The most efficient way to reduce the complexity of real-time watermarking algorithms is to avoid computationally demanding operations by exploiting the compression format of the host video data. An advantage of this approach is that the watermark automatically becomes video content dependent. Since lossy compression algorithms discard the video information to which the human visual system is less sensitive and only encode visual important information, the watermark is only embedded in visual important areas. A disadvantage of closely following a compression standard and applying the constraint that the compressed video stream may not increase in size is that the number of locations to embed watermark information is limited significantly. The distortions caused by a watermark that is applied on a compressed video stream differ also from the distortions caused by a watermark applied on an uncompressed video stream. Due to block-based transformations and motion compensated frame prediction, distortions may spread over blocks and accumulate over the consecutive frames.

In this chapter we discussed four low complexity watermarking algorithms. The first correlation-based algorithm only uses the DC-coefficients. Although the algorithm can be performed completely in the coefficient domain, the low frequency watermark causes too many visible artefacts. The second correlation-based method does not only take into account the DC-coefficients, but also the AC-coefficients and applies drift compensation to prevent that the watermark becomes visible. Since it utilizes more locations to embed the watermark energy, the watermark is more robust. However, adding a drift compensation signal and extracting the watermark information cannot be performed in the coefficient domain, since a full MPEG decoding operation is required. The algorithm is therefore more complex than an algorithm that can be applied completely in the coefficient domain. The third LSB-modification method that we proposed fully operates in the bit domain, and is therefore the method that is computationally most efficient, but the least robust. Other advantages of this method are the enormous payload and the invisibility of the watermark. The fourth method extends the LSB-modification method and achieves a higher robustness by reducing the payload of the watermark.

There are two important differences between the correlation-based methods and the LSB-modification methods. A watermark embedded by a correlation-based method can still be extracted from the decoded raw video, since the watermarking procedure adds a spatial noise pattern to the pixel values. If the pixel values are available in raw format or in another compressed format the watermark can still be detected. Once a video stream watermarked by the LSB-modification methods is decoded, the watermark is lost, because the watermark embedding and extraction procedures are completely dependent on the MPEG structure of the video. This structure disappears or changes when the video is decoded or re-encoded at another bit rate. Since full MPEG decoding and encoding is a task that is computationally quite demanding, this is not really an issue for consumer applications requiring moderate robustness. Furthermore, correlation-based methods and LSB-modification methods differ considerably in complexity. LSB-modification methods are computationally far more efficient since they can operate on the lowest level in the bit domain.

For real-time applications that require the same level of robustness as the correlation-based methods but do not have enough computational power to perform full MPEG decoding for drift compensation and watermark detection, we have developed a completely new watermarking concept, which is presented in Chapters 10 and 11.

Chapter 10

Differential Energy Watermarks (DEW)

10.1 Introduction

In Chapter 9 we noticed that correlation-based watermarking techniques have the advantage that watermarks can be extracted from decoded or re-encoded video streams. However, in order to embed or detect an invisible correlation-based watermark, a full MPEG decoding operation is required. This might be computationally too demanding. On the other hand, we have seen that the Least Significant Bit (LSB) based algorithms are computationally highly efficient. But watermarks embedded by these algorithms cannot be extracted from decoded or re-encoded video streams. For real-time consumer applications that require the same level of robustness as the correlation-based methods and the same computational efficiency as the LSB-based methods, we therefore developed the Differential Energy Watermarking (DEW) concept [Lan97a], [Lan97b], [Lan98a] and [Lan99b]. As can be seen in Figure 10.1.1 the DEW concept can be applied directly on MPEG/JPEG compressed video as well as on raw video.

In the case of MPEG/JPEG encoded video data, the DEW embedding and extracting procedures can be performed completely in the coefficient domain (see Section 9.2). The encoding parts of the coefficient-domain watermarking concept can even be omitted. This means that the complexity of the DEW algorithm is only slightly higher than the LSB-based methods discussed in Section 9.4, but its complexity is considerably lower than the correlation-based method with drift compensation discussed in Section 9.3.

The application of the DEW concept is not limited to MPEG/JPEG coded video only; it is also suitable for video data compressed using other coders, for instance embedded zero-tree wavelet coders [Sha93]. The DEW algorithm embeds label bits by selectively discarding high frequency coefficients in certain video frame regions. The label bits of the watermark are encoded in the pattern of energy differences between DCT blocks or hierarchical wavelet trees.

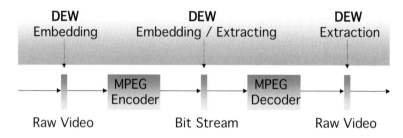

Figure 10.1.1: *DEW embedding / extracting in compressed and raw video.*

In Section 10.2 the general DEW concept for MPEG/JPEG coders is explained, followed by a more detailed description in Section 10.3. In Section 10.4 the DEW concept is evaluated for MPEG compressed video. Section 10.5 explains the general DEW concept for embedded zero-tree wavelet coded video. Finally the results are discussed in Section 10.6.

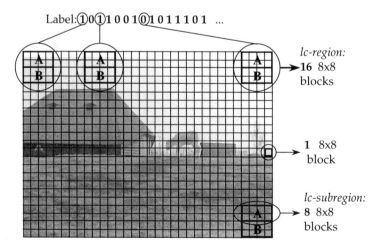

Figure 10.2.1: *Label bit positions and region definitions in a frame.*

10.2 The DEW concept for MPEG/JPEG encoded video

The Differential Energy Watermarking (DEW) method embeds a watermark consisting of *l* label bits b_j ($j = 0, 1, 2,..., l-1$) in a JPEG image or in the I-frames of an MPEG video stream. Each bit out of the label bit string has its own label-bit-carrying region, the *lc-region*, consisting of *n* 8x8 DCT luminance blocks.

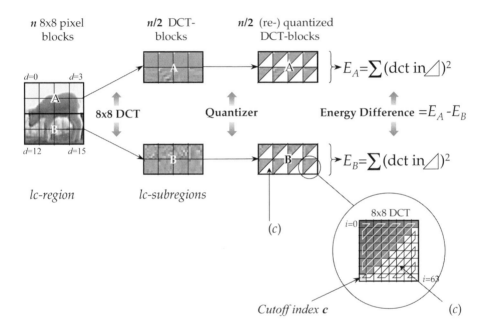

Figure 10.2.2: *Energy definitions in an lc-region of n=16 8x8 DCT blocks.*

For instance the first label bit is located in the top-left-corner of the image or I-frame in an lc-region of $n=16$ 8x8 DCT blocks, as illustrated in Figure 10.2.1. The size of this lc-region determines the label bit rate. The higher *n*, the lower the label bit rate. In case the video data is not DCT compressed, but in raw format, the DEW algorithm requires a block-based DCT transformation as preprocessing step.

A label bit is embedded in an lc-region by introducing an "energy" difference *D* between the high frequency DCT coefficients of the top half of the lc-region (denoted by *lc-subregion A*) and the bottom half (denoted by *B*). The energy in an lc-subregion equals the squared sum of a particular subset of DCT coefficients in this lc-subregion. This subset is denoted by $S(c)$, and is illustrated in Figure 10.2.2 by the white triangularly shaped areas in the DCT-blocks.

We define the total energy in $S(c)$, computed over the $n/2$ blocks in *subregion A* as:

$$E_A(c,n,Q_{jpeg}) = \sum_{d=0}^{n/2-1} \sum_{i \in S(c)} ([\theta_{i,d}]_{Q_{jpeg}})^2 \qquad (10.2.1)$$

Here $\theta_{i,d}$ denotes the non-weighted zigzag scanned DCT coefficient with index i in the d-th DCT block of the lc-subregion A under consideration. The notation $[]_{Q_{jpeg}}$ indicates that, prior to the calculation of E_A, the DCT coefficients of JPEG compressed video are optionally re- or pre-quantized using the standard JPEG quantization procedure [Pen93] with quality factor Q_{jpeg}. For embedding labels bits into MPEG compressed I-frames a similar approach can be followed, but here we restrict ourselves to the JPEG notation without loss of generality. The pre-quantization is done only in determining the energies, but is *not* applied to the actual video data upon embedding the label. The energy in lc-subregion B, denoted by E_B, is defined similarly.

$S(c)$ is typically defined according to a *cutoff index c* in the zigzag scanned DCT coefficients.

$$S(c) = \{h \in \{1,63\} \mid (h \geq c)\} \qquad (10.2.2)$$

The selection of suitable cutoff indices for lc-regions is very important for the robustness and the visibility of the label bits and will be discussed in the next section. First we focus on how the watermarking procedure works, assuming that we have available suitable cutoff indices c for each lc-region. The energy difference D between top and bottom half of an lc-region is defined as:

$$D(c,n,Q_{jpeg}) = E_A(c,n,Q_{jpeg}) - E_B(c,n,Q_{jpeg}) \qquad (10.2.3)$$

In Figure 10.2.2 the complete procedure to calculate the energy difference D of an lc-region $(n=16)$ is illustrated.

We now define the label bit value as the sign of the energy difference D. Label bit "0" is defined as $D>0$ and label bit "1" as $D<0$. The watermark embedding procedure must therefore adapt E_A and E_B to manipulate the energy difference D. If label bit "0" must be embedded, all energy after the cutoff index in the DCT-blocks of lc-subregion B is eliminated by setting the corresponding DCT coefficients to zero, so that:

$$D = E_A - E_B = E_A - 0 = +E_A \qquad (10.2.4)$$

If label bit "1" must be embedded, all energy after the cutoff index in the DCT-blocks of lc-subregion A must be eliminated, so that:

$$D = E_A - E_B = 0 - E_B = -E_B \qquad (10.2.5)$$

There are several reasons for computing this energy difference over the *triangularly shaped areas*. The most important reason is that it is easy to calculate the difference in energy and to change E_A and E_B accordingly in the compressed stream. All DCT coefficients needed for the calculation of E_A or E_B are conveniently located at the end of the compressed 8x8 DCT-block after zigzag ordering. The coefficients can be forced to zero to adapt the energy without re-encoding the stream by shifting the end of block marker (EOB) towards the DC-coefficient. Figure 10.2.3 illustrates the procedure of calculating E in a single compressed DCT-block and changing E by removing DCT coefficients located at the end of the zigzag scan (i.e. high frequency DCT coefficients).

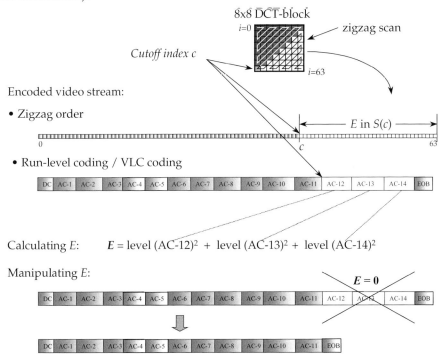

Figure 10.2.3. *Calculating and adapting energy in an 8x8 compressed DCT-block.*

Adding a watermark by removing coefficients has two advantages. Since no coefficients are adapted or added to the stream, the encoding parts of the coefficient domain watermarking concept can be omitted, as illustrated in Figure 10.2.4. This means that the DEW algorithm has only half the complexity of other coefficient domain watermarking algorithms.

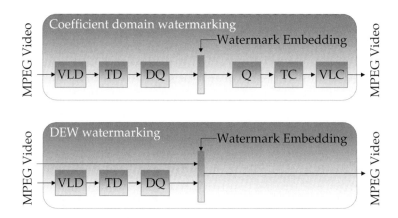

Figure 10.2.4. *Complexity difference between the DEW algorithm and other Coefficient domain watermarking algorithms.*

Furthermore, removing coefficients will always reduce the size of the watermarked compressed video stream compared to the unwatermarked video stream. If it is necessary that the watermarked compressed video stream keeps its original size, stuffing bits can be inserted before each macro block.

10.3 Detailed DEW algorithm description

The energies present in lc-subregions A and B defined by Equations 10.2.1 and 10.2.2 play a central role in the watermark embedding and extraction process. The values of E_A and E_B are determined by 4 factors:

- the spatial content of the lc-subregions A and B
- the number of blocks n per lc-region
- the pre- or re-quantization JPEG quality factor Q_{jpeg}
- the size of subset $S(c)$ (i.e. the *triangularly shaped areas*)

If the spatial content of an lc-region is very smooth and only coded by de DC-DCT coefficients, the AC-energy will be zero. The energy will be larger for regions containing a lot of texture or edges. The more DCT-blocks are taken to form the lc-region, the higher the energy will be, since the energy is the sum of the energies in all individual DCT-blocks in the lc-region.

The optional pre- or re-quantization JPEG quality factor Q_{jpeg} controls the robustness of the watermark against re-encoding attacks. In a re-encoding attack the watermarked video data is partially or fully decoded and subsequently re-encoded at a lower bit rate. Our method anticipates the re-encoding at lower bit rates up to a certain minimal rate. The smaller Q_{jpeg} is chosen, the more robust the watermark is against re-encoding attacks. However, the smaller Q_{jpeg} is chosen, the smaller the energies E_A and E_B will be, since most high frequency coefficients are quantized to zero, and can no longer contribute to the energy.

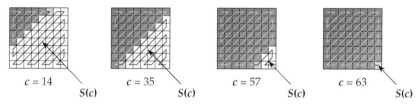

(a) Subset $S(c)$ of DCT coefficients defined by zigzag scan and cutoff index

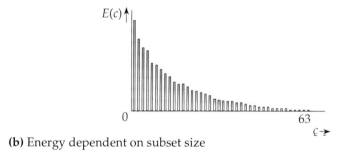

(b) Energy dependent on subset size

Figure 10.3.1: *(a) Examples of subsets and (b) energies for several cutoff indices.*

The size of subset $S(c)$ (Equation 10.2.2) is determined by the standard zigzag scan and a cutoff index c. If the zigzag scanned DCT coefficients are numbered from 0 to 63, where the coefficient with index 0 represents the DC-component and the coefficient with index 63 represents the highest frequency component. This subset consists of the DCT coefficients with indices $c...63$ ($c>0$). In Figure 10.3.1 some examples are shown of subsets defined by increasing cutoff indices. The corresponding experimentally determined energies are plotted below. This figure shows that increasing the cutoff index decreases the energy.

To enforce an energy difference, the watermark embedding process has to discard all DCT coefficients in the subset $S(c)$ in lc-subregion A or B. Since discarding coefficients introduces visual distortion, the number of discarded DCT coefficients has to be minimized. This means that the watermark embedding algorithm has to find a suitable cutoff index for each lc-region that defines the smallest subset $S(c)$ for which the energy in both lc-subregions A and B exceeds the desired energy difference. To find the cutoff index that defines the desired subset, we first calculate the energies $E_A(c,n,Q_{jpeg})$ and $E_B(c,n,Q_{jpeg})$ for all possible cutoff indices $c = 1...63$. If D is the energy difference that is needed to represent a label bit in an lc-region, the cutoff index c is found as the *largest* index of the DCT coefficients for which (10.2.1) gives an energy *larger* than the required difference D in *both* subregions A and B.

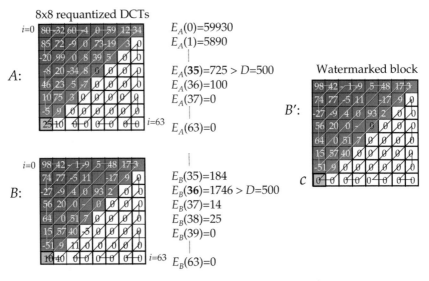

Figure 10.3.2: *Embedding label bit $b_0=0$ in an lc-region of $n=2$ DCT blocks.*

In controlling the visual quality of the watermarked video data, we wish to avoid the situation that the important low frequency DCT coefficients are discarded. For this reason, only cutoff indices larger than a certain minimum c_{min} may be selected. Mathematically, this gives the following expression for determining c:

$$c(n,Q_{jpeg},D,c_{min}) = \max\{c_{min}, \max\{g \in \{1,63\} \mid (E_A(g,n,Q_{jpeg}) > D) \wedge (E_B(g,n,Q_{jpeg}) > D)\}\} \quad (10.3.1)$$

Figure 10.3.2 shows an example of the embedding of label bit $b_0=0$ with an energy difference of $D=500$ in an lc-region consisting of $n=2$ DCT blocks. The maximum cutoff index for which the energy E_A exceeds $D=500$ is 35, while for E_B a cutoff index of 36 is sufficient. This means that the algorithm must select a cutoff index c of 35 so that both lc-subregions A and B have sufficient energy. Since the label bit to be embedded is zero, a positive energy difference has to be enforced by setting E_B to zero (Equation 10.2.4). This is done by discarding all non-zero DCT coefficients with indices 35...63 in lc-subregion B.

To extract a label bit from an lc-region we have to retrieve the cutoff index that was used for that lc-region during the embedding process. We therefore first calculate the energies $E_A(c,n,Q_{jpeg})$ and $E_B(c,n,Q_{jpeg})$ for all possible cutoff indices $c = 1...63$. Since either in lc-subregion A or lc-subregion B several DCT coefficients have been eliminated during the watermark embedding, we first find the *largest* index of the DCT coefficients, for which Equation 10.2.1 gives an energy *larger* than a threshold $D' \leq D$ in either of the two lc-subregions. The actually used cutoff index is then found as the maximum of these two numbers:

$$c^{(extract)}(n,Q'_{jpeg},D') = \max \{ \max\{g \in \{1,63\} \mid E_A(g,n,Q'_{jpeg}) > D'\},$$
$$\max\{g \in \{1,63\} \mid E_B(g,n,Q'_{jpeg}) > D'\} \} \quad (10.3.2)$$

In the above procedure, the parameters D' and Q'_{jpeg} can be chosen equal to the parameters D and Q_{jpeg}, which are used in the embedding phase. The detection threshold D' influences the determination of the cutoff index. This value must be smaller than the enforced energy difference D, but larger than 0. If $D' = 0$ the label can be extracted correctly, but only if the video stream is not affected by processing like adding noise, filtering or re-encoding. However, if a small amount of noise is introduced in the highest DCT coefficients, cutoff indices will be detected which are higher than the originally enforced ones. D' determines which amount of energy will be seen as noise. The re-quantization step can also be omitted ($Q'_{jpeg}=100$) without significantly influencing the reliability of the label bit extraction. Since Q_{jpeg} and D are not fixed parameters but may vary per image, the label extraction procedure must be able to determine suitable values for Q'_{jpeg} and D' itself. The most reliable way for doing this is to start the label bit string with several fixed label bits, so that during the label extraction those values for Q'_{jpeg} and D' can be chosen that result in the fewest errors in the known label bits.

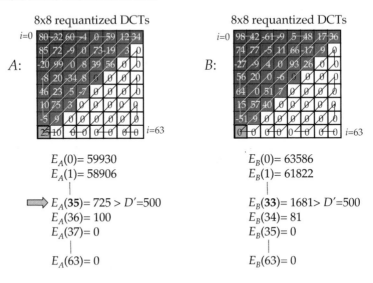

Figure 10.3.3: *Extracting label bit b_0 from an lc-region of n=2 DCT blocks.*

In Figure 10.3.3 an example is given of the extraction of label bit b_0 from the lc-region consisting of $n=2$ DCT blocks watermarked in Figure 10.3.2. For the extraction $D'=D=500$ is used. The maximum cutoff index for which the energy E_A exceeds $D'=500$ is 35; for E_B this cutoff index is 33. This means that the watermark embedding algorithm has used a cutoff index of 35. The energy difference $E_A(35)-E_B(35)=+725$. Since the energy difference is positive, the value zero is assigned to label bit b_0.

The algorithm applied in this form is heavily dependent on the video content. Figure 10.3.4 shows several examples of this content dependency. In Figure 10.3.4a an lc-region is depicted in which the lc-subregions A and B both contain edges, smooth and textured areas. These are typical examples of regions with average energy in the AC DCT coefficients. In this case, the watermark embedding procedure will select a subset $S(c)$ with a cutoff index somewhere in the middle of the range 1...63. This means that some coefficients in the highest and middle frequency bands are discarded. If the amount of energy that is discarded in these frequency bands is limited, the label bit will not be noticeable. Since re-quantization by re-encoding at a lower bit rate will not seriously affect the energy difference in the middle frequency band, the label bit will survive a re-encoding attack.

DIFFERENTIAL ENERGY WATERMARKS (DEW)

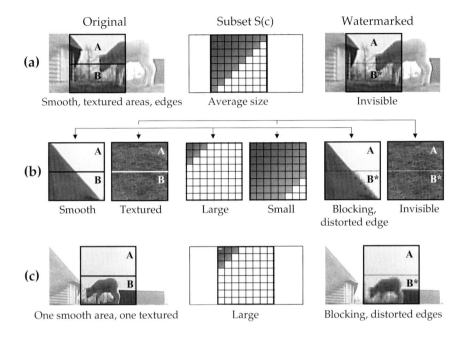

Figure 10.3.4: *Examples of subset sizes depending on video content.*

In Figure 10.3.4b two lc-regions are presented in which the lc-subregions are both very smooth or both very textured. If there is not much energy in a smooth lc-region, a very large subset $S(c)$ has to be chosen. This means that low frequency DCT coefficients are discarded. The human eye is quite sensitive to these, so block artefacts and distorted edges are the result. If there is much energy in a textured lc-region, a very small subset $S(c)$ is sufficient to find the required energy difference. Since here only the highest frequency components are discarded, the label bit will not be noticeable. However, since re-quantization by re-encoding at a lower bit rate will seriously affect the energy difference in the highest frequency bands, the label bit will not survive a re-encoding attack.

The worst-case situation is depicted in Figure 10.3.4c, where one lc-subregion is completely smooth, while the other one is textured and contains sharp edges. If a positive energy difference $D = E_A - E_B$ must be generated in this lc-region, all AC DCT coefficients in lc-subregion B must be eliminated by selecting an extremely large subset $S(c)$ to make $E_A > E_B$. The presence of the label bit obviously becomes clearly visible in lc-subregion B.

From these situations we conclude that it is not desirable to select very small subsets $S(c)$ defined by high cutoff indices, since energy differences embedded in the highest frequency bands do not survive re-encoding attacks. Furthermore, the selection of large subsets defined by low cutoff indices should be avoided, since energy differences enforced in the lowest frequency bands cause visible artefacts like blocking and distortion of sharp edges.

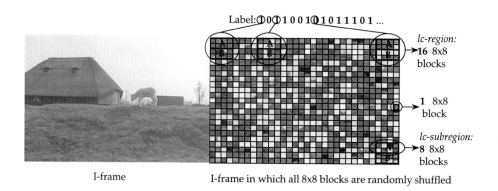

Figure 10.3.5: *Label bit positions and region definitions in a shuffled frame.*

In order to avoid the use of an extremely high or low cutoff index, we pseudorandomly shuffle all DCT-blocks in the image or I-frame using a secret key prior to embedding the label bits, as illustrated in Figure 10.3.5.

This does not pose any problems when we use MPEG or JPEG streams in practice, because effectively we now select randomly DCT-blocks from the compressed stream to define an *lc-region*, instead of spatially neighboring blocks. As a result of the shuffling operation, smooth 8x8 DCT-blocks and textured 8x8 DCT-blocks will alternate in the lc-subregions. The energy is now distributed more equally over all lc-regions, significantly diminishing the chance of a completely smooth or a completely textured lc-subregion. Another major advantage of the shuffle operation is that each label bit is scattered over the image or frame, which makes it impossible for an attacker to localize the lc-subregions. The complete watermark embedding and extraction procedures are shown in Figure 10.3.6.

DIFFERENTIAL ENERGY WATERMARKS (DEW)

Watermark embedding procedure:

- Shuffle all 8x8 DCT luminance blocks of an image or I-frame pseudorandomly
- FOR all label bits b_j in label string L DO

  - Select lc-subregion A consisting of $n/2$ 8x8 DCT-blocks,
    Select lc-subregion B consisting of $n/2$ other blocks (Fig. 10.3.5)
  - Calculate cutoff index c:

  $$c(n, Q_{jpeg}, D, c_{min}) = \max\{c_{min}, \max\{g \in \{1,63\} \mid (E_A(g,n,Q_{jpeg}) > D) \wedge (E_B(g,n,Q_{jpeg}) > D)\}\}$$

  where $E_{A,B}(c, n, Q_{jpeg}) = \sum_{d=0}^{n/2-1} \sum_{k \in S(c)} ([\theta_{i,d}]_{Q_{jpeg}})^2$

  $$S(c) = \{h \in \{1,63\} \mid (h \geq c)\}$$

  - IF ($b_j = 0$) THEN discard coefficients of area B in $S(c)$
    IF ($b_j = 1$) THEN discard coefficients of area A in $S(c)$
- Shuffle all 8x8 DCT luminance blocks back to their original locations

Watermark extraction procedure:

- Shuffle all 8x8 DCT luminance blocks of an image or I-frame pseudorandomly
- FOR all label bits b_j in label string L DO
  - Select lc-subregion A consisting of $n/2$ 8x8 DCT-blocks,
    Select lc-subregion B consisting of $n/2$ other blocks (Fig. 10.3.5)
  - Calculate cutoff index c:

  $$c^{(extract)}(n, Q'_{jpeg}, D') = \max\{\max\{g \in \{1,63\} \mid E_A(g,n,Q'_{jpeg}) > D'\},$$
  $$\max\{g \in \{1,63\} \mid E_B(g,n,Q'_{jpeg}) > D'\}\}$$

  where $E_{A,B}(c, n, Q_{jpeg}) = \sum_{d=0}^{n/2-1} \sum_{k \in S(c)} ([\theta_{i,d}]_{Q_{jpeg}})^2$

  $$S(c) = \{h \in \{1,63\} \mid (h \geq c)\}$$

  - Calculate energy difference:

  $$D = E_A(c^{(extract)}, n, Q'_{jpeg}) - E_B(c^{(extract)}, n, Q'_{jpeg})$$

  IF ($D > 0$) THEN $b_j = 0$
                ELSE $b_j = 1$

Figure 10.3.6: *Complete procedure for watermark embedding and extraction.*

10.4 Evaluation of the DEW algorithm for MPEG video data

10.4.1 Payload of the watermark

To evaluate the effect of the label bit rate on the visual quality of the video stream we applied the DEW algorithm to the "sheep sequence" coded at different bit rates. The label bit rate is fixed and determined by n, the number of 8x8 DCT-blocks per lc-region. In the experiments we omitted the optional re-quantization stage (Q_{jpeg}=100). Over a wide range of sequences we have found a reasonable setting for the energy difference $D = 20$ and the detection threshold $D' = 15$. The cutoff indices c for each label bit are allowed to vary in the range from 6 to 63 (c_{min}=6). Informal subjective tests show that the watermark, embedded with $n = 32$, is not noticeable in video streams coded at 8 and 6 Mbit/s. If MPEG streams coded at a lower bit rate are labeled with $n = 32$, blocking artefacts appear around edges of smooth objects. By increasing n further to 64 we make these artefacts disappear in the MPEG stream coded at 4 Mbit/s. At a rate of 1.4 and 2 Mbit/s the compression artefacts always dominate the additional degradations due to watermarking.

Video bit rate	n	Discarded bits	% Bit errors	Label bit rate
1.4 Mbit/s	64	1.6 kbit/s	24.6	0.21 kbit/s
2.0 Mbit/s	64	4.6 kbit/s	0.1	0.21 kbit/s
4.0 Mbit/s	64	3.8 kbit/s	0.0	0.21 kbit/s
6.0 Mbit/s	32	7.2 kbit/s	0.0	0.42 kbit/s
8.0 Mbit/s	32	6.6 kbit/s	0.0	0.42 kbit/s

Table 10.4.1: *Number of 8x8 DCT-blocks per bit, number of bits discarded by the watermarking process, percentage label bit errors and label bit rate for the "sheep sequence" coded at different bit rates.*

In Table 10.4.1 the results of the experiments are listed. The third column shows the number of bits discarded by the watermark embedding process. The fourth column presents the percentage bit errors found by extracting the label L' from the watermarked stream and comparing L' with the originally embedded one, L. Bit errors occur when the embedding algorithm selects cutoff indices below c_{min}. In this case the energy difference cannot be enforced. It appears that not enough high frequency coefficients exist in the compressed stream coded at 1.4Mbit/s to create the energy differences D for the label bits, since only 75% of the extracted label bits are correct.

10.4.2 Visual impact of the watermark

In Figure 10.4.1a the original I-frame of the MPEG-2 coded "sheep sequence" is represented. The sequence is MPEG-2 encoded at 8 Mbit/s. Figure 10.4.1b shows the corresponding watermarked I-frame. In Figure 10.4.1c the strongly amplified difference between the original I-frame and the watermarked frame is presented. Figure 10.4.1d shows the difference between the original I-frame coded at 4Mbit/s and the corresponding watermarked frame.

(a) Unwatermarked frame I (8 Mbit/s)   (b) Watermarked Frame I_w (8 Mbit/s)

(c) Difference $W(x,y)=I-I_w$ (8 Mbit/s) label bit rate 0.42 kbit/s

(d) Difference $W(x,y)=I-I_w$ (4 Mbit/s) label bit rate 0.21 kbit/s

Figure 10.4.1: *DEW watermarking by discarding DCT coefficients.*

It appears that all degradations are located in DCT-blocks with a relatively large number of high frequency DCT-components, textured blocks and blocks with edges. If we compare Figure 10.4.1 with Figure 9.4.3, we see that the DEW watermarking method causes fewer differences per frame than the LSB-based method described in Section 9.4, although the differences per block are larger. If the Bit Domain Labeling method is used, a DCT coefficient is only altered by one quantization level, while here the DCT coefficients are completely discarded.

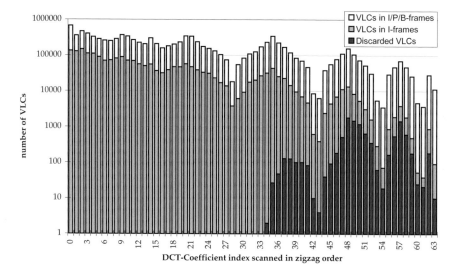

Figure 10.4.2: *Number of VLCs coding non-zero DCT coefficients in 10 s MPEG-2 video coded at 8 Mbit/s vs. number of VLCs discarded by the watermark.*

In Figure 10.4.2 a histogram is shown of the "sheep sequence" coded at 8 Mbit/s. The number of all VLCs (including the fixed length codes) that code non-zero DCT coefficients, the number of all VLCs in the I-frames and the number of discarded VLCs are plotted along the logarithmic vertical axis. The DCT coefficient index scanned in the zigzag order ranging from 0 to 63 is shown on the horizontal axis. Figure 10.4.2 shows that only high frequency DCT coefficients with an index above 33 are discarded for this particular parameter setting.

The histograms of the cutoff indices in the "sheep sequence" coded at 1.4 and 8 Mbit/s are plotted in Figure 10.4.3. The minimum cutoff index for the "sheep sequence" coded at 8 Mbit/s is 33; for a stream coded at 1.4 Mbit/s the minimum is equal to the minimum cutoff index $c_{min}=6$. The lower the bit rate is, the lower the cutoff indices have to be because of the lack of high energy components in the compressed video stream.

The visual impact of the labeling will be much smaller if the degradations introduced by discarding DCT coefficients are distributed more or less uniformly over the frame. Removing all VLCs from a few textured blocks will cause visible artefacts.

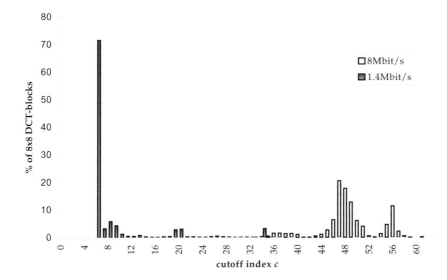

Figure 10.4.3: *Histograms of the cutoff indices in an MPEG-2 sequence coded at 1.4 and 8 Mbit/s, label bit rates are respectively 0.21 kbit/s and 0.42 kbit/s.*

In Figure 10.4.4 a histogram is shown of 10 seconds of the watermarked "sheep sequence" coded at 8 Mbit/s. On the vertical axis the number of discarded VLCs per 8x8 DCT-block is shown. The number of 8x8 DCT-blocks that contain this amount of discarded VLCs is plotted along the logarithmic horizontal axis.

It appears that 95% of all coded 8x8 blocks in the I-frames are not affected by the DEW algorithm. From an *lc-region* only the DCT coefficients above a certain cutoff index in the half, an *lc-subregion*, are eliminated. This means that only a few 8x8 blocks from an *lc-subregion* (average 10%) have energy above the cutoff index.

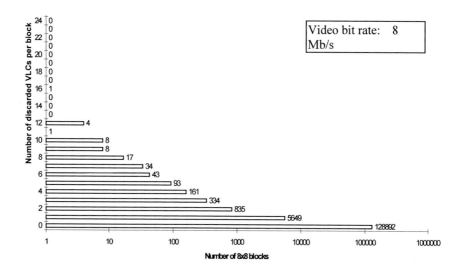

Figure 10.4.4: *Number of discarded VLCs per 8x8 DCT-block.*

Like in the Bit Domain watermarking algorithm described in Section 9.4, per 8x8 block a limit T_m can be set on the number of VLCs that are discarded during the watermarking process. Whereas in the Bit Domain watermarking algorithm this limit decreases the label bit rate, the DEW algorithm has a fixed label bit rate. Instead, setting a limit T_m affects the robustness of the label. If some DCT coefficients in one 8x8 block of an *lc-subregion* are not eliminated because the limit T_m prohibits it, in the worst case one label bit error can occur if the label extracted from this stream is compared with the originally embedded one. However, since each label bit is dependent on n 8x8 blocks, the likelihood that this error occurs is relatively small.

T_m =Max. number of discarded VLCs per block	Worst case % bit errors
2	17%
3	9%
4	5%
5	3%
6	2%
Unlimited	0%

Table 10.4.2. *Worst case % label bit errors introduced by limit T_m, the maximum number of discarded VLCs per 8x8 block (Video bit rate 8 Mbit/s, Label bit rate 0.42 kbit/s).*

In Table 10.4.2 the worst case percentages of bit errors introduced in the label of the "sheep sequence" coded at 8 Mbit/s are listed for several values of T_m. By applying proper error correcting codes on the label stream, we can greatly reduce the number of VLCs to be removed. In this way we obtain a better visual quality without significantly affecting the label retrieval.

10.4.3 Drift

Since P- and B-frames are predicted from I- and P-frames, the degradations in the I-frames introduced by watermarking also appear in the predicted frames. Because the P- and B-frames are only partially predicted from other frames and partially intracoded, the degradations will fade out. No degradations are introduced in the intracoded parts of the predicted frames by the labeling. The error fade-out can clearly be seen in Figure 10.4.5, where the difference $MSE_l - MSE_u$ is plotted. The MSE_u is the Mean-Square-Error (MSE) per frame between the original uncoded "sheep sequence" and the "sheep sequence" coded at 8 Mbit/s. The MSE_l is the MSE per frame between the uncompressed sequence and the watermarked sequence coded at 8 Mbit/s.

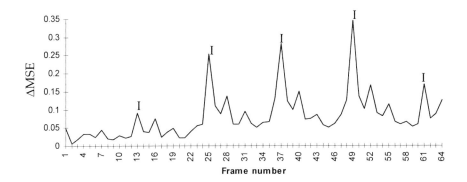

Figure 10.4.5: *ΔMSE of the watermarked "sheep sequence" coded at 8 Mbit/s with a label bit rate of 0.42 kbit/s.*

The average *PSNR* between the MPEG-compressed original and the uncompressed original is 37dB. If the watermarked video stream coded at 8 Mbit/s is compared with the original compressed stream, the ΔMSE causes an average $\Delta PSNR$ of 0.06dB and a maximum $\Delta PSNR$ of 0.3dB. It appears that this method has less impact on the average $\Delta PSNR$ and more impact on the maximum $\Delta PSNR$ than the method described in Section 9.4. From the $\varDelta PSNR$ values we conclude that no drift compensation signal is required.

10.4.4 Robustness

Unlike the watermark embedded by means of the LSB-based methods described in Section 9.4, the watermark embedded by the DEW algorithm cannot be removed by watermarking the video stream again using another watermark if another pseudorandom block shuffling is used. Other, more time-consuming methods, which are computationally more demanding and require a larger memory (disk) have to be applied to the watermarked compressed video stream in an attempt to remove the watermark. For simple filtering techniques the compressed stream must be decoded and completely re-encoded. A less complex method requiring lesser disk space, but which is still computationally highly demanding would be transcoding. To see if the watermark is resistant to transcoding or re-encoding at a lower bit rate, we performed the following experiment. The "sheep sequence" is MPEG-2 encoded at 8 Mbit/s and this compressed stream is watermarked (n = 32). Hereafter, the watermarked video sequence is transcoded at different lower bit rates.

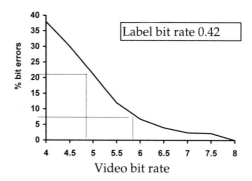

Figure 10.4.6: *% Bit errors after transcoding a watermarked 8 Mbit/s MPEG-2 sequence at a lower bit rate.*

The label bit strings are extracted from the transcoded video streams and each label bit string is compared with the originally embedded label bit string. If 50% bit errors are made the label is completely removed. The bit errors introduced by decreasing the bit rate are represented in Figure 10.4.6. We see that if the video bit rate is decreased by 25%, only 7% label bit errors are introduced. Even if the video bit rate is decreased by 38%, 79% of the label bit stream can be extracted correctly. Error correcting codes can further improve this result.

To embed a label bit in an lc-region, the DEW algorithm removes some high frequency DCT coefficients in one of the lc-subregions. This can be seen as locally applying a low-pass filter to an lc-subregion. To detect the label bit, the amount of high frequency components in the two lc-subregions is compared. If small geometrical distortions are applied to the video data, e.g. shifting, there is a mismatch between the lc-regions chosen during the embedding phase and the lc-regions chosen during the detection phase. Parts of the lc-region chosen during the embedding phase are in the detection phase replaced by adjacent lc-regions. Although the adjacent lc-regions introduce high frequency components in the low-pass filtered lc-subregions, the difference in high frequency components is still measurable if the geometrical distortions are relatively small. The DEW algorithm should therefore exhibit some degree of resistance to geometrical distortions like line shifting. The experiments performed in the next chapter show that the DEW algorithm is resistant to line shifts up to 3 pixels.

10.5 Extension of the DEW concept for EZW-coded images

The DEW concept is not only suitable for MPEG/JPEG compressed video data, but can also be applied to video compressed using embedded zero-tree wavelets [Sha93]. For an explanation about wavelet-based compression the reader is referred to [Aka96], [Bar94] and [Vet95]. In MPEG/JPEG compressed video data the natural starting point for computing energies and creating energy differences is the DCT-block. In embedded zero-tree wavelet compressed video data the natural starting point is the hierarchical tree structure. Instead of embedding a label bit by enforcing an energy difference between two lc-subregions of DCT-blocks, we now enforce energy differences between two sets of hierarchical trees. Figure 10.5.1 shows a typical tree structure that is used in the wavelet compression of images or video frames.

As can be seen in Figure 10.5.1, a tree in this 3-level wavelet decomposed image or video frame starts with a root Discrete Wavelet Transform (DWT) coefficient in the LL_3 band and counts 64 DCT coefficients. Unlike in the DCT situation, where the discarding of high frequency DCT coefficients is implicitly restricted by the zigzag scan order, in wavelet compressed video data different ways of pruning the hierarchical trees can be envisioned.

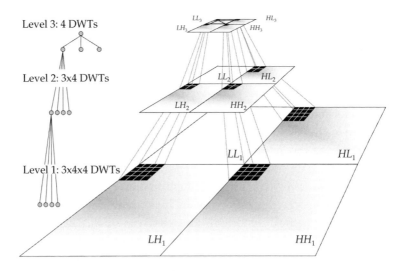

Figure 10.5.1: *Hierarchical tree structure of a DWT 3-level decomposition.*

The simplest way of removing energy is to truncate the trees below the hierarchical levels. A scheme in which trees are pruned coefficient-by-coefficient allows for fine-tuning the energy difference and for minimization of the visual impact. Therefore we have numbered the DWT coefficients of the hierarchical tree and defined a pseudo zigzag scan order, as illustrated in Figure 10.5.2.

This pseudo zigzag order is not the only possible way to order the DWT coefficients. More sophisticated orderings are possible, which take the human visual system into account. The advantage of using the straightforward numbering defined by Figure 10.5.2 is that we can use the same scheme as we used for the DCT situation. Only two minor changes are required. First, the quantization step in the energy definitions has to be adapted, as the DWT coefficients are now optionally re- or pre-quantized using a uniform quantizer instead of the standard JPEG quantization procedure. Second, not the 8x8 blocks are shuffled, but the roots of the hierarchical trees are pseudorandomly shuffled.

The complete procedure to calculate the energy difference in an lc-region is illustrated in Figure 10.5.3. In Figure 10.5.4a an example is given of the DEW algorithm applied to the embedded zero-tree wavelet coded "Lena image" using a 3-level wavelet decomposition. Here a label bit string of 64 label bits is embedded, for which end lc-regions of 64 hierarchical trees are used. One can see clearly that the watermark in this variation of the DEW algorithm also adapts to the image content.

DIFFERENTIAL ENERGY WATERMARKS (DEW)

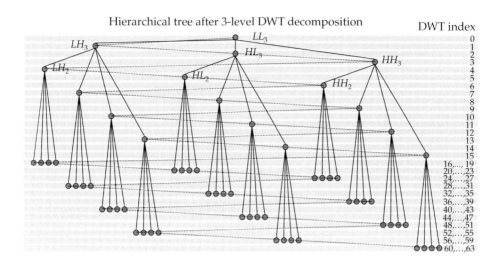

Figure 10.5.2: *DWT coefficient numbering and pseudo zigzag scan order.*

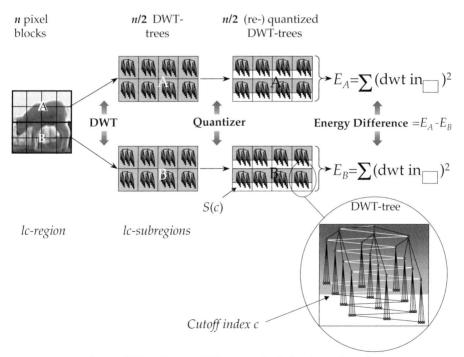

Figure 10.5.3: *Energy difference calculation in an lc-region.*

(a) Watermarked image   (b) Difference $W(x,y)=I(x,y)-I_w(x,y)$

Figure 10.5.4: *Level 3 EZW coded image watermarked using the DEW concept.*

10.6 Conclusions

In this chapter we introduced the Differential Energy Watermarking (DEW) concept. Unlike the correlation-based method with drift compensation described in Section 9.3.2, the DEW embedding and extraction algorithm can completely be performed in the coefficient domain and does not require a drift compensation signal. The encoding parts of the coefficient domain watermarking concept can even be omitted. The complexity of the DEW watermarking algorithm is therefore only slightly higher than the LSB methods described in Section 9.4. Furthermore, the DEW label bit rate is about 25 times higher than the label bit rate of the correlation-based methods described in Section 9.3. Like these correlation-based methods, a watermark embedded with the DEW concept can also be embedded and extracted from raw video data and the label string is resistant to re-labeling. Besides the low complexity and the much higher label bit rate the advantages the DEW concept has over other methods are that it provides a parameter Q_{jpeg} to anticipate to re-encoding attacks, that it exhibits some degree of resistance to geometrical distortions like line shifting and that it is directly applicable to video data compressed using other coders, for instance embedded zero-tree wavelet coders. Since many parameters are involved in the watermark embedding process of the DEW algorithm (n, Q_{jpeg}, D and c_{min}) heuristically determining optimal parameter settings is a quite elaborate task. Therefore in the next chapter a statistical model is derived that can be used to find these optimal parameter settings for DCT based coders.

Chapter 11

Finding Optimal Parameters for the DEW Algorithm

11.1 Introduction

The performance of the DEW algorithm proposed in the previous chapter greatly depends on the four parameters used in the watermark embedding phase. All parameters involved in the watermarking process are presented in Figure 11.1.

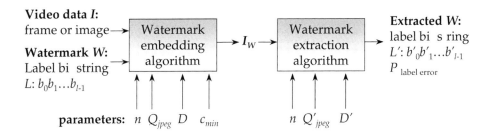

Figure 11.1: *Parameters involved in the DEW watermarking process.*

The first parameter is the number of 8×8 DCT blocks n used to embed a single information bit of the label bit string. The larger n is chosen, the more robust the watermark becomes against watermark-removal attacks, but the fewer information bits can be embedded into an image or a single frame of a video sequence.

The second parameter controls the robustness of the watermark against re-encoding attacks. In a re-encoding attack the watermarked image or video is partially or fully decoded and subsequently re-encoded at a lower bit rate. Our method anticipates the re-encoding at lower bit rates up to a certain minimal rate. Without loss of generality we will elaborate on the re-encoding of JPEG compressed images, where we express the anticipated re-encoding bit rate by the JPEG quality factor setting Q_{jpeg}. The smaller Q_{jpeg}, the more robust the watermark is against re-encoding attacks. However, for decreasing Q_{jpeg} increasingly more (high to middle frequency) DCT coefficients have to be removed upon embedding of the watermark, which leads to an increasing probability that artifacts become visible due to the presence of the watermark.

The third parameter is the energy difference D that is enforced to embed a label bit. This parameter determines the number of DCT coefficients that are discarded. Therefore, it directly influences the visibility and robustness of the label bits. Increasing D increases the probability that artifacts become visible and increases the robustness of the label.

The fourth parameter is the so-called minimal *cutoff index* c_{min}. This value represents the smallest index – in zigzag scanned fashion – of the DCT coefficient that is allowed to be removed from the image data upon embedding the watermark. The smaller c_{min} is chosen, the more robust the watermark becomes but at the same time, image degradations due to removing high frequency DCT coefficients may become apparent. For a given c_{min} there is a certain probability that a label bit cannot be embedded. Consequently, sometimes a *random* information bit will be recovered upon watermark detection, which is denoted as a *label bit error* in this chapter. Clearly, the objective is to make the probability that label bit errors occur as small as possible.

In order to optimize the performance of the DEW watermark technique, the settings of the above-mentioned parameters have to be determined. In the previous chapter we have used experimentally determined settings for these parameters. For a given watermark and image or video frame this is, however, an elaborate process. In this chapter we will show that it is possible to derive an expression for the label bit error probability P_{be} as a function of the parameters n, Q_{jpeg} and c_{min}. The relations that we derive analytically describe the behavior of the watermarking algorithm, and they make it possible to select suitable values for the three parameters (n, Q_{jpeg}, c_{min}), as well as suitable error correcting codes for dealing with label bit errors [Lan99b] and [Lan99c]. Although the expressions in this chapter are derived and validated for JPEG compressed images, they are also directly applicable to MPEG compressed I-frames.

FINDING OPTIMAL PARAMETERS FOR THE DEW ALGORITHM

In Section 11.2, we derive an analytical expression for the probability mass function (PMF) of the cutoff indices. In Section 11.3, this PMF is verified with real-world data. After deriving and validating the obtained PMF, we use the PMF to find the probability that a label string cannot be recovered correctly in Section 11.4 and the optimal parameter settings (n, Q_{jpeg}, c_{min}) in Section 11.5. Subsequently in Section 11.6, we experimentally validate the results from Section 11.5. The chapter concludes with a discussion on the DEW watermarking technique and its optimization in Section 11.7.

11.2 Modeling the DEW concept for JPEG compressed video

When operating the DEW algorithm, different values for the cutoff index are obtained. Insight in the actually selected cutoff indices is important since the cutoff indices determine the quality and robustness of the DEW. Therefore, in this section we will derive the probability mass function (PMF) for the cutoff index based on a stochastic model for DCT coefficients. This PMF depends only on the parameters Q_{jpeg} and n.

11.2.1 PMF of the cutoff index

The optimal cutoff index varies per label bit that we wish to embed. Therefore, it can be interpreted as a stochastic variable that depends on n, Q_{jpeg}, D, and c_{min}, i.e. $C(n, Q_{jpeg}, D, c_{min})$. Mathematically, this gives the following expression for determining C (see Sections 10.2 and 10.3):

$$C(n, Q_{jpeg}, D, c_{min}) = \max\{c_{min}, \max\{g \in \{1,63\} \mid (E_A(g,n,Q_{jpeg}) > D) \wedge (E_B(g,n,Q_{jpeg}) > D)\}\} \quad (11.2.1a)$$

where
$$E_A(c, n, Q_{jpeg}) = \sum_{d=0}^{n/2-1} \sum_{i \in S(c)} ([\theta_{i,d}]_{Q_{jpeg}})^2 \quad (11.2.1b)$$

$$S(c) = \{h \in \{1,63\} \mid (h \geq c)\} \quad (11.2.1c)$$

In order to be able to compute the PMF of the cutoff index, we assume that the energy difference D in Equation 11.2.1a is chosen in the range $[1, D_{max}(Q_{jpeg})]$. Here $D_{max}(Q_{jpeg})$ indicates the maximum of the range of energies defined by Equation 11.2.1b that do *not* occur in quantized DCT blocks because of the JPEG or MPEG compression process.

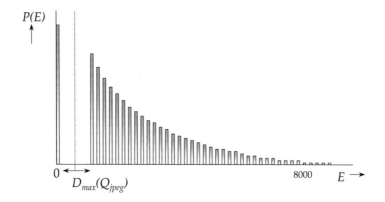

Figure 11.2.1: *Energy histogram of $E_{A,B}$ for a wide range of parameters (c,n,Q_{jpeg}).*

Figure 11.2.1 illustrates this effect; it is a histogram of the energy $E(c,n,Q_{jpeg})$ for a wide range of values of c, n, and Q_{jpeg}. We notice a clear "gap" in the histogram for smaller energies, because DCT blocks with that small amount of energy can no longer exist after compression.

In general the maximum $D_{max}(Q_{jpeg})$ depends on the extent to which the image has been compressed, i.e. it depends on Q_{jpeg}. The smaller Q_{jpeg}, the larger $D_{max}(Q_{jpeg})$. Mathematically this relation is given by:

$$D_{max}(Q_{jpeg}) = \left(F(Q_{jpeg}) \min_i(W_i) \right)^2$$

$$F(Q_{jpeg}) = \begin{cases} 50/Q_{jpeg} & Q_{jpeg} < 50 \\ (100 - Q_{jpeg})/50 & Q_{jpeg} \geq 50 \end{cases} \quad (11.2.2)$$

where $F(Q_{jpeg})$ denotes the coarseness of the quantizer used, and W_i is the i-th element ($i \in [c_{min}, 63]$) of the zigzag scanned standard JPEG luminance quantization table [Pen93].

Theorem I:

If the enforced energy difference D is chosen in the range $[1, D_{max}(Q_{jpeg})]$, where $D_{max}(Q_{jpeg})$ is defined by Equation 11.2.2, and if we do not constrain the cutoff index by c_{min}, the PMF of the cutoff index is given by:

FINDING OPTIMAL PARAMETERS FOR THE DEW ALGORITHM 255

$$P[C(n,Q_{jpeg})=c] = P[E(c,n,Q_{jpeg}) \neq 0]^2 - P[E(c+1,n,Q_{jpeg}) \neq 0]^2 \qquad (11.2.3)$$

where $E(c,n,Q_{jpeg})$ is defined in Equation 11.2.1b. Observe that in this theorem $C(n,Q_{jpeg})$ – besides being not constrained by c_{min} – is no longer dependent on D due to the wide range of values in which D can be selected.

Proof:

We first rewrite the definition of the cutoff index in Equation 11.2.1a to avoid the maximum operators as follows:

$$P[C(n,Q_{jpeg},D)=c] = P[\ \{(E_A(c,n,Q_{jpeg})>D) \land (E_B(c,n,Q_{jpeg})>D)\} \land$$
$$\{(E_A(c+1,n,Q_{jpeg})<D) \lor (E_B(c+1,n,Q_{jpeg})<D)\}\ ] \qquad (11.2.4)$$

In the following, we will drop the dependencies on n and Q_{jpeg} of the energies for notational simplicity. To calculate Equation 11.2.4, we need to have an expression for probabilities of the form $P[E_A(c)>D]$. As illustrated by Figure 11.2.1, the histogram of $E_A(c)$ is zero for small $E_A(c)$s because the quantization process maps many small DCT coefficients to zero. As a consequence, the energy defined in Equation 11.2.1b is either equal to 0 (for instance for large values of c), or the energy has a value larger than the smallest non-zero squared *quantized* DCT coefficient in the lc-subregion under consideration. This value has been defined as $D_{max}(Q_{jpeg})$ in Equation 11.2.2. Since we always choose the value of D smaller than $D_{max}(Q_{jpeg})$, Equation 11.2.4 can be simplified to:

$$P[C(n,Q_{jpeg})=c] = P[\ \{(E_A(c) \neq 0) \land (E_B(c) \neq 0)\} \land \{(E_A(c+1)=0) \lor (E_B(c+1)=0)\}\ ] \qquad (11.2.5)$$

Due to the random shuffling of the positions of the DCT blocks, we can now assume that $E_A(c)$ and $E_B(c)$ are mutually independent. Following several standard probability manipulations, Equation 11.2.5 can then be rewritten as follows:

$$\begin{aligned}P[C(n)=c] &= P[(E_A(c) \neq 0) \land (E_B(c) \neq 0) \land (E_A(c+1)=0)] \\ &\quad + P[(E_A(c) \neq 0) \land (E_B(c) \neq 0) \land (E_B(c+1)=0)] + \\ &\quad - P[(E_A(c) \neq 0) \land (E_B(c) \neq 0) \land (E_A(c+1)=0) \land (E_B(c+1)=0)] \\ &= P[(E_A(c) \neq 0) \land (E_A(c+1)=0)]\ P[E_B(c) \neq 0]\ ] \\ &\quad + P[(E_B(c) \neq 0) \land (E_B(c+1)=0)]\ P[E_A(c) \neq 0] \\ &\quad - P[(E_A(c) \neq 0) \land (E_A(c+1)=0)]\ P[(E_B(c) \neq 0) \land (E_B(c+1)=0)]\end{aligned} \qquad (11.2.6)$$

We first expand the first term of Equation 11.2.6 using conditional probabilities:

$P[(E_A(c) \neq 0) \wedge (E_A(c+1)=0)]$
$= 1 - P[(E_A(c+1)=0) \wedge (E_A(c)=0)] - P[(E_A(c+1) \neq 0) \wedge (E_A(c) \neq 0)]$
$\quad - P[(E_A(c+1) \neq 0) \wedge (E_A(c)=0)]$
$= 1 - P[E_A(c+1)=0 \,/\, E_A(c)=0]\, P[E_A(c)=0]$
$\quad - P[E_A(c) \neq 0 \,/\, E_A(c+1) \neq 0]\, P[E_A(c+1) \neq 0]$
$\quad - P[E_A(c)=0 \,/\, E_A(c+1) \neq 0]\, P[E_A(c+1) \neq 0]$ (11.2.7)

From the definition in Equation 11.2.1b we know that $E_A(c)$ is a strictly non-increasing function. Therefore, if there is no energy above cutoff index c, i.e., $E_A(c)=0$, there is also no energy above $c+1$, i.e. $E_A(c+1)=0$. This yields $P[E_A(c+1)=0 \,/\, E_A(c)=0] = 1$. On the other hand, if there is energy above cutoff index $c+1$, the same amount of energy or more must be present above cutoff index c, therefore $P[E_A(c) \neq 0 \,/\, E_A(c+1) \neq 0] = 1$ and $P[E_A(c)=0 \,/\, E_A(c+1) \neq 0] = 0$. Substitution of these conditional probabilities into Equation 11.2.7 gives the following result:

$P[(E_A(c) \neq 0) \wedge (E_A(c+1)=0)] = 1 - P[E_A(c)=0] - P[E_A(c+1) \neq 0]$
$\quad = P[E_A(c) \neq 0] - P[E_A(c+1) \neq 0]$ (11.2.8)

A similar approach can be followed to simplify the other terms in Equation 11.2.6. This results in the following expression:

$P[C(n)=c] = (P[E_A(c) \neq 0] - P[E_A(c+1) \neq 0])\, P[E_B(c) \neq 0]$
$\quad + (P[E_B(c) \neq 0] - P[E_B(c+1) \neq 0])\, P[E_A(c) \neq 0]$
$\quad + (P[E_A(c) \neq 0] - P[E_A(c+1) \neq 0])\, (P[E_B(c) \neq 0] - P[E_B(c+1) \neq 0])$
$= P[E_A(c) \neq 0]\, P[E_B(c) \neq 0] - P[E_A(c+1) \neq 0]\, P[E_B(c+1) \neq 0]$ (11.2.9)

Since the lc-subregions are both built-up from block-shuffled image data, we can assume that the probabilities in Equation 11.2.9 do not depend on the actual lc-subregion for which they are calculated, i.e. $P[E_A(c) \neq 0] = P[E_B(c) \neq 0] = P[E(c) \neq 0]$. Substitution of this equality results in Equation 11.2.3.

11.2.2 Model for the DCT-based energies

Theorem II:

If the PDF of the DCT coefficients is modeled as a generalized Gaussian distribution with shape parameter γ, then the probability that the energy $E_A(c,n,Q_{jpeg})$ is not equal to zero is given by:

FINDING OPTIMAL PARAMETERS FOR THE DEW ALGORITHM

$$P[E(c,n,Q_{jpeg}) \neq 0] = 1 - \left(\prod_{i=c}^{63} \left(1 - e^{-(\psi_i Q_i)^\gamma} \cdot \left(\sum_{h=0}^{\gamma^{-1}-1} \frac{(\psi_i Q_i)^{h\gamma}}{h!} \right) \right) \right)^{\frac{n}{2}} \quad (11.2.10)$$

where

$$\gamma^{-1} = 1, 2, 3, \ldots \quad (11.2.11a)$$

$$\psi_i Q_i = \frac{w_i F(Q_{jpeg})}{2\sigma_i} \sqrt{\frac{(3 \cdot \gamma^{-1} - 1)!}{(\gamma^{-1} - 1)!}} \quad (11.2.11b)$$

Further, $F(Q_{jpeg})$ denotes the coarseness of the quantizer as defined in Equation 11.2.2, σ_i^2 represents the variance of the i-th DCT coefficient (in zigzag scanned fashion), and w_i represents the corresponding element of standard JPEG luminance quantization table.

Proof:

The expression for $P[E_A(c) \neq 0]$ can be derived using Equation 11.2.1b. First we need a probability model for the DCT coefficients θ_i. Following the usual course of action taken in the literature at this point, we use the generalized Gaussian distribution [Mul93] and [Var89] with shape parameter γ:

$$P(\theta_i) = \xi_i e^{-|\psi_i \theta_i|^\gamma} \quad (11.2.12a)$$

where

$$\xi_i = \frac{\psi_i \cdot \gamma}{2(\gamma^{-1} - 1)!} \quad \text{and} \quad \psi_i = \frac{1}{\sigma_i} \sqrt{\frac{(3 \cdot \gamma^{-1} - 1)!}{(\gamma^{-1} - 1)!}} \quad \text{for } \gamma^{-1} = 1,2,3,\ldots \quad (11.2.12b)$$

This PDF has zero mean and variance σ_i^2. Typically, the shape parameter γ takes on values between 0.10 and 0.50. In a more complicated model, the shape parameter could be made to depend on the index of the DCT coefficient. We will, however, use a constant shape parameter for all DCT coefficients. Using Equation 11.2.12 we can now calculate the probability that a DCT coefficient is quantized as zero:

$$P[\hat{\theta}_i = 0] = \int_{-Q_i}^{Q_i} \xi_i \cdot e^{-|\psi_i \theta_i|^\gamma} d\theta_i = 1 - e^{-(\psi_i Q_i)^\gamma} \cdot \left(\sum_{h=0}^{\gamma^{-1}-1} \frac{(\psi_i Q_i)^{h\cdot\gamma}}{h!} \right) \quad (11.2.13)$$

where Q_i is the coarseness of the quantizer applied to the DCT coefficients. The probability that $E_A(c,n,Q_{jpeg})$ is equal to zero is now given by the probability that all quantized DCT coefficients with an index larger than c in all $n/2$ DCT blocks are equal to zero:

$$P[E(c)=0] = \left(\prod_{i=c}^{63} P[\hat{\theta}_i = 0] \right)^{\frac{n}{2}} \quad (11.2.14)$$

Equations 11.2.13 and 11.2.14 use the quantizer parameter Q_i. In JPEG this parameter is determined by the parameter w_i and the function $F(.)$, which depends on the user parameter Q_{jpeg} via Equation 11.2.2. Taking into account that JPEG implements quantization through rounding operations yields:

$$Q_i = \tfrac{1}{2} w_i F(Q_{jpeg}) \quad (11.2.15)$$

Combining Equations 11.2.12 - 11.2.15 yields Equation 11.2.10.

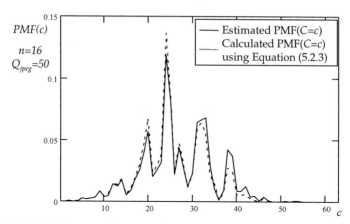

Figure 11.3.1: *Probability mass function of the cutoff index $P[C(n,Q_{jpeg})=c]$ as a function of c, calculated as a normalized histogram directly from watermarked images (solid line), and calculated using the derived Theorem I (dotted line).*

11.3 Model validation with real-world data

We validate Theorem I as follows. From a wide range of images we calculate the normalized histogram of $P[E(c,n,Q_{jpeg}) \neq 0]$ as a function of c. As an example we show the situation where $Q_{jpeg}=50$ and $n=16$. Using this histogram we evaluate Equation

11.2.3 to get an estimate of the PMF $P[C(n,Q_{jpeg})=c]$. The resulting PMF is shown in Figure 11.3.1 (dotted line). Using the same test data, we then directly calculate the histogram of $P[C(n,Q_{jpeg})=c]$ as a function of c. The resulting (normalized) histogram is also shown in Figure 11.3.1 (solid line). The figure shows that both curves fit well, which validates the correctness of the assumptions made in the derivation of Theorem I.

For the validation of Theorem II, we first need a reasonable estimate of the shape parameter γ and the variance σ_i^2 of the DCT coefficients. In fitting the PDF of the DCT coefficient we concentrated on obtaining a correct fit for the more important low frequency DCT coefficients, and obtained $\gamma=1/7$.

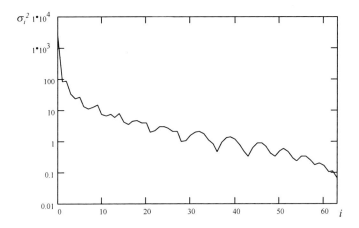

Figure 11.3.2: *Measured variances of the unquantized DCT coefficients as a function of the coefficient number along the zigzag scan.*

The variances of the DCT coefficients were measured over a large set of images, yielding Figure 11.3.2. For the time being, we will use these experimentally determined variances, but later on we will replace these with a fitted polynomial function.

In Figure 11.3.3a normalized histograms of the energy $E(c,n,Q_{jpeg})\neq 0$ are plotted for $n=16$ and several values of Q_{jpeg} as a function of c. In Figure 11.3.3b the probabilities $P[E(c,n,Q_{jpeg})\neq 0]$ are shown as calculated with Equation 11.2.10 from Theorem II using the measured variances of the DCT coefficients. Comparing the Figures 11.3.3a and 11.3.3b, we see that the estimated and calculated probabilities match quite well. There are some minor deviations for very small values of Q_{jpeg}

($Q_{jpeg}<15$), which is the result of the imperfect model for the DCT coefficients of real image data. We consider these deviations insignificant since they occur only at very high image-compression factors. We conclude that the models underlying Theorem II give results for $P[E(c,n,Q_{jpeg})\neq 0]$ that are sufficiently close to the actually observed data.

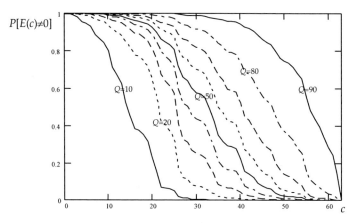

(a) $P[E(c,n,Q_{jpeg})\neq 0]$ calculated as normalized histogram directly from watermarked images

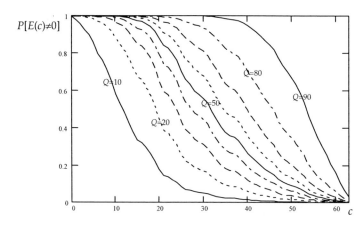

(b) $P[E(c,n,Q_{jpeg})\neq 0]$ calculated using Theorem II

Figure 11.3.3: *The probabilities $P[E(c,n,Q_{jpeg})\neq 0]$ as functions of c for n=16.*

FINDING OPTIMAL PARAMETERS FOR THE DEW ALGORITHM

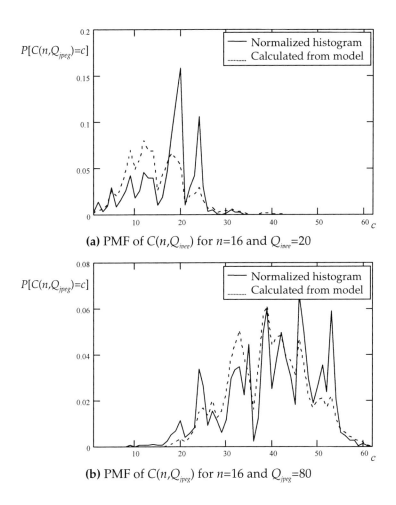

Figure 11.3.4: *Probability mass function of $C(n,Q_{jpeg})$, calculated as the normalized histogram directly from watermarked image data (solid line), and calculated using Equations 11.2.3 and 11.2.10.*

By combining Theorem I and II, we can derive PMFs of the cutoff index as a function of the parameters n and Q_{jpeg} based merely on the variances of the DCT coefficients. To validate the combined theorems we compared the PMFs calculated using Equations 11.2.3 and 11.2.10 with the normalized histograms directly calculated on a wide range of images. In Figure 11.3.4 two examples of the PMFs are plotted. In these examples, the solid lines represent the normalized histograms of $C(n,Q_{jpeg})$

calculated from watermarked image data, while the dotted lines represent the PMF $P[C(n,Q_{jpeg})=c]$ calculated using Equations 11.2.3 and 11.2.10. The greatly varying behavior of these curves as a function of c is mainly due to the zigzag scanning order of the DCT coefficients. We observe that an acceptable fit between the two curves is obtained with some deviations for higher cutoff indices. Since the PMF $P[C(n,Q_{jpeg})=c]$ will be used for calculating the probability of a label bit error, i.e. the probability that the watermarking procedure selects a cutoff index smaller than the minimum allowed values c_{min}, slight deviations at higher values for the cutoff index are not relevant to the purpose of this chapter.

The final step is to use the relation (11.2.3) and (11.2.10) to *analytically* estimate the PMF $P[C(n,Q_{jpeg})=c]$ of the cutoff index for different values of the parameters Q_{jpeg} and n. In this final step we rid ourselves of the erratic behavior of the curves in Figure 11.3.2 and 11.3.4 due to the zigzag scan order of the DCT coefficients by approximating the variances of the DCT coefficients in Figure 11.3.2 by a second-order polynomial function. The overall effect of using a polynomial function for the DCT coefficients is the smoothing of the PMF $P[C(n,Q_{jpeg})=c]$.

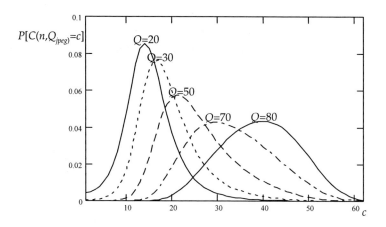

Figure 11.3.5: *Analytically calculated PMF $P[C(n,Q_{jpeg})=c]$ using Theorem I and II for various values of Q_{jpeg} and $n=16$.*

In Figures 11.3.5 and 11.3.6 the analytically calculated PMFs are shown. These curves are computed using Theorems I and II with only the shape parameter γ and the fitting parameters of the DCT variances as input. In Figure 11.3.5 $P[C(n,Q_{jpeg})=c]$ is shown as a function of Q_{jpeg} where n is kept constant, and in Figure 11.3.6

$P[C(n,Q_{jpeg})=c]$ is shown as a function of n where Q_{jpeg} is kept constant. It can clearly be seen that decreasing n or Q_{jpeg} leads to an increased probability of lower cutoff indices. This complies with our earlier experiments in Section 10.4.1, which showed that watermarks embedded with small values for n yield visible artifacts due to the removal of high frequency DCT coefficients.

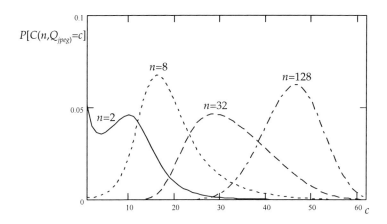

Figure 11.3.6: *Analytically calculated PMF $P[C(n,Q_{jpeg})=c]$ using Theorem I and II for various values of n and $Q_{jpeg}=50$.*

11.4 Label error probability

In the analysis of the DEW algorithm, we have seen that depending on the parameter settings (n,Q_{jpeg}) certain cutoff indices are more likely to occur than others. In this analysis, however, the selection of the cutoff index by the watermarking algorithm has been carried out irrespective of the visual impact on the image data. To ensure an invisible watermark, the cutoff indices must be larger than a certain minimum c_{min}. Consequently, it may happen in certain lc-regions that a label bit cannot be embedded. This random event is typically the case in lc-(sub)regions that contain insufficient high frequency detail.

Using Theorems I and II, we are able to derive the probability that this undesirable situation occurs, and obtain an expression for the *label bit error probability* P_{be} that depends on Q_{jpeg}, n and c_{min}. If a label bit cannot be embedded because of the minimally required value of the cutoff index c_{min}, there is a probability of 0.5 that during the extraction phase a random bit is extracted which equals the original label bit. We assume that due to the random shuffling of DCT blocks, the occurrence of a label bit error can be considered as a random event, independent of other label bit errors. The probability that a random error occurs in a label bit can therefore be computed as follows:

$$P_{be}(n,Q_{jpeg},c_{min}) = 0.5 \, P[\, C(n,Q_{jpeg}) < c_{min}] = 0.5 \sum_{c=0}^{c_{min}} P[C(n,Q_{jpeg}) = c] \tag{11.4.1}$$

Using this relation, we can calculate the label bit error probability for each value of c_{min} as a function of Q_{jpeg} and n. As an example Figure 11.4.1 shows the analytically computed label bit error probability $P_{be}(n,Q_{jpeg},c_{min})$ as a function of Q_{jpeg} and n for $c_{min}=3$. From this example it is immediately clear that for a given c_{min} certain (Q_{jpeg}, n) combinations must be avoided in practice because they lead to unacceptably high label bit error probabilities.

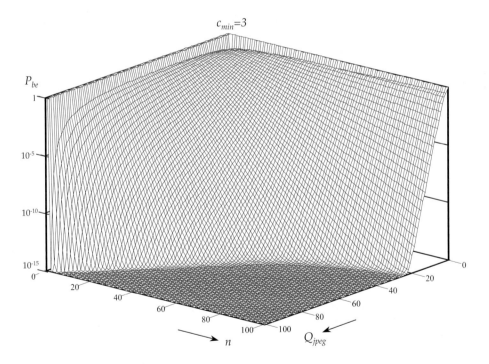

Figure 11.4.1: *The bit error probability P_{be} as a function of Q_{jpeg} and n for $c_{min}=3$.*

Using the label *bit* error probability in Equation 11.4.1, we can now derive the *label* error probability P_e, which is here defined as the probability that one or more label *bit* errors occur in the embedded information bit string. Assuming image dimensions of N×M, the number of information bits l that the image can contain is given by

$$l(N, M, n) = \text{int}\left(\frac{N \cdot M}{64 \cdot n}\right) \tag{11.4.2a}$$

with which the label error probability can be calculated as:

$$P_e(n, Q_{jpeg}, c_{min}, N, M) = 1 - (1 - P_{be})^{l(N,M,n)} \tag{11.4.2b}$$

Let us consider one particular numerical example. If, for instance in a broadcast scenario, one incorrect label is accepted per month in a continuous 10 Mbit/s video stream, the label bit error rate should be smaller than 10^{-7}. For selecting the optimal setting for Q_{jpeg} and n that comply with this label bit error rate, we use the curves of the combinations Q_{jpeg} and n for which P_e equals 10^{-7} shown in Figure 11.4.2. Different curves refer to different values of c_{min}. Further we have assumed the image dimensions N×M = 1024 × 768.

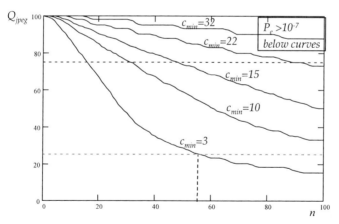

Figure 11.4.2: *Combinations of Q_{jpeg} and n for which $P_e = 10^{-7}$.*

11.5 Optimal parameter settings

Using results such as the ones shown in Figure 11.4.2, we can now select optimal settings for Q_{jpeg} and n for specific situations. We consider three different cases, namely:

- optimization of re-encoding robustness;
- optimization of the number of information bits l;
- optimization of the watermark invisibility.

In all cases the parameter D must be chosen in the range $[1, D_{max}(Q_{jpeg})]$ to ensure the validity of the models in Theorem I and II and the validity of the analytical results obtained from these models.

If we tune the DEW watermark such that it is optimized for maximum re-encoding robustness, typical choices are to anticipate re-encoding up to a JPEG quality factor of $Q_{jpeg}=25$, and to allow a minimal cutoff index of $c_{min}=3$. In this case – using Figure 11.4.2 – we need at least $n=54$ DCT blocks per label bit (which directly determines the number of information bits that can be stored in an image) to achieve the required label error probability of 10^{-7}.

If we require a large label and robustness against re-encoding attacks is not an issue, we can store more than 3 times as many bits in a label with the same label error probability of 10^{-7}. A typical parameter setting would be for instance, $Q_{jpeg}=75$, $n=16$ and $c_{min}=3$, as can be seen from Figure 11.4.2.

If visual quality is the most important factor, we must make sure that the minimal cutoff index is sufficiently large. For instance we choose $c_{min}=15$. Clearly, to obtain the same label bit error probability more DCT blocks per label bit are required since the allowed minimal cutoff index is larger than the one in the previous example. Using Figure 11.4.2, we find the optimal settings $Q_{jpeg}=75$ and $n=48$.

The performance of any watermarking system can be improved by applying error-correcting codes (ECCs). Since we know that the label bit errors occur randomly and independently of other label bit errors, we can compute the probability for *label error* in case an ECC is used that can correct up to R label bit errors, namely

$$P_e^{ECC(R)}(n, Q_{jpeg}, c_{min}, N, M) = 1 - \sum_{j=0}^{R} \binom{l(N,M,n)}{j} P_{be}^j (1-P_{be})^{l(N,M,n)-j} \qquad (11.5.1)$$

with the label bit error probability P_{be} given by Equation 11.4.1.

In Figure 11.5.1 the label error probability $P_e^{ECC(R)}$ is shown as a function of the number of DCT blocks used to embed a single label bit (n) for R=0, 1, 2, Q_{jpeg}=25 and c_{min}=3. We had already found that for a watermark optimized for robustness without error correcting codes, the optimal value is n=54 for a required bit error probability of P_e<10^{-7}. From Figure 11.5.1 we see that the same label error probability can be obtained using smaller values for n if we apply error correcting codes.

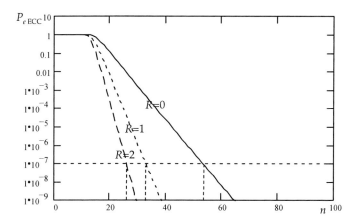

Figure 11.5.1: *Label error probability with (R=1,2) and without (R=0) error correcting codes for Q_{jpeg}=25 and c_{min}=3.*

For instance, by using an ECC that can correct one error, n can be decreased from 54 to 33. Obviously the use of ECCs introduces some redundant bits. However, this overhead is small compared to the increase in capacity due to the use of a smaller value of n. Table 11.5.1 gives some examples of the effective length of labels that can be embedded for $N \times M$ = 1024 × 768. In this table standard BCH codes [Rhe89] are used that can correct one or two errors.

ECC-Type	R	n	Parity-check bits ECC	Label size corrected for extra parity-check bits
no-ECC	0	54	0	227
BCH (511,502)	1	33	9	363
BCH (511,493)	2	27	18	437

Table 11.5.1: *Effective number of bits per label that can be embedded into an image of size $N \times M$= 1024 × 768, with required performance parameters c_{min}=3, Q_{jpeg}=25 and $P_e^{ECC(R)}$<10^{-7}.*

11.6 Experimental results

In this section, we will compare the robustness of labels embedded using settings optimized for maximum label size, namely $c_{min}=3$, $n=16$, $Q_{jpeg}=75$, and $D=25$ with labels embedded using settings optimized for robustness, namely $c_{min}=3$, $n=64^{\clubsuit}$, $Q_{jpeg}=25$, and $D=400$.

(a) Watermarked image **(b)** Difference $W(x,)=I-I_W$

Figure 11.6.1: *DEW watermarking using optimal settings for maximum label size.*

 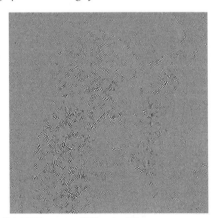

(a) Watermarked image **(b)** Difference $W(x,)=I-I_W$

Figure 11.6.2: *DEW watermarking using optimal settings for robustness.*

$^{\clubsuit}$ Our software implementation choices require that $n=16 \cdot k^2$, where k=1,2,3.... We therefore selected $n=64$ instead of the optimal value $n=54$.

FINDING OPTIMAL PARAMETERS FOR THE DEW ALGORITHM

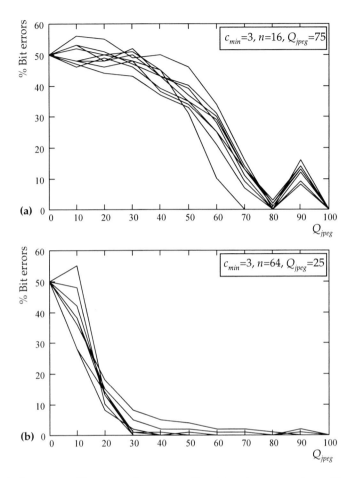

Figure 11.6.3: *Percentage bit errors after re-encoding (a) using parameter settings optimized for label size; (b) using parameter settings optimized for robustness.*

The "Lena image" watermarked with the DEW algorithm using settings optimized for maximum label size and the corresponding strongly amplified watermark are presented in Figure 11.6.1. Figure 11.6.2 shows the same images resulting from the DEW algorithm using settings optimized for robustness.

We will first check the robustness against re-encoding. Images are JPEG compressed with quality factor of 100. From these JPEG compressed images two watermarked versions are produced; one for each parameter setting. Next, the images are re-encoded using a lower JPEG quality factor. The quality factor of the re-encoding process is variable. Finally, the watermark is extracted from the re-encoded images and compared bit by bit with the originally embedded watermark. For the labels embedded using settings optimized for maximum label size, we used the extraction parameters $D'=40$ and $Q'_{jpeg}=75$. For the labels embedded using settings optimized for robustness, we used the extraction parameters $D'=400$ and $Q'_{jpeg}=80$. With this experiment, we find the percentages of label bit errors due to re-encoding as a function of the re-encoding quality factor. In Figure 11.6.3 the resulting label bit error curves are shown for nine different images.

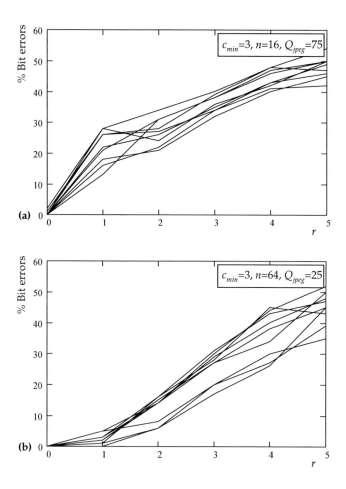

Figure 11.6.4: *Percentage bit errors after shifting over r pixels (a) using parameter settings optimized for label size; (b) using parameter settings optimized for robustness.*

Although we expect that the percentages of label bit errors are very small for JPEG quality factors between 75 and 100, because the parameter Q_{jpeg} is set to 75, we see in Figure 11.6.3a a small increase in bit errors for images re-encoded using a JPEG quality factor of 90. This effect is caused by the two consecutive quantization steps using JPEG quality factors 90 and 75, which are performed before the energy differences are calculated. These quantization steps introduce minor differences in the DCT coefficients. If these minor differences are squared and accumulated over 16 DCT blocks, the energy differences can significantly differ from the originally enforced small energy differences ($D=25$). This effect can be canceled by omitting the optional quantization step ($Q'_{jpeg}=100$) during the watermark extraction phase, or by increasing the enforced energy difference D.

Comparing Figure 11.6.3a (parameter setting optimized for label length using $c_{min}=3$, $n=16$, $Q_{jpeg}=75$, and $D=25$) and Figure 11.6.3b (parameter setting optimized for label robustness using $c_{min}=3$, $n=64$, $Q_{jpeg}=25$, and $D=400$), we see an enormous gain in robustness. In Figure 11.6.3b, we see a breakpoint around $Q_{jpeg}=25$. For higher re-encoding qualities, the percentage label bit errors is below 10%.

In the previous chapter we noticed that the DEW watermarking technique is slightly resistant to line shifting. To investigate the effect of the parameter settings optimized for robustness on the resistance to line shifting, we carry out the following experiment. Images are JPEG compressed with a quality factor of 85. These JPEG images are watermarked using the parameter settings optimized for label size or for robustness. Next the images are decompressed, shifted to the right over r pixels and re-encoded using the same JPEG quality factor. Finally, a watermark is extracted from these re-encoded images and compared bit by bit with the originally embedded watermark. Consequently, we find the percentages of bit errors due to line shifting. In Figure 11.6.4 the bit error curves are shown for nine different images. As in the previous experiment, we see an improvement in robustness between Figure 11.6.4a and Figure 11.6.4b. Using the parameter settings optimized for robustness, the DEW watermark becomes resistant to line shifts of up to 3 pixels.

11.7 Conclusions

In this chapter we have derived, experimentally validated, and exploited a statistical model for our DCT-based DEW watermarking algorithm. The performance of the DEW algorithm has been defined as its robustness against re-encoding attacks, the label size, and the visual impact. We have analytically shown how the performance is controlled by three parameters, namely Q_{jpeg}, n and c_{min}. The derived statistical model gives us an expression for the label bit error probability as a function of these three parameters Q_{jpeg}, n and c_{min}. Using this expression, we can optimize a watermark for robustness, size or visibility and add adequate error correcting codes.

The obtained expressions for the probability mass function of the cutoff indices can also be used for other purposes. For instance, with this PMF an estimate can be made of the variance of the watermarking "noise" that is added to an image by the DEW algorithm. This measure, possibly adapted to the human visual perception, can be used to carry out an overall optimization of the watermark embedding procedure using the (perceptually weighted) signal-to-noise ratio as optimization criterion.

Chapter 12

Benchmarking the DEW Watermarking Algorithm

12.1 Introduction

In literature many watermarking algorithms have been presented in recent years. Most authors claim that the watermark embedded by their algorithm is robust and invisible. However, they all use different robustness criteria and quality measures. Furthermore, the term "robustness" is hard to define and it is even questionable if it can be defined formally. A watermark that is fully resistant to lossy compression techniques may be very vulnerable to a dedicated attack, which may consist of some low complexity processing steps like concatenated filtering. Besides robustness and visibility, the payload and complexity of the embedding and extraction procedure may play an important role. Also the weighting of these performance factors varies significantly for different applications. This makes the comparison of the performance of the different algorithms a difficult task. In spite of this, we attempt in this chapter to derive a fair benchmark for the DEW algorithm by taking into account attacks known from the literature and by weighting the performance factors according to the requirements imposed by the application.

In Section 12.2 two watermark benchmarking approaches from literature are discussed. In Section 12.3 two dedicated watermark attacks are presented, which can be part of a benchmarking process. The performance of the DEW algorithm is compared to the real-time spread spectrum method of Hartung and Girod [Har98] and the basic spread spectrum method of Smith and Comiskey [Smi96] in Section 12.4. Section 12.5 concludes the chapter.

12.2 Benchmarking methods

In literature two watermark benchmarking methods are proposed, namely [Fri99a] and [Kut99]. The authors of both methods notice that the robustness is dependent on the payload and the visibility of the watermark. Therefore, to allow a fair comparison between different watermarking schemes, watermarks are embedded in a pre-defined video data set with the highest strength, which does not introduce annoying effects according to a pre-defined visual quality metric. Subsequently processing techniques and attacks are applied to the watermarked data and the percentages bit errors are measured to estimate the performance of the watermarking schemes.

The two benchmarking methods differ in the choice of the payload of the watermark, the visual quality metric and the processing techniques. In [Fri99] the payload of the watermark is fixed to 1 or 60 bits. To evaluate the visual quality of the watermarked video data, the spatial masking model of Girod [Gir89] is used. This model is based on the human visual system and accurately describes the visibility of artefacts around edges and flat areas in video data. The watermark strength is adjusted in such a way that Girod's model indicates less than one percent of pixels with visible changes. Subsequently, the watermarked data is subject to the processing operations listed in Table 12.2.1 and the bit error rate is measured as a function of the corresponding parameters.

Operation	Parameter
JPEG compression	Quality factor
Blurring	Kernel size
Noise addition	Noise amplitude (SNR)
Gamma correction	Gamma exponent
Permutation of pixels	Kernel size
Mosaic filter	Kernel size
Median filter	Kernel size
Histogram equalization	-

Table 12.2.1: *List of processing operations to which the robustness of a watermarking method is tested.*

The authors [Fri99a] do not claim that this list is exhaustive; other common lossy compression techniques, such as wavelet compression should probably be included.

Using the benchmarking approach described in [Kut99] the payload of the watermark is fixed to 80 bits. To evaluate the visual quality of the watermarked video data, the distortion metric proposed by Van den Branden Lambrecht and Farrell [Bra96] is used. This perceptual quality metric exploits the contrast sensitivity and masking phenomena of the Human Visual System and is based on a multi-channel model of the human spatial vision. The unity for this metric is given in *units above threshold*, also referred to as *Just Noticeable Difference* (JND). In [Kut99] this quality metric is normalized using the ITU-R Rec. 500 quality rating [ITU95]. In Table 12.2.2 the ratings and the corresponding visual perception and quality are listed.

Rating	Impairment	Quality
5	Imperceptible	Excellent
4	Perceptible, not annoying	Good
3	Slightly annoying	Fair
2	Annoying	Poor
1	Very annoying	Bad

Table 12.2.2: *ITU-R Rec. 500 quality ratings on a scale from 1 to 5.*

The ITU-R quality rating Q_{ITU} is computed as follows:

$$Q_{ITU} = \frac{5}{1 + CN + MD} \tag{12.2.1}$$

where *MD* is the measured distortion according to the model of Van den Branden Lambrecht and Farrell and *CN* is a normalization constant. *CN* is usually chosen such that a known reference distortion maps to the corresponding quality rating. The results generated by the model cannot be used to determine if for instance an image with quality rating Q_{ITU}=4.5 looks better than an image with quality rating Q_{ITU}=4.6. The results should be interpreted in combination with a threshold: images with quality ratings above Q_{ITU}=4 may only contain perceptible, and not annoying artefacts.

The watermark strength is adjusted in such a way that the quality rating is at least 4. Subsequently, the watermarked data is subject to a list of processing operations, including lossy JPEG compression, geometric transformations and filters. Most of these processing operations are implemented in one single program called StirMark, which is described in the next section. Instead of applying each processing operation listed in Table 12.2.1 to the watermarked data, only StirMark is applied to the data, which has the same effect as performing the transformations separately with various parameters. Finally, the error rate for the retrieved bits is measured.

12.3 Watermark attacks

12.3.1 Introduction

Watermarks are vulnerable to processing techniques. Therefore, every processing technique that does not significantly impair the perceptual quality of the watermarked data can be considered as an intentional or unintentional watermark attack. In [Har99] the watermark attacks are classified in four groups:

A. "Simple attacks" are conceptually simple attacks that attempt to impair the embedded watermark by manipulations of the whole watermarked data, without attempting to identify and isolate the watermark. Examples include linear and general non-linear filtering, lossy compression techniques like JPEG and MPEG compression, noise addition, quantization, D/A conversion and gamma correction.

B. "Detection-disabling attacks" are attacks that attempt to break the correlation and to make the recovery of the watermark impossible for the watermark detector, mostly by geometrical distortions like scaling, shifting in spatial or temporal direction, rotation, shearing, cropping and removal or insertion of pixel clusters. A typical property of this type of attacks is that the watermark remains in the attacked data and can still be recovered with increased intelligence of the watermark detector.

C. "Ambiguity attacks" are attacks that attempt to confuse by producing fake original data or fake watermarked data. This attack is only useful for copyright purposes and therefore outside the scope of this book. An example of this attack is the inversion attack described in [Cra96], which attempts to discredit the authority of the watermark by embedding additional watermarks so that it is unclear which was the first watermark and who the legitimate copyright owner is.

D. "Removal attacks" are attacks that attempt to analyze the watermarked data, estimate the watermark or the host data, and separate the watermark from the watermarked data to discard the watermark.

The authors [Har99] note that the distinction between the groups is sometimes vague, since some attacks belong to two or more groups. In Section 12.3.2 the StirMark attack is discussed, which belongs to groups A and B. A removal attack on spatial spread spectrum watermarking techniques belonging to group D is presented in Section 12.3.3.

12.3.2 Geometrical transforms

StirMark is a watermark removal attack that is based on the idea that although many watermarking algorithms can survive simple video processing operations, they cannot survive combinations of them [Pet98b] and [Pet99]. In its simplest form StirMark emulates a resampling process. It applies minor geometrical distortions by slightly stretching, shearing, shifting and/or rotating an image or video frame by an unnoticeable random amount and then resampling the video data using either bi-linear or Nyquist interpolation. In addition, a transfer function that introduces a small and smooth distributed error into all sample values is applied. This emulates the small non-linear analog/digital converter imperfection typically found in scanners and display devices. In Figure 12.3.1b an example is given of how StirMark resamples the data. The distortions are exaggerated for viewing purposes. As can be seen the distortion of each pixel is the greatest at the borders of the video data and almost zero at the center.

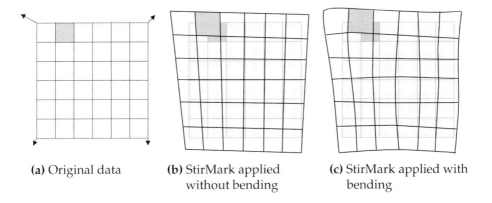

(a) Original data   (b) StirMark applied without bending   (c) StirMark applied with bending

Figure 12.3.1: *Exaggerated example of distortions applied by StirMark.*

In addition to this procedure StirMark can also apply global bending to the video data. This results in an additional slight deviation for each pixel, which is greatest at the center of the video data and almost zero at the borders. The bending process is depicted in Figure 12.3.1c. Finally the resulting data is compressed with the lossy JPEG algorithm using a quality factor for medium visual quality.

In Figure 12.3.2b an example is shown of the "Lena image" after applying StirMark. Figure 12.3.2c shows the difference between the original image and the StirMarked image. It can be seen that although some pixels are shifted over more than 3 pixels, the image quality is not affected seriously.

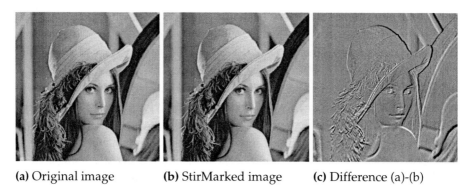

(a) Original image   **(b)** StirMarked image   **(c)** Difference (a)-(b)

Figure 12.3.2: *Example of an image after applying StirMark.*

The StirMark attack confuses most watermarking schemes available on the market [Pet98b]. Only watermarking schemes with a very low payload can survive this kind of attack.

12.3.3 Watermark estimation

12.3.3.1 Introduction

The spatial spread spectrum watermarking methods described in Chapter 8 basically add a pseudorandom pattern to an image in the spatial domain to embed a watermark. This watermark can be detected by correlating with the same pattern or by applying other statistics to the watermarked image. In this section two attacks are discussed to estimate the pseudorandom spread spectrum watermark from the watermarked image only. If a nearly perfect estimation of the watermark can be found, this estimated watermark can be subtracted from the watermarked image. In this way the watermark is removed without affecting the quality of the image [Lan98b] and [Lan98c].

For our initial experiments we use the basic spread spectrum implementation of Smith and Comiskey [Smi96]. If we apply this method to an image I, a random pattern W consisting of the constants $-k$ and $+k$ is added to obtain the watermarked image $I_w = I+W$, where k is a positive integer value. The watermark energy resides in all frequency bands. Compression and other degradations may remove signal energy from certain parts of the spectrum, but since the energy is distributed all over the spectrum, some of the watermark remains. The random pattern W is uncorrelated with image I, but correlated with I_w:

$$\text{cov}(W, I+W) = \text{var}(W) + \text{cov}(I, W) \approx \text{var}(W) + 0$$

$$\rho(W, I+W) = \frac{\text{cov}(W, I+W)}{\sqrt{\text{var}(W)}\sqrt{\text{var}(I+W)}} \approx \sqrt{\frac{\text{var}(W)}{\text{var}(I+W)}} \quad (12.3.1)$$

$$\rho(W, I+W) \approx \frac{k}{\sqrt{\text{var}(I+W)}}$$

Evaluation of Equation 12.3.1 for typical images yields the conclusion that ρ ranges from 0.02 to 0.05. However, if the watermarked images are compressed using the JPEG algorithm or distorted, the approximation in Equation 12.3.1 does not hold. Indeed, the correlation coefficients decrease by a factor 2, while the variance of $(I+W)$ nearly equals the variance of the JPEG compressed version of $(I+W)$.

If an arbitrary random pattern W_x is used, the correlation coefficient will be very small:

$$\text{cov}(W_x, I+W) = \text{cov}(W_x, W) + \text{cov}(W_x, I) \approx 0 + 0$$

$$\rho = \frac{\text{cov}(W_x, I+W)}{\sqrt{\text{var}(W_x)}\sqrt{\text{var}(I+W)}} \approx 0 \quad (12.3.2)$$

This holds only if W and W_x are orthogonal and W_x is not correlated with I. Typical values for correlation coefficients between I_w and arbitrary random watermark W_x are a factor 10^2 smaller than $\rho(W, I_w)$.

A simple estimation attack would be to search for all possible random patterns and take the one with the highest correlation value as the possible watermark pattern. This approach has several disadvantages. In the first place the search space is huge. Even if the watermark pattern consisting of the integers [-1,1] should meet the requirement that the number of -1s and the number of +1s need to be equal, more than 4×10^{306} possible patterns have to be checked for a 32x32 pixel watermark.

As a first step, we carried out experiments with a genetic algorithm to search the random pattern with the highest correlation coefficient with $I_w = I+W$. In some cases the genetic algorithm found a pattern with a relatively high correlation (0.3) with I_w and no correlation with W (10^{-5}). This means that the pattern is adapted to the image contents and not to the watermark. To avoid that the genetic algorithm finds random patterns that have higher correlation coefficients than the embedded watermark we must adapt our optimization criterion. From the properties of spread spectrum watermarks we know the following about W:

- $\rho(W, I_w) \in [0.01 .. 0.05]$
- $\rho(W, I) \approx 0$
- W is pseudorandom and has a flat spectrum

If the image is distorted by compression, $\rho(W, I_w)$ is unknown. Too many patterns meet the requirement $\rho(W, I) \approx 0$. The additional information that W is random and has a flat spectrum is not enough either to create a suitable optimization criterion function. If we have several different images with the same watermark to our disposal, there are several possibilities (e.g. collusion attacks). A fitness function for the genetic algorithm dependent on all images can be used, and if there are enough images, the average of the images can be taken as estimation of the watermark.

In [Kal98b] and [Lin98] the watermark is estimated by analyzing the watermark detector. However, if different watermarks are used for each image and the watermark detector is not available, we have to follow other approaches that estimate the watermark from only the watermarked data. In [Mae98] an approach is proposed to estimate spatial spread spectrum watermarks by histogram analysis. The results of this approach depend very much on the content of the images. Watermarks can be estimated quite accurately for images with peaky histograms, but the results for images with a smooth histogram are poor. In the next subsection we propose a watermark estimation approach which is based on non-linear filtering.

12.3.3.2 Watermark estimation by non-linear filtering

In general, a watermark can be regarded as an enforced distortion in the image that is perceptually invisible. In most cases, this distortion is not correlated to the image contents. If we could apply a nearly perfect image model to the watermarked image $I_w = I+W$, we could predict the image content $\hat{I}$ and estimate the watermark: $\hat{W} = I_w - \hat{I}$. Because perfect image models and perfect noise filters do not exist, $\hat{I}$ will be different from I and $\hat{W}$ will be different from W. Our objective is to separate I_w

$= \hat{I} + \hat{W}$ in such a way that the watermark is totally removed from $\hat{I}$ and resides completely in $\hat{W}$ [Lan98b] and [Lan98c]. This means that image contents may remain in the predicted watermark.

We tested the performance of the following separation operations to divide I_w in $\hat{I}$ and $\hat{W}$: an AR-model, linear smoothing filters (3x3 and 5x5), Kuwahara filters [Kuw76] (several sizes), non-linear region based filters and filters based on thresholding in the DCT-domain (coring). In some cases, the watermark can be retrieved from both $\hat{I}$ and $\hat{W}$, while $\hat{I}$ has still a reasonable quality and $\hat{W}$ does not contain any image information. In other cases the watermark can only be retrieved from $\hat{W}$, but the quality of $\hat{I}$ is significantly affected and the image contents, especially the edges, remain in $\hat{W}$.

We select some candidates from the separation operations that totally destroy the watermark in $\hat{I}$, $\rho(W, \hat{I}) \approx 0$. From these candidates we select the operation that has the highest correlation coefficient $\rho(\hat{W}, W)$ in a test set of 9 images. In Table 12.3.1 the correlation coefficients for several separation operations are listed.

Separation Operation	$\rho(\hat{W}, W)$
Misc. Noise Reduction Filters	0.08-0.12
Autoregressive Model	0.10-0.17
Median 3x3	0.13-0.22

Table 12.3.1: *Correlation coefficients $\rho(\hat{W}, W)$ using different separation operations.*

The 3x3 median filter turns out to be the best separation operation and is used for the rough estimation of $\hat{W} = I_w\text{-med}_{3x3}(I_w)$. However, correlation coefficients $\rho(\hat{W}, W)$ between 0.13 and 0.22 are still too low and $\hat{W}$ must be refined further by using information about the watermark properties.

The estimate $\hat{W}$ does still contain edge information. To protect the edges in I_w we limit the range of $\hat{W}$ from [-128..128] to [-2..2] before we subtract $\hat{W}$ from I_w. In Figure 12.3.3 the modulus of the Fourier Transform of the truncated $\hat{W}$ is presented.

The horizontal, vertical and diagonal patterns in Figure 12.3.3 clearly indicate that some dominating low frequency components are present in the spectrum. Since a spread spectrum watermark should not contain such dominating components, these come certainly from the image content. To remove these components a 3x3 linear high-pass filter is applied to the non-truncated $\hat{W}$. Truncating the filtered $\hat{W}$ to the range [-2,2] results in the Fourier spectrum as presented in Figure 12.3.4. The correlation coefficients between the high-pass filtered $\hat{W}$ and W, $\rho(\hat{W}, W)$ now increase to values around 0.4.

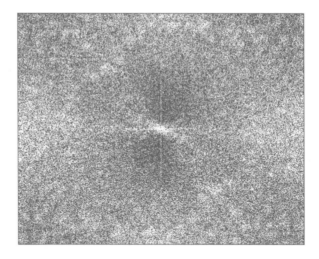

Figure 12.3.3: *Power density spectrum of* $\hat{W}_{[-2..2]}$.

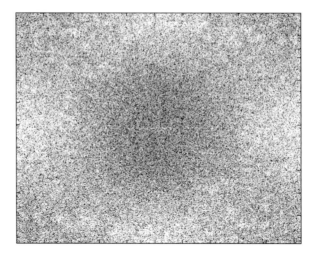

Figure 12.3.4: *Power density Spectrum of high-pass filtered* $(\hat{W})_{[-2..2]}$.

If this watermark $\hat{W}$ is subtracted from the watermarked image I_w the watermark is not completely removed. This is not surprising, since we are not able to predict the low frequency components of the watermark. These components are discarded during the high-pass filtering stage of $\hat{W}$ or are left in $\hat{I}$ by the median filter. The low frequency components, which cannot be estimated properly, give a positive contribution to correlation of the watermark detector, while subtracting the estimated watermark $\hat{W}$, which mainly consists of high frequency components, gives a negative correlation contribution. By amplifying the estimated watermark $\hat{W}$ with a certain gain factor G before subtraction, the overall correlation of the watermark detector can be forced to zero. The complete scheme for removing a watermark is represented in Figure 12.3.5.

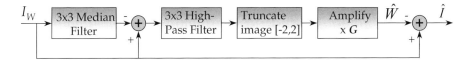

Figure 12.3.5: *The complete watermark removing scheme (WRS).*

The value of G is dependent on the image content and the amount of energy in the embedded watermark. If G is chosen too high, the watermark inverts and one can still retrieve it from $\hat{I}$ by inverting the image before retrieving the watermark.

The value G is experimentally determined. A watermark is added to an image using the method of Smith and Comiskey [Smi96], 32x32 pixels are used to store one bit of watermark information and the watermark carrier consists of the integers {-2,2}. The watermark removing scheme is applied to the watermarked image with several values for G. The percentage watermark bit errors is plotted as a function of G in Figure 12.3.6. If 50% bit errors are made, the watermark is removed; if 100% bit errors are made, the watermark is totally inverted. According to Figure 12.3.6 the gain factor G should have a value between 2 and 3 to remove the watermark from this image. The values of the gain factor vary for different kinds of images but are typically in the range from 2 to 3. We therefore fixed the gain factor G to 2.5 for all images.

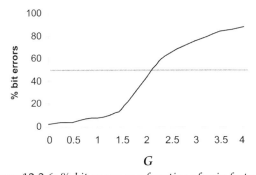

Figure 12.3.6: *% bit errors as a function of gain factor G.*

We tested the watermark removing scheme (WRS) represented in Figure 12.3.5 on a set of 9 true color images. Informal subjective tests were performed to determine the quality of the images. Some images hardly contain any textured areas and sharp edges, some contain many sharp edges and much detail, others contain both smooth and textured areas. First, the WRS ($G=2.5$) is applied to the methods of Bender *et al* [Ben95] and Pitas and Kaskalis [Pit95]. The watermarks in the 9 test images are all removed while the quality of the images is not reduced significantly.

Subsequently the WRS ($G=2.5$) is applied to the more robust watermarking method of Smith and Comiskey [Smi96]. The watermarks are added using P pixels per bit and a gain factor of k, where $k=1$ or 2. If higher gain factors k are used, the watermark becomes visible. For the values P=8x8, 16x16, 32x32, 64x64 and k=1,2 the watermarks can be removed without affecting the visual quality significantly. An example is given in Figure 12.3.7. An image is watermarked using the parameters k=2, P=32x32. To remove the watermark completely (about 44% bit errors) using the JPEG compression algorithm, we have to use a quality factor Q=10. The result of this compression operation is presented in Figure 12.3.7a. If we apply the WRS to the watermarked image, the watermark is completely removed (>50% bit errors) and we obtain the image which is shown in Figure 12.3.7b. This image is hardly distorted. If the number of pixels per bit P is increased further to 128x128 or 256x256, the watermark is fully removed in smooth images, but only partially in textured images.

(a) Removal by JPEG compression **(b)** Removal by the WRS scheme

Figure 12.3.7: *Removing a watermark from a watermarked image.*

Finally, the WRS ($G=2.5$) is applied to the method of Langelaar *et al* [Lan97a]. This watermarking method determines the gain factor k for each watermark bit automatically. Therefore only the number of pixels per bit P can be changed. All watermarks added with this method can be removed for P=8x8, 16x16, 32x32. For P=64x64, 128x128, ... the watermarks are only partially removed. In this case the watermark information is only removed from the smooth regions of the images, but remains in the more textured regions, since here the watermark estimate is not accurate enough.

Some methods (e.g. [Wol96]) first subtract the original image from the watermarked image and apply the watermark retrieval operation on this difference image. However, the WRS also removes the watermarks in this case. Other methods using a similar approach as [Smi96] are not tested, but we expect that such watermarks will be affected in the same way as those in [Smi96], since these methods use the same basic principle.

12.4 Benchmarking the DEW algorithm

12.4.1 Introduction

In this section the DEW algorithm is compared to other watermarking methods known from literature. Section 12.4.2 discusses the performance factors on which the comparison is based. In Section 12.4.3 the real-time DEW algorithm for MPEG compressed video is compared to the basic spread spectrum technique of Smith and Comiskey [Smi96], which operates on raw video data, and to other real-time watermarking algorithms that operate directly on the compressed data. In this comparison the emphasis is on the real-time aspect. This holds for both the watermarking procedures and the watermark removal attacks. The attacks are therefore limited to transcoding operations.

In Section 12.4.4 the DEW algorithm for JPEG compressed and uncompressed still images is compared to the basic spread spectrum method of Smith and Comiskey [Smi96]. Since the latter method is not specially designed for real-time operation on compressed data, the real-time aspect is neglected in this comparison. In our evaluation we follow the guidelines of the benchmarking methods from literature described in Section 12.2.

12.4.2 Performance factors

To evaluate the performance of the DEW algorithm we have to compare it to other watermarking algorithms with respect to complexity, payload, impact on the visual quality, and robustness. Of these performance factors, the impact on the visual quality is most important. A watermark may not introduce annoying effects; in that case, watermarking algorithms will not be accepted as protection techniques by the users, who expect excellent quality of digital data. The weighting of the other performance factors depends heavily on the application of the watermarking method.

As already mentioned in Section 7.4, Part II of this book focuses mainly on the class of watermarking algorithms which can, for instance, be used in fingerprinting and copy protection systems for home-recording devices for the consumer market. For this class of watermarking algorithms the complexity of the watermark embedding and extraction procedures is an important performance factor for two reasons. On one hand, because the algorithms have to operate in real-time and on the other hand, because the algorithms have to be inexpensive for the use in consumer products.

Another performance factor is the payload of the watermark. For fingerprinting applications and protection of intellectual property rights a label bit rate of at least 60 bits per second is required to store one identification number per second that is similar to the one used for ISBN or ISRC [Kut99]. For copy protection purposes, a label bit rate of one bit per second is usually sufficient to control digital VCRs.

The last performance factor is the robustness of the watermark. The robustness is closely related to the payload of a watermark. The robustness can be increased by decreasing the payload and visa versa. Sections 12.2 and 12.3 gave an overview of processing techniques to which watermarks are vulnerable. Most of these processing techniques require that the compressed video stream is decoded and completely re-encoded. This task is computationally demanding and requires a lot of storage space. The most obvious way to intentionally remove a watermark from a compressed video stream is to circumvent these MPEG decoding and re-encoding steps. This can be done, for instance, by transcoding the video stream.

12.4.3 Evaluation of the DEW algorithm for MPEG compressed video

To evaluate the DEW algorithm for MPEG compressed video we compare it with the real-time watermarking algorithms known from literature as described in Chapter 9.

Since the bit domain methods do not survive MPEG decoding and re-encoding, we restrict ourselves here to the correlation-based methods described in Section 9.3. Because the method described in [Wu97] decreases the visual quality of the video stream drastically, the method described in [Har98] is the only comparable real-time watermarking method that operates directly on compressed video and keeps the video bit rate constant.

The authors [Har98] report that the complexity of their watermark embedding process is much lower than the complexity of a decoding process followed by watermarking in the spatial domain and re-encoding, but that it is somewhat higher than the complexity of a full MPEG decoding operation. Since the DEW algorithm adds a watermark only by removing DCT coefficients and no DCT, IDCT or full decoding steps are involved, the complexity of the DEW algorithm is less than half the complexity of a full MPEG decoding operation.

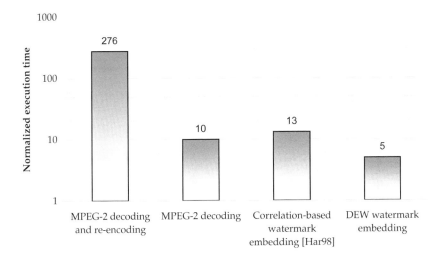

Figure 12.4.1: *Normalized execution times of software MPEG-2 re-encoding and decoding operations in comparison to two real-time watermarking techniques.*

In Figure 12.4.1 an indication is given of the execution times of the following operations on 60 frames of MPEG-2 encoded video. The first bar represents the execution time of a full software MPEG decoding step followed by an MPEG re-encoding step. These steps are necessary if we want to embed a watermark to the compressed video data, for instance using the method of [Smi96]. The second bar represents the execution time of a full software MPEG decoding step. This step is required to extract a watermark from the compressed video data, for instance by means of the method of [Smi96]. The third and fourth bars represent the execution times of the fastest software implementation of the correlation-based watermarking algorithm described in Section 9.3 [Har98] and the DEW algorithm. The execution times are normalized such that the execution time of MPEG-2 decoding 60 frames equals 10.

Concerning the payload of the watermark, the DEW algorithm clearly outperforms the real-time correlation-based method. The authors [Har98] report maximum watermark label bit rates of only a few bytes per second, while the DEW algorithm has a watermark label bit rate of up to 52 bytes per second (see Table 10.4.1).

Since no experimental results about robustness against transcoding are reported in literature for the real-time correlation-based method [Har98], we compare the DEW algorithm with the basic spread spectrum method of Smith and Comiskey [Smi96]. Although the real-time method of [Har98] uses the same basic principles as the method of [Smi96], the latter method can embed 100% of the watermark energy instead of 0.5-3% and has a much higher payload, since it is not limited by the constraint that the watermark embedding process must take place in the compressed domain.

To evaluate the resistance to transcoding or re-encoding at a lower bit rate, we performed the following experiments. The "sheep sequence" described in Section 9.4.2.1 is MPEG-2 encoded at 8 Mbit/s. This compressed stream is directly watermarked with the DEW algorithm using 3 different parameter settings:

- $n = 32, D = 20, c_{min}=6, D' = 15$, without pre-quantization (0.42kbit/s)
- $n = 64, D = 20, c_{min}=6, D' = 15$, without pre-quantization (0.21kbit/s)
- $n = 64, D = 20, c_{min}=6, D' = 15$, with pre-quantization in the embedding stage (0.21kbit/s)

Pre-quantization means here that, prior to the calculation of the energies (Equation 10.2.1), the DCT coefficients of MPEG compressed video are pre-quantized using the default MPEG intrablock quantizer matrix [ISO96]. The DCT coefficients are divided by this matrix, rounded and multiplied by the same matrix.

Next, the "sheep sequence" encoded at 8Mbit/s is watermarked with the spatial spread spectrum method [Smi96] (Section 8.2.2) by subsequently decoding, watermarking the I-frames and re-encoding the video stream. For the watermarking procedures the following settings are used:

- $k=1$, $P=64\times64$, without pre-filter in the detector (0.21kbit/s)
- $k=1$, $P=64\times64$, with pre-filter in the detector (0.21kbit/s)
- $k=2$, $P=64\times64$, without pre-filter in the detector (0.21kbit/s)
- $k=2$, $P=64\times64$, with pre-filter in the detector (0.21kbit/s)

As pre-filter a 3x3 edge-enhance filter is applied to the pixels of the I-frames before the correlation is calculated. The convolution kernel of the filter is given by Equation 8.2.5. Hereafter, the watermarked video sequences are transcoded at different lower bit rates. The label bit strings are extracted from the transcoded video streams and each label bit string is compared with the originally embedded label bit string. If 50% bit errors are made, the label is completely removed. The percentages label bit errors that are introduced when the bit rate is decreased, are represented in Figure 12.4.2.

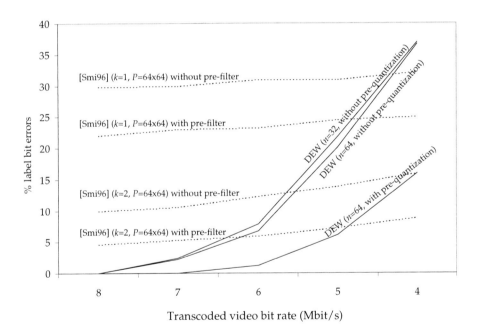

Figure 12.4.2: *% Bit errors after transcoding a watermarked 8 Mbit/s MPEG-2 sequence at a lower bit rate.*

From this figure several conclusions can be drawn. First, with respect to the DEW algorithm, increasing the number of 8x8 DCT blocks per label bit does not significantly increase the robustness to transcoding. This shows that *n* only needs to be increased if the watermarking process results in visual artefacts; otherwise *n* should be chosen as low as possible and error correcting codes should be used to improve the robustness.

Second, the robustness of the DEW algorithm increases drastically if pre-quantization is used during the embedding stage. We take a closer look at the results of the video stream transcoded to 5Mbit/s. Instead of using the averages over 21 frames (Figure 12.4.2), we look at the percentages of label bit errors of each separate frame (Figure 12.4.3). It now becomes clear that in some frames still no errors occur after transcoding (frame numbers: 1,2,7,20). However, in some other frames the percentage of label bit errors is quite high (frame numbers: 12,13). This is due to the fact that for the experiments a fixed pre-quantization level is used for each frame. This is not an optimal solution, since in MPEG coded video streams the quantization levels vary not only temporally, but also spatially, depending on the video bit rate, the video content and the buffer space of the encoder. The robustness of the DEW algorithm can therefore be improved further by locally adapting the pre-quantization.

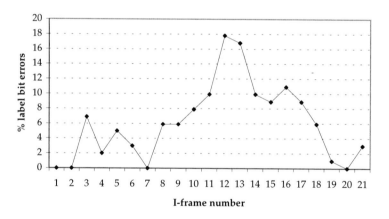

Figure 12.4.3: *% Bit errors after transcoding an 8 Mbit/s MPEG-2 sequence water-marked using the DEW algorithm (n=64, with pre-quantization) at 5 Mbit/s.*

The third and last conclusion that can be drawn from Figure 12.4.2 is that the DEW algorithm outperforms the correlation-based method [Smi96] with respect to the transcoding attack for bit rates between 8 and 5 Mbit/s.

Since due to the bit rate constraint, the real-time correlation-based version described in [Har98] is only capable of embedding 0.5...3% of the total watermark energy, which is embedded using [Smi96], it can be expected that this method performs less than the method of [Smi96] and the DEW algorithm concerning the transcoding attack.

12.4.4 Evaluation of the DEW algorithm for still images

To evaluate the DEW algorithm for JPEG compressed and uncompressed still images we compare it to the basic spread spectrum method of Smith and Comiskey [Smi96]. For all experiments in this section we use the parameter settings optimized for robustness for the DEW algorithm, namely $c_{min}=3$, $n=64$, $Q_{jpeg}=25$ and $D=400$. For the watermark extraction the parameters $n=64$ and $D'=400$ are used. Since the detector results are significantly influenced by the pre-quantization stage in the detector, a value for Q'_{jpeg} is chosen out of the set [25, 80, 99] such that the error rate of the detector is minimized. This process can be automated by, for instance, starting the label bit string with several fixed label bits, so that during the extraction the value Q'_{jpeg} can be chosen that results in the fewest errors in the known label bits.

For all experiments in this section with the method of [Smi96], $P=64 \times 64$ pixels are used to store each label bit, while the watermark carrier consists of the integers {-2,2} ($k=2$). This means that the watermarks embedded with both methods have the same payload. Since we noticed in the previous section that pre-filtering significantly improves the performance of the correlation-based method [Smi96], we apply a 3x3 edge-enhance filter to the watermarked images before calculating the correlation. The convolution kernel of the filter is given by Equation 8.2.5.

We watermarked a set of twelve images with the two watermarking methods using the parameter settings described above. First we calculate the ITU-R Rec. 500 quality ratings of the watermarked images using the approach described in Section 12.2 (Equation 12.2.1) and test the robustness of the watermarks against the attacks described in Section 12.3. In Table 12.4.1 the results of these experiments are listed for the DEW algorithm. For the StirMark attack version 1.0 is used, using the default parameter settings. In this version only the geometrical distortions are performed as described in Section 12.3.2, the final JPEG compression step is not implemented.

For the images watermarked with the method of [Smi96] the ITU-R Rec. 500 quality ratings are in the range of 4.7…4.8, the percentages of label bit errors after the StirMark attack exceed 40% for all images and the percentages of label bit errors after the watermark removal attack by non-linear filtering exceed 30% for all images. From Table 12.4.1 it can be concluded that the DEW algorithm affects the visual quality marginally more than the correlation-based method. However, the ITU-R quality ratings are far above the required minimum of 4. Further it can be concluded that for both watermark removal attacks the DEW algorithm clearly outperforms the correlation-based method.

Image name	Size	ITU-R Rec. 500 rating	% Label bit errors	
			StirMark Attack[Pet98b]	WRS [Lan98b]
Bike	720x512	4.3	34%	7%
Bridge	720x512	4.5	16%	17%
Butterfly	720x512	4.6	11%	7%
Flower	720x512	4.5	15%	5%
Grand Canyon	720x512	4.4	24%	13%
Lena	512x512	4.6	17%	6%
Parrot	720x512	4.7	28%	8%
Rafter	720x512	4.3	24%	7%
Red Square	720x512	4.6	15%	7%
Sea	720x512	4.4	15%	4%
Temple	720x512	4.6	17%	5%
Tree	720x512	4.3	9%	13%

Table 12.4.1: *ITU-R Rec. 500 quality ratings and percentages label bit errors for the DEW algorithm after applying the StirMark attack based on geometrical distortions (Q'_{jpeg}=99) and the Watermark Removing Scheme (WRS) based on watermark estimation (Q'_{jpeg}=25).*

To evaluate the robustness of both algorithms against common simple processing techniques we further tested the robustness against re-encoding, linear and non-linear filtering, noise addition, simple geometrical transformations, gamma correction, dithering and histogram equalization.

A set of twelve images is watermarked with both watermarking methods. The images are re-encoded using a lower JPEG quality factor. The quality factor of the re-encoding process is made variable. Finally, the watermarks are extracted from the re-encoded images and compared bit by bit with the original embedded watermarks. From this experiment, we find the percentages of label bit errors caused by re-encoding as a function of the re-encoding quality factor. In Figure 12.4.4 the resulting label bit error curves are shown for twelve different images.

As can be seen in Figure 12.4.4 the DEW algorithm is slightly more robust to re-encoding attacks than the correlation-based method. To test the robustness against non-linear filtering we filtered the test set of twelve images watermarked with both watermarking methods using a median filter with a kernel size of 3x3. To test the robustness against linear filtering we first filtered the watermarked images with a

3x3 smoothing filter F_{smooth} and subsequently with an edge-enhance filter F_{edge}, where F_{smooth} and F_{edge} are given by the following convolution kernels:

$$F_{smooth} = \begin{pmatrix} 1 & 1 & 1 \\ 1 & 5 & 1 \\ 1 & 1 & 1 \end{pmatrix} / 13 \quad \text{and} \quad F_{edge} = \begin{pmatrix} -1 & -1 & -1 \\ -1 & 10 & -1 \\ -1 & -1 & -1 \end{pmatrix} / 2 \quad (12.4.1)$$

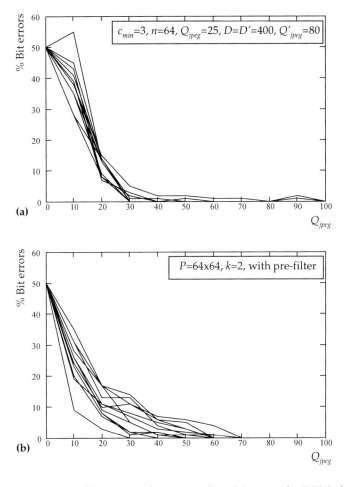

Figure 12.4.4: *Percentages of bit errors after re-encoding (a) using the DEW algorithm; (b) using the correlation-based method of [Smi96].*

Image name	% Label bit errors			
	Median Filtering 3x3		Linear Filtering	
	DEW	Corr.-based	DEW	Corr.-based
Bike	16%	3%	10%	0%
Bridge	9%	8%	0%	0%
Butterfly	15%	3%	0%	0%
Flower	9%	0%	0%	0%
Grand Canyon	18%	4%	2%	0%
Lena	5%	2%	0%	0%
Parrot	22%	3%	1%	0%
Rafter	15%	2%	1%	0%
Red Square	14%	10%	0%	0%
Sea	10%	7%	0%	0%
Temple	13%	8%	0%	0%
Tree	21%	15%	0%	0%

Table 12.4.2: *Percentages of label bit errors for the DEW algorithm ($Q'_{jpeg}=99$) and for the correlation-based method of [Smi96] after non-linear and linear filters were applied to the watermarked images.*

Table 12.4.2 presents The percentages of label bit errors in the labels extracted from the non-linear and linear filtered images. Table 12.4.2 shows that both methods are more vulnerable to non-linear filtering than to linear filtering. The correlation-based method is slightly more robust to filtering than the DEW algorithm. The reason for this is that the energy of the DEW algorithm is located more or less in a middle frequency band, and the energy of the correlation-based method is distributed uniformly over the spectrum. If some frequency bands are affected by filtering operations, there is enough energy left in other frequency bands in the case of the correlation-based method.

Correlation-based methods are quite resistant to uncorrelated additive noise. Experiments show that uniformly distributed noise in the range from -25 to 25 added to images watermarked with the method of [Smi96] does not introduce label bit errors in the extracted labels (0%). To investigate the robustness of the DEW algorithm against additive noise, we add noise to the watermarked images, where the noise amplitude $[-N_a, N_a]$ varies between 0 and 25. The results of this experiment are shown in Figure 12.4.5. This figure shows that the DEW algorithm is also quite insensitive to additive noise.

Robustness against geometrical distortions is very important, since shifting, scaling and rotating are very simple processing operations that hardly introduce visual quality loss. We already tested the robustness of the DEW algorithm against line shifting followed by lossy JPEG compression in Section 11.6 and the resistance to minor geometrical distortions applied by StirMark at the beginning of this section. Nevertheless here we perform some additional experiments to check the robustness against scaling and rotating. We enlarge the watermarked images 1% and crop them to their original size. Next, we rotate the watermarked images 0.5 degree and crop them to their original size. Finally, the watermark labels are extracted and compared bit by bit with the original embedded ones. The percentages of label bit errors in the labels extracted with the DEW algorithm from the scaled and rotated images are presented in Table 12.4.3. It appears that these geometrical transformations, line shifting, scaling and rotating, completely remove the watermarks embedded by the correlation-based method (percentages of bit errors > 40). From Table 12.4.3 and the experiments performed in Section 11.6 we can conclude that the DEW algorithm clearly outperforms the correlation-based method concerning geometrical transformations.

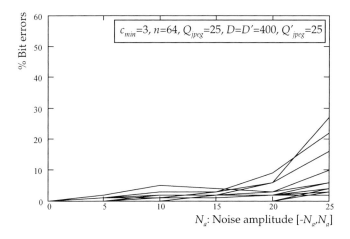

Figure 12.4.5: *Percentages of label bit errors after labels have been extracted from images affected with additive noise using the DEW algorithm.*

Both the DEW algorithm and the correlation-based method are insensitive to gamma correction and histogram equalization. Even quantization of the color channels from 256 levels to 32, 16 or 8 levels followed by dithering does not affect the watermarks embedded by the DEW algorithm ($Q'_{jpeg}=25$) or by the correlation-based method.

Image name	% Label bit errors	
	Zoom 1% and crop	Rotate 0.5° and crop
Bike	14%	17%
Bridge	3%	5%
Butterfly	0%	7%
Flower	7%	9%
Grand Canyon	17%	6%
Lena	0%	6%
Parrot	11%	10%
Rafter	10%	6%
Red Square	10%	3%
Sea	10%	10%
Temple	7%	8%
Tree	3%	0%

Table 12.4.3: *Percentages of label bit errors for the DEW algorithm ($Q'_{jpeg}=99$) after scaling or rotating and cropping the watermarked images.*

12.5 Conclusions

Benchmarking watermarking algorithms is a difficult task. Performance factors like visibility, robustness, payload and complexity have to be taken into account, but the weighting of these factors is application dependent. Furthermore it is questionable if robustness can be defined formally.

In this chapter we discussed two benchmarking approaches for watermarking methods and two dedicated watermark removal attacks. The benchmarking approaches discussed here only give some general guidelines on how watermarking methods can be evaluated. More research and standardization is necessary to derive more sophisticated benchmarking systems. Also the attacks discussed here are just examples to show that robustness against simple standard image processing techniques is not enough to call a watermarking method robust. Other simple processing techniques exist or may be developed that do not significantly affect the image quality, but can defeat most watermarking schemes.

The attacks presented here can be counterattacked by increasing the complexity of the watermark detectors. But the attacks can also be improved by taking these changes of the detectors into account. For instance, the watermark removal technique presented in Section 12.3.3 can be counterattacked by applying a special low-pass pre-filter in the detector [Har99]. However, by replacing the 3x3 high-pass filter in the removal scheme by a filter with a larger kernel and appropriate coefficients, this counterattack can be rendered useless. Furthermore an attack can be improved by combining it with a different attack, for instance, combining a watermark estimation attack with a geometrical transformation attack will defeat any watermarking scheme.

In spite of the problems mentioned above, we evaluated the DEW algorithm in this chapter taking into account the benchmarking approaches and attacks mentioned in the literature. We found that of all real-time watermarking algorithms for MPEG compressed video known from literature, the correlation-based method described in [Har98] is the only algorithm that can directly be compared with the DEW algorithm. In this comparison it turned out that the DEW algorithm has only less than half the complexity of this correlation-based method. Furthermore, the payload of the DEW algorithm is up to 25 times higher and the DEW algorithm is more robust against transcoding attacks than the correlation-based methods in the spatial domain. The robustness of the DEW algorithm can even be improved further by making the pre-quantization step variable.

For still images we compared the DEW algorithm to the basic spread spectrum method of [Smi96], which is not designed for real-time watermarking in the compressed domain.

In this comparison it turned out that the DEW algorithm and the correlation-based method perform equally well concerning the robustness against linear filtering, histogram equalization, gamma correction, dithering and additive noise. The DEW algorithm clearly outperforms the correlation-based method where it concerns the dedicated watermark removal attacks, geometrical transformations and re-encoding attacks using lossy JPEG compression.

Bibliography - Part II

[Aka96]   Ali N. Akansu, Mark J.T. Smith eds., "Subband and Wavelet Transforms: Design and Applications", Kluwer Academic Publishers, 1996

[And98]   Ross J. Anderson, Fabien A.P. Petitcolas, "On the limits of steganography. IEEE Journal of Selected Areas in Communications", 16(4):474-481, May 1998, Special Issue on Copyright & Privacy Protection, ISSN 0733-8716

[Aur95]   Tuomas Aura, "Invisible Communication", Proceedings of the HUT Seminar on Network Security '95, Espoo, Finland, November 6, 1995

[Aur96]   Tuomas Aura, "Practical invisibility in digital communication", Proceedings of the Workshop on Information Hiding, Cambridge, England, May 1996, Lecture Notes in Computer Science 1174, Springer Verlag 1996

[Bar94]   M. Barlaud ed., "Wavelets in Image Communication", Advances in image communication 5, Elsevier, ISBN:0444892818, 1994

[Bar98]   F. Bartolini, M. Barni, V. Cappellini, A. Piva, "Mask building for perceptually hiding frequency embedded watermarks", Proceedings of 5th IEEE International Conference on Image Processing ICIP'98, Chicago, Illinois, USA, Vol I, pp. 450-454, October 4-7, 1998

[Bar99]   Mauro Barni, Franco Bartolini, Vito Cappellini, Alessandro Lippi, Alessandro Piva, "A DWT-based technique for spatio-frequency masking of digital signatures", Proceedings of the SPIE/IS&T International Conference on Security and Watermarking of Multimedia Contents, vol. 3657, San Jose, CA, January 25-27, 1999

[Bas98] Patrick Bas, Jean-Marc Chassery, Franck Davoine, "Self-similarity based image watermarking", IX European Signal Processing Conference, Island of Rhodes, Greece, September 8-11, 1998

[Bas99] Patrick Bas, Jean-Marc Chassery, Franck Davoine, "A Geometrical and Frequential Watermarking Scheme Using Similarities", Proceedings of SPIE Electronic Imaging '99, Security and Watermarking of Multimedia Contents, San Jose (CA), USA, January 1999

[Ben95] W. Bender, D. Gruhl, N. Morimoto, "Techniques for Data Hiding", Proceedings of the SPIE, Vol. 2420, pp. 165-173, Storage and retrieval for image and Video Databases III, San Jose CA, USA, February 9-10, 1995

[Bol95] F.M. Boland, J.J.K. Ó Ruanaidh and C. Dautzenberg, "Watermarking Digital Images for Copyright Protection", IEE Int. Conf. on Image Processing and Its Applications, pp. 326-330, Edinburgh, Scotland, July 1995

[Bor96a] A.G. Bors, I.Pitas, "Embedding Parametric Digital Signatures in Images", EUSIPCO-96, Trieste, Italy, vol. III, pp. 1701-1704, September 1996

[Bor96b] A.G. Bors, I.Pitas, "Image Watermarking Using DCT Domain Constraints", IEEE International Conference on Image Processing (ICIP'96), Lausanne, Switzerland, vol. III, pp. 231-234, September 16-19, 1996

[Bra96] C.J. van den Branden Lambrecht and J.E. Farrell, "Perceptual quality metric for digitally coded color images", EUSIPCO-96, pp. 1175-1178, Trieste, Italy, September 1996

[Bra97] G.W. Braudaway, "Protecting Publicly-Available Images with an Invisible Watermark", Proceedings of ICIP 97, IEEE International Conference on Image Processing, Santa Barbara, California, October 1997

[Bur98] S.Burgett, E.Koch, J.Zhao, "Copyright Labeling of Digitized Image Data", IEEE Communications Magazine, pp. 94-100, March 1998

[Car95] G. Caronni, "Assuring Ownership Rights for Digital Images", Proceedings of Reliable IT Systems, pp. 251-263, VIS '95, Vieweg Publishing Company, Germany, 1995

[Che99] Brian Chen, Gregory W Wornell, "An Information-Theoretic Approach to the Design of Robust Digital Watermarking Systems", Proceedings ICASSP' 99, Vol 4., Phoenix, Arizona, USA, March 15-19, 1999

[Col99] Dinu Coltuc, Philippe Bolon, "Watermarking by histogram specification", Proceedings of SPIE ELECTRONIC IMAGING '99, Security and Watermarking of Multimedia Contents, January 1999, San Jose (CA), USA

[Cox95] I.J. Cox, J. Kilian, T. Leighton, T. Shamoon, "Secure Spread Spectrum Watermarking for Multimedia", Technical Report 95 - 10, NEC Research Institute, Princeton, NJ, USA, 1995

[Cox96a] I.J. Cox, J. Kilian, T. Leighton and T. Shamoon, "A Secure, Robust Watermark for Multimedia", Preproceedings of Information Hiding, an Isaac Newton Institute Workshop, Univ. of Cambridge, May 1996

[Cox96b]   I.J. Cox, J.Kilian, T.Leighton and T.Shamoon, "Secure spread spectrum watermarking for images, audio and video", Proceedings of the 1996 International Conference on Image Processing, vol. 3, pp. 243-246, Lausanne, Switzerland, September 16-19, 1996

[Cox97]   Ingemar J. Cox, and Matt L. Miller "A review of watermarking and the importance of perceptual modeling", Proceedings of SPIE Electronic Imaging '97, Storage and Retrieval for Image and Video Databases V, San Jose (CA), USA, February 1997

[Cra96]   S. Craver, N. Memon, B.-L. Yeo, and M. Yeung, "Can invisible watermarks resolve rightful ownerships?", IBM Research Report RC 20509, 25 July 1996

[Dav96]   Paul Davern, Michael Scott, "Fractal Based Image Steganography", Preproceedings of Information Hiding, An Isaac Newton Institute Workshop, pp. 245-256, University of Cambridge, UK, May 1996

[DCM98]   U.S. Copyright Office Summary, "The Digital Millennium Copyright Act of 1998", http://lcweb.loc.gov/copyright/legislation/dmca.pdf, December 1998

[Dep98]   G. Depovere, T. Kalker, J.-P. Linnartz, "Improved watermark detection using filtering before correlation", Proceedings of 5th IEEE International Conference on Image Processing ICIP'98, Chicago, Illinois, USA, Vol I, pp. 430-434, October 4-7, 1998

[Fle97]   D.J. Fleet, D.J. Heeger, "Embedding Invisible Information in Color Images", Proceedings of ICIP 97, IEEE International Conference on Image Processing, Santa Barbara, California, October 1997

[Fri99a]   Jiri Fridrich, Miroslav Goljan, "Comparing robustness of watermarking techniques", Electronic Imaging '99, The International Society for Optical Engineering, Security and Watermarking of Multimedia Contents, vol 3657, San Jose, CA, USA, January 25-27, 1999

[Fri99b]   Jiri Fridrich, "Robust Bit Extraction From Images", submitted to IEEE ICMCS'99 Conference, Florence, Italy, June 7-11, 1999

[Fri99c]   Jiri Fridrich, Miroslav Goljan, "Protection of Digital Images Using Self Embedding", submitted to The Symposium on Content Security and Data Hiding in Digital Media, New Jersey Institute of Technology, March 16, 1999

[Gir87]   B. Girod, "The efficiency of motion-compensating prediction for hybrid coding of video sequences", IEEE Journal on Selected Areas in Communications, Vol. 5, pp. 1140-1154, August 1987

[Gir89]   B. Girod, "The information theoretical significance of spatial and temporal masking in video signals", Proceedings of the SPIE Human Vision , Visual Processing, and Digital Display, Vol. 1077, pp. 178-187, 1989

[Gof97]   F. Goffin, J.-F. Delaigle, C. De Vleeschouwer, B.Macq and J.-J. Quisquater, "A low cost perceptive digital picture watermarking method", Proceedings of SPIE Electronic Imaging '97, Storage and Retrieval for Image and Video Databases V, San Jose (CA), USA, February 1997

[Han96]   A. Hanjalic, G.C. Langelaar, R.L. Lagendijk, M. Ceccarelli, M. Soletic, "Report on Technical Possibilities and Methods for Security of SMASH and for Fast Visual Search on Compressed/encrypted Data", Deliverable #5, AC-018, SMASH, SMS-TUD-648-1, November 1996

[Har96]   F. Hartung and B. Girod: "Digital Watermarking of Raw and Compressed Video", Proceedings SPIE 2952: Digital Compression Technologies and Systems for Video Communication, pp. 205-213, October 1996 (Proc. European EOS/SPIE Symposium on Advanced Imaging and Network Technologies, Berlin, Germany)

[Har97a]   F. Hartung and B. Girod, "Watermarking of MPEG-2 Encoded Video Without Decoding and Re-encoding", Proceedings Multimedia Computing and Networking 1997 (MMCN 97), San Jose, CA, February 1997

[Har97b]   F. Hartung and B. Girod, "Digital Watermarking of MPEG-2 Coded Video in the Bitstream Domain", Proceedings ICASSP 97, Volume 4, pp. 2621-2624, Munich, Germany, April 21-24, 1997

[Har97c]   F. Hartung and B. Girod, "Copyright Protection in Video Delivery Networks by Watermarking of Pre-Compressed Video", in: S. Fdida, M. Morganti (eds.), "Multimedia Applications, Services and Techniques - ECMAST '97", Springer Lecture Notes in Computer Science, Vol. 1242, pp. 423-436, Springer, Heidelberg, 1997

[Har98]   F. Hartung and B. Girod, "Watermarking of Uncompressed and Compressed Video", Signal Processing, Vol. 66, no. 3, pp. 283-301, (Special Issue on Copyright Protection and Control, B. Macq and I. Pitas, eds.), May 1998

[Har99]   F. Hartung, J.K. Su and B. Girod, "Spread Spectrum Watermarking: Malicious Attacks and Counterattacks", Proceedings of SPIE Electronic Imaging '99, Security and Watermarking of Multimedia Contents, San Jose (CA), USA, January 1999

[Her72]   Herodotus, "The Histories (trans. A. de Selincourt), Middlesex, England: Penguin, 1972

[Her98a]   Alexander Herrigel, Holger Petersen, Joseph O' Ruanaidh, Thierry Pun, Pereira Shelby, "Copyright Techniques for Digital Images Based On Asymmetric Cryptographic Techniques", Workshop on Information Hiding, Portland, Oregon, USA, April 1998

[Her98b]   Alexander Herrigel, Joe J. K. Ó Ruanaidh, Holger Petersen, Shelby Pereira and Thierry Pun, "Secure copyright protection techniques for digital images", In David Aucsmith ed., Information Hiding, pp. 169-190, Vol. 1525 of Lecture Notes in Computer Science, Springer, Berlin, 1998

[Hir96]   K. Hirotsugu, "An Image Digital Signature System with ZKIP for the Graph Isomorphism", Proceedings ICIP-96, IEEE International Conference on Image Processing, Volume III pp. 247-250, Lausanne, Switzerland, September 16-19, 1996

[Hsu96]   C.-T. Hsu, J.-L. Wu, "Hidden Signatures in Images", Proceedings ICIP-96, IEEE International Conference on Image Processing, Volume III, pp. 223-226, Lausanne, Switzerland, September 16-19, 1996

[IEC958]   Digital audio interface, International Standard, IEC 958

[Int97]	International Federation of the Phonographic Industry, "Request for Proposals", Embedded Signalling Systems Issue 1.0. 54 Regent Street, London W1R 5PJ, June 1997
[ISO96]	ISO/IEC 13818-2:1996(E), "Information Technology – Generic Coding of Moving Pictures and Associated Audio Information", Video International Standard, 1996
[ISO95]	Recommendation ITU-R BT.500-7, "Methodology for the subjective assessment of the quality of television pictures", International Telecommunication Union, Broadcasting Service (Television), 1995 BT Series Fascicle, Radiocommunication assembly, Geneva, 1995
[Jac92]	A.E. Jacquin, "Image coding based on a fractal theory of iterated contractive image transformations", IEEE transactions on Image Processing, Vol. 2, No. 1, pp. 18-30, January 1992
[Jai81]	Anil K. Jain, "Image Data Compression: A Review", Proceedings IEEE, Vol. 69, no. 3, pp. 349-389, March 1981
[Joh98]	Neil F. Johnson, Sushil Jajodia, "Exploring Steganography: Seeing the Unseen", IEEE Computer, Vol. 31, no 2, pp. 26-34, February 1998
[Kah67]	D. Kahn, "The Codebreakers", New York: MacMillan, 1967
[Kal98a]	Ton Kalker, Jean-Paul Linnartz, Geert Depovere, Maurice Maes, "On the Reliability of Detecting Electronic Watermarks in Digital Images", IX European Signal Processing Conference, Island of Rhodes, Greece, September 8-11, 1998
[Kal98b]	Ton Kalker, Jean-Paul Linnartz, Marten van Dijk, "Watermark estimation through detector analysis", Proceedings of 5th IEEE International Conference on Image Processing ICIP'98, Chicago, Illinois, USA, October 4-7, 1998
[Kal99]	Ton Kalker, Geert Depovere, Jaap Haitsma, Maurice Maes, "A Video Watermarking System for Broadcast Monitoring", Proceedings of SPIE ELECTRONIC IMAGING '99, Security and Watermarking of Multimedia Contents, San Jose (CA), USA, January 1999
[Kob97]	M. Kobayashi, "Digital Watermarking: Historical Roots", IBM Research Report, RT0199, Japan, April 1997
[Koc94]	E. Koch, J. Rindfrey, J. Zhao, "Copyright Protection for Multimedia Data", Proceedings of the International Conference on Digital Media and Electronic Publishing, Leeds, UK, December 6-8, 1994
[Koc95]	E. Koch, J. Zhao, "Towards Robust and Hidden Image Copyright Labeling", Proceedings IEEE Workshop on Non-lineair Signal and Image Processing, pp. 452-455, Neos Marmaras (Thessaloniki, Greece), June, 1995
[Kun97]	D. Kundur, D. Hatzinakos, "A robust digital image watermarking scheme using wavelet-based fusion," Proceedings of ICIP 97, IEEE International Conference on Image Processing, Santa Barbara, California, October 1997

[Kun98] D. Kundur, D. Hatzinakos, "Digital watermarking using multi resolution wavelet decomposition", Proceedings IEEE International Conference on Acoustics, Speech and Signal Processing, Seattle, Washington, Vol. 5, pp. 2969-2972, May 1998

[Kut97] M. Kutter, F. Jordan, F. Bossen, "Digital Signature of Color Images using Amplitude Modulation", Proceedings of SPIE Electronic Imaging '97, Storage and Retrieval for Image and Video Databases V, San Jose (CA), USA, February 1997

[Kut98] M. Kutter "Watermarking resistant to rotation, translation and scaling", Proceedings of SPIE, Boston, USA, November 1998

[Kut99] M. Kutter, F.A.P. Petitcolas, "A fair benchmark for image watermarking systems", Electronic Imaging '99. Security and Watermarking of Multimedia Contents, vol. 3657, The International Society for Optical Engineering, San Jose, CA, USA, January 25-27, 1999,

[Kuw76] M. Kuwahara, K. Hachimura, S. Eiho, M. Kinoshita, "Processing of RI-angiocardiographic images", in Digital Processing of Biomedical Images, K. Preston and M. Onoe, Editors, pp. 187-203, Plenum Press, New York, 1976

[Lan96a] G.C. Langelaar, J.C.A. van der Lubbe, J. Biemond, "Copy Protection for Multimedia Data based on Labeling Techniques", 17th Symposium on Information Theory in the Benelux, Enschede, The Netherlands, May 30-31, 1996

[Lan96b] G.C. Langelaar, "Feasibility of security concept in hardware", AC-018, SMASH, SMS-TUD-633-1, August 1996

[Lan97a] G.C. Langelaar, J.C.A. van der Lubbe, R.L. Lagendijk, "Robust Labeling Methods for Copy Protection of Images", Proceedings of SPIE Electronic Imaging '97, Storage and Retrieval for Image and Video Databases V, San Jose (CA), USA, February 1997

[Lan97b] G.C. Langelaar, R.L. Lagendijk, J. Biemond "Real-time Labeling Methods for MPEG Compressed Video", 18th Symposium on Information Theory in the Benelux, Veldhoven, The Netherlands, May 15-16, 1997

[Lan98a] G.C. Langelaar, R.L. Lagendijk, J. Biemond, "Real-time Labeling of MPEG-2 Compressed Video", Journal of Visual Communication and Image Representation, Vol 9, No 4, December, pp. 256-270, 1998, ISSN 1047-3203

[Lan98b] G.C. Langelaar, R.L. Lagendijk, J. Biemond, "Watermark Removal based on Non-linear Filtering", ASCI'98 Conference, Lommel, Belgium, June 9-11, 1998

[Lan98c] G.C. Langelaar, R.L. Lagendijk, J. Biemond, "Removing Spatial Spread Spectrum Watermarks by Non-linear Filtering", IX European Signal Processing Conference, Island of Rhodes, Greece, September 8-11, 1998

[Lan99a] G.C. Langelaar, "Conditional Access to Television Service", Wireless Communication, the interactive multimedia CD-ROM, 3rd edition 1999, Baltzer Science Publishers, Amsterdam, ISSN 1383 4231

[Lan99b] G.C. Langelaar, R.L. Lagendijk, J. Biemond, "Watermarking by DCT Coefficient Removal: A Statistical Approach to Optimal Parameter Settings", Proceedings of SPIE Electronic Imaging '99, Security and Watermarking of Multimedia Contents, San Jose (CA), USA, January 1999

[Lan99c] G.C. Langelaar, R.L. Lagendijk, "Optimal Differential Energy Watermarking (*DEW*) of DCT Encoded Images and Video", submitted to the IEEE Transactions on Image Processing, 1999

[Lea96] T. Leary, "Cryptology in the 15th and 16th century", Cryptologica v XX, no 3, pp. 223-242, July 1996

[Lin98] J.-P. M.G. Linnartz, M. van Dijk, "Analysis of the sensitivity attack against electronic watermarks in images", Workshop on Information Hiding, Portland, Oregon, USA, April 1998

[Mac95] B.M. Macq, J.-J. Quisquater, "Cryptology for Digital TV Broadcasting", Proceedings of the IEEE Vol. 83 no. 6, pp. 944-957, June 1995

[Mae98] Maurice Maes, "Twin Peaks: The histogram attack to fixed depth image watermarks", Workshop on Information Hiding, Portland, Oregon, USA, April 1998

[Mul93] F. Muller, "Distribution Shape of Two-Dimensional DCT Coefficients of natural Images", *Electronic Letters*, Vol. 29, no. 22, pp. 1935-1936, 1993

[Ng99] K.S. Ng, L.M. Cheng, "Selective block assignment approach for robust digital image watermarking", Proceedings of the SPIE/IS&T International Conference on Security and Watermarking of Multimedia Contents, vol. 3657, San Jose, CA, January 25-27, 1999

[Nik96] N. Nikolaidis and I.Pitas, "Copyright protection of images using robust digital signatures", Proceedings of IEEE International Conference on Acoustics, Speech and Signal Processing (ICASSP-96), vol. 4, pp. 2168-2171, Atlanta, USA, May 1996

[Pen93] W.B. Pennebakker, J.L. Mitchell, "The JPEG Still Image Data Compression Standard", Van Nostrand Reinhold, New York, 1993

[Per99] Shelby Pereira, Joe J. K. Ó Ruanaidh, Frédéric Deguillaume, Gabriella Csurka and Thierry Pun, Template based recovery of Fourier-based watermarks using log-polar and log-log maps, In IEEE Multimedia Systems 99, International Conference on Multimedia Computing and Systems, Florence, Italy, June 7-11, 1999

[Pet98a] Fabien A.P. Petitcolas and Ross J. Anderson, "Weaknesses of copyright marking systems", Multimedia and Security Workshop at ACM Multimedia '98. Bristol, UK, September 1998

[Pet98b] Fabien A.P. Petitcolas, Ross J. Anderson and Markus G. Kuhn, "Attacks on copyright marking systems", in David Aucsmith (Ed.), Proceedings LNCS 1525, Springer-Verlag, ISBN 3-540-65386-4, pp. 219-239, Information Hiding, Second International Workshop, IH'98, Portland Oregon, USA, April 15-17, 1998

[Pet99]   Fabien A.P. Petitcolas and Ross J. Anderson, "Evaluation of copyright marking systems", Proceedings of IEEE Multimedia Systems (ICMCS '99), Florence, Italy, June 7-11, 1999

[Pit95]   I. Pitas, T.H. Kaskalis, "Applying signatures on digital images", Proceedings of IEEE Workshop on Nonlinear Signal and Image Processing, pp. 460-463, Neos Marmaras, Greece, June 20-22, 1995

[Pit96a] I. Pitas, "A method for Signature Casting on Digital Images", Proceedings ICIP-96, IEEE International Conference on Image Processing, Volume III pp. 215-218, Lausanne, Switzerland, September 16-19, 1996

[Piv97]   A. Piva, M. Barni, F. Bartolini, V. Cappellini, "DCT-based Watermark Recovering without Resorting to the Uncorrupted Original Image", Proceedings of ICIP 97, IEEE International Conference on Image Processing, Santa Barbara, California, October 1997

[Pod97]   C. Podilchuk, W. Zeng, "Perceptual Watermarking of Still Images", Proceedings of 1997 IEEE First Workshop on Multimedia Signal Processing, pp. 363-368, Princeton, New Jersey, USA, June 23-25, 1997

[Por93]   Giovanni Baptista Porta, "De occultis literarum notis", Facsimil of edition from 1593, Cryptography chair Univerisity of Zaragoza, Spain 1996

[Pua96]   J. Puate, F. Jordan, "Using fractal compression scheme to embed a digital signature into an image", Proceedings of SPIE Photonics East Symposium, Boston, USA, November 18-22, 1996

[Ren96]   J.-L. Renaud, "PC industry could delay DVD", Advanced Television Markets, Issue 47, May 1996

[Rhe89]   M.Y. Rhee, "Error Correcting Coding Theory", McGraw-Hill Publishing Company, New York, 1989

[Ron99]   P.M.J. Rongen, M.J.J.J.B. Maes, C.W.A.M. van Overveld, "Digital Image Watermarking by Salient Point Modification", Proceedings of SPIE Electronic Imaging '99, Security and Watermarking of Multimedia Contents, San Jose (CA), USA, January 1999

[Rua96a] J.J.K. Ó Ruanaidh, F.M. Boland, O. Sinnen, "Watermarking Digital Images for Copyright Protection", Electronic Imaging and the Visual Arts 1996, Florence, Italy, February 1996

[Rua96b] J.J.K. Ó Ruanaidh, W.J. Dowling, F.M. Boland, "Watermarking Digital Images for Copyright Protection", IEE Proceedings Vision, Image- and Signal Processing, 143(4) pp. 250-256, August 1996

[Rua96c] J.J.K. Ó Ruanaidh, W.J. Dowling, F.M. Boland, "Phase Watermarking of Digital Images", Proceedings of the IEEE International Conference on Image Processing, Volume III pp. 239-242, Lausanne, Switzerland, September 16-19, 1996

[Rua97]   Joseph J. K. Ó Ruanaidh, Thierry Pun, "Rotation, scale and translation invariant digital image watermarking" Proceedings of ICIP 97, IEEE International Conference on Image Processing, pp. 536-539, Santa Barbara, CA, October 1997

[Rua98a]   Joe J. K. Ó Ruanaidh, Shelby Pereira, "A secure robust digital image watermark" Electronic Imaging: Processing, Printing and Publishing in Colour, SPIE Proceedings, (SPIE/IST/Europto Symposium on Advanced Imaging and Network Technologies), Zürich, Switzerland, May 1998

[Rua98b]   Joe J. K. Ó Ruanaidh, Thierry Pun, "Rotation, scale and translation invariant spread spectrum digital image watermarking", Signal Processing, Vol. 66, no. 3, pp. 303-317, (Special Issue on Copyright Protection and Control, B. Macq and I. Pitas, eds.), May 1998

[Rup96]   S. Rupley, "What's holding up DVD?", PC Magazine, vol. 15, no. 20, pp. 34, November 19, 1996

[Sam91]   P. Samuelson, "Legally Speaking: Digital Media and the Law", Communications of ACM, vol. 34, no. 10, pp. 23-28, October 1991

[Sch94]   R.G. van Schyndel, A.Z. Tirkel, C.F. Osborne, "A Digital Watermark", Proceedings of the IEEE International Conference on Image Processing, volume 2, pages 86-90, Austin, Texas, USA, November 1994

[Sha49]   C.E. Shannon, W.W. Weaver, "The Mathematical Theory of Communications", The University of Illinois Press, Urbana, Illinois, 1949

[Sha93]   J. M. Shapiro, "Embedded image coding using zerotrees of wavelet coefficients", IEEE Transactions on Signal Processing, 41(12), pp. 3445-3462, December 1993

[Smi96]   J.R. Smith, B.O. Comiskey, "Modulation and Information Hiding in Images", Preproceedings of Information Hiding, an Isaac Newton Institute Workshop, University of Cambridge, UK, May 1996

[Swa96a]   M.D. Swanson, B. Zhu, A.H. Tewfik, "Transparant Robust Image Watermarking", Proceedings of the IEEE International Conference on Image Processing, Volume III pp. 211-214, Lausanne, Switzerland, September 16-19, 1996

[Swa96b]   Mitchell D. Swanson, Bin Zhu, Ahmed H. Tewfik, "Robust Data Hiding for Images", 7th IEEE Digital Signal Processing Workshop, pp. 37-40, Loen, Norway, September 1996

[Swa98]   Mitchell D. Swanson, Mei Kobayashi, Ahmed H. Tewfik, "Multimedia Data-Embedding and Watermarking Technologies", Proceedings of the IEEE, Vol. 86(6) pp. 1064-1087, June 1998

[Tan90]   K. Tanaka, Y. Nakamura, K. Matsui, "Embedding Secret Information into a Dithered Multi-level Image", Proceedings of the 1990 IEEE Military Communications Conference, pp. 216-220, September 1990

[Tao97]   Bo Tao and Bradley Dickinson, "Adaptive Watermarking in the DCT Domain", IEEE Internatational Conference on Acoustics, Speech and Signal Processing, April 1997

[Tay97]   J. Taylor, "DVD Demistified : the Guidebook for DVD-Video and DVD-Rom", McGraw Hill Text, 1997

[Var89]   M.K. Varansai and B. Aazhang, "Parametric generalized Gaussian density estimation", J.Acoust.Soc.Amer., vol.86, no. 4, pp. 1404-1415, October 1989

[Vet95]   Martin Vetterli, Jelena Kovacevic, "Wavelets and subband coding", Prentice Hall Signal Processing Series, New Jersey, ISBN 0-13-097080-8, 1995

[Voy96]   G. Voyatzis and I. Pitas, "Applications of Toral Automorphisms in Image Watermarking", Proceedings ICIP-96, IEEE International Conference on Image Processing, Volume II, pp. 237-240, Lausanne, Switzerland, September 16-19, 1996

[Voy98]   G. Voyatzis, N. Nikolaides, I. Pitas, "Digital Watermarking: An Overview", Proceedings of IX European Signal Processing Conference (EUSIPCO), pp. 13-16, Island of Rhodes, Greece, September 8-11, 1998

[Wan95]   Brian A. Wandell, "Foundations of Vision", Sinauer Associates, Inc., Sunderland, Massachusetts, USA, 1995

[Wol96]   R.B. Wolfgang and E. J. Delp, "A Watermark for Digital Images", Proceedings of the IEEE International Conference on Image Processing, Volume III, pp. 219-222, September 16-19, 1996, Lausanne, Switzerland

[Wol97]   R.B. Wolfgang and E.J. Delp, "A Watermarking Technique for Digital Imagery: Further Studies," Proceedings of the International Conference on Imaging Science, Systems, and Technology, Las Vegas, USA, June 30 - July 3, 1997

[Wol98]   R. B. Wolfgang and E. J. Delp, "Overview of Image Security Techniques with applications in Multimedia Systems," Proceedings of the SPIE Conference on Multimedia Networks: Security, Displays, Terminals, and Gateways, Vol. 3228, pp. 297-308, Dallas, Texas, USA, November 2-5, 1997

[Wol99a]   Raymond B. Wolfgang, Edward J. Delp, "Fragile Watermarking Using the VW2D Watermark" Electronic Imaging '99, The International Society for Optical Engineering, Security and Watermarking of Multimedia Contents, Vol 3657, San Jose, CA, USA, January 25-27, 1999

[Wol99b]   R. B. Wolfgang, C. I. Podilchuk, E. J. Delp, "Perceptual Watermarks for Digital Images and Video", Proceedings of the SPIE/IS&T International Conference on Security and Watermarking of Multimedia Contents, vol. 3657, San Jose, CA, January 25-27, 1999

[Wol99c]   R. B. Wolfgang, C. I. Podilchuk, E. J. Delp, "Perceptual Watermarks for Digital Images and Video", Proceedings of IEEE, May 1999

[Wu97]   T.L. Wu, S.F. Wu, "Selective encryption and watermarking of MPEG video", International Conference on Image Science, Systems, and Technology, CISST'97, June 1997

[Xia97]   X.-G. Xia, C.G. Boncelet, G.R. Arce, "A Multiresolution Watermark for Digital Images", Proceedings of ICIP 97, IEEE International Conference on Image Processing, Santa Barbara, California, October 1997

[Zen97] W.Zeng, B. Liu, "On resolving Rightful Ownerships of Digital Images by Invisible Watermarks", Proceedings of ICIP 97, IEEE International Conference on Image Processing, Santa Barbara, California, October 1997

[Zha95] J. Zhao, E. Koch, "Embedding Robust Labels into Images for Copyright Protection", Proceedings of the International Congress on Intellectual Property Rights for Specialized Information, Knowledge and New Technologies, Vienna, Austria, August 21-25, 1995

Image and Video Databases

Part III: Retrieval

Chapter 13

Information Retrieval: An Introduction

13.1 Information retrieval systems: From needs to technical solutions

There is hardly a better way to describe the development stage of our civilization at the end of the second millennium than as the *information era*, and this has a quite obvious reason: never before was the impact of information on the human lifestyle and way of thinking as enormous as it is in the second half of the 20th century. People are not only exposed to information all the time, this experience also becomes more intensive, which greatly contributes to broadening their views; they acquire knowledge and awareness about the environment and the world in general. This process globalizes society, and as such, creates new living and educational standards.

We can explain such an impact mainly as a consequence of an overwhelming *digital revolution*, which started some decades ago and has continuously gained in strength. On the one hand, the digital way of representing information opened completely new perspectives for further developments in information technology. It became possible to compress information, which resulted in a strong reduction of the time and channel capacity required for its transmission and of the space required for its storage. Information can be transmitted or manipulated without quality loss and it is possible to combine and transmit or process different types of information together, like audio, visual or textual: *multimedia* was born. On the other hand, digital hardware technology has rapidly developed and grown in the last decades, so that the performance-versus-price ratio of various digital systems, storage and transmission media steadily increased. All this has led to continuous advances in the quality of transmitted and received audiovisual information [Hua99a], in digital telecommunication networks providing high-speed information transfer ("information superhighway"), in fast digital signal processors and in compact high-capacity storage media like Digital Versatile Disc (DVD), which is seen by many as "the epitome of the digital age" [TNO97]. In view of such technological growth, it is not difficult to understand that an average information consumer easily raises his expectations regarding the amount, variety and technical quality of the received information, as well as of the systems for information receiving, processing, storage and re- or display. It will soon become quite usual that each household is equipped with receivers for Digital Video and Audio Broadcasting (DVB [ETS94] and DAB [ETS97]) providing together hundreds of high-quality audiovisual channels, accompanied by a broadband Internet connection, which gives access to countless on-line information archives all over the world.

However, it is beyond human capabilities to digest all the received information in an on-line manner. Large volumes of digital information obtained from digital TV/radio channels, Internet etc. will need to be stored temporarily, or if they are of long-term value, permanently. In this sense, we witness a strong development of *home digital multimedia archives* [SMA]. And, naturally, with an increasing information production even larger digital multimedia archives appear at service providers (e.g. TV and radio broadcasters, Internet providers, etc.). Thus, the issue of digital information storage steadily becomes more and more interesting and we can talk about emerging *digital libraries*. This term stands for a (large-scale) collection of stored digital information of any type (e.g. audio, visual, textual), made for either professional or consumer environments; examples of this are digital museum archives, Internet archives, video libraries available to commercial service providers and private information collections in the home, all of them being characterized by a quickly increasing capacity and content variety.

The development of digital libraries is not only related to technological advances in high-capacity storage media. The issue of efficiently retrieving the information stored in these libraries becomes of utmost importance as larger data volumes are stored. Actually, it can be said that the missing possibility to quickly access stored information degrades the high technological value of new high-capacity storage media and seriously jeopardizes the usability of the stored information. As nicely formulated in the preface of [Sme97], "anyone who has surfed the Web has exclaimed at one point or another that there is so much information available, so much to search and so much to keep up with". This citation describing a particular problem of finding a desired information on the World-Wide Web (WWW) can analogously be applied to a digital library of any type:

If information of interest is not easily accessible within a large digital library, that information can be of no use, in spite of its value and the fact that it is present in that library.

Manually searching through GBytes of unorganized stored data is a tedious and time-consuming task. Consequently, with increasing information volumes the need grows for shifting the information retrieval to automated systems. There, algorithms are applied capable of performing any information retrieval task with the same reliability and with the same or even higher efficiency as when the retrieval is done manually.

Realizing this shifting in practice is not a trivial problem, especially in the case of images or video[*]. To explain this, we here analyze some characteristic retrieval tasks, such as "find me an image with a bird", "find me the movie scene where Titanic hits the iceberg", "find me the CNN business news report from 15 November 1999", "find me a 'western' movie in the database", "classify all the images according to the place where they were taken" or "find me all images showing Paris". These retrieval tasks are formulated on a *cognitive level*, according to the human capability of understanding the information content and analyzing it in terms of objects, persons, sceneries, meaning of speech fragments or the context of a story in general. Opposed to this, the only feasible analysis of a video or an image at the algorithmic or *system level* can be in terms of their *features*, such as color, texture, shape, frequency components, audio and speech signal characteristics, and using the algorithms operating on these features. Such algorithms are, for instance, image segmentation, detection of moving objects, extraction of textures and shapes, recognition of color compositions, determination of relations among different objects or analysis of the frequency spectrum of the audio or speech signal. These algorithms can be developed using the state-of-the-art in image and audio analysis and processing, computer vision, statistical signal processing, machine intelligence, pattern recognition and other related areas.

As illustrated in Figure 13.1, we can understand an automated feature-based content analysis as a system-level parallel to the cognition-based analysis. There the features are chosen and algorithms are developed in the way that the retrieval

---

[*] Within the context of this book we refer to *video* as to a program in its entirety, consisting of an image sequence and the eventual accompanying audio/speech stream.

results are similar at the end of each branch of the scheme. Experience shows, however, that the parallelism in Figure 13.1 is not viable in all cases. We can explain this with the example of searching for an image containing a bird. While such a search performed by a human will always succeed, this cannot be said for the feature-based image analysis, simply because a complicated and large feature set describing the characteristics of a bird in general is required as well as complex algorithms operating on that feature set, which would enable the system to recognize the appearance of any arbitrary bird, in any possible pose and also in cases where parts of a bird are occluded. Finding a suitable feature set and developing related algorithms for such a retrieval task is very difficult. Consequently, the development of feature-based content analysis algorithms for the scheme in Figure 13.1 has not been directed to enable queries on the highest semantic level, such as the above example with a bird, but mainly towards extracting certain semantic aspects of the information which would allow for a reduction of the overall large search space. This tendency can be recognized in numerous algorithms proposed in recent literature, many of which will be explained in detail in further chapters of this book.

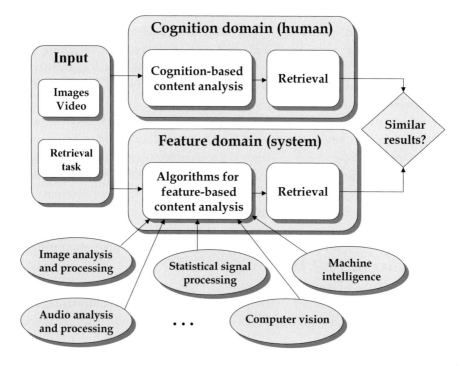

Figure 13.1: *Cognitive versus feature-based retrieval*

For instance, an algorithm in [Vai98] is able to classify with high accuracy images showing a city versus those showing a landscape. Further in [Vai99], a Bayesian framework is presented for semantic classification of outdoor vacation images. There, landscape images can be classified into those showing a sunset, a forest or mountains. Similar examples can also be found in the area of digital video libraries. The algorithms proposed in [Han99b] and [Yeu97] provide the possibility to detect episode or scene boundaries of a TV broadcast (movie, situation comedy, etc.). A methodology for detecting commercial breaks in a TV news program is presented in [Liu98]. There, the audio track of the broadcast is analyzed and commercial breaks are efficiently separated from the rest of the program because of their specific audio characteristics. An approach to detect commercial breaks in an arbitrary TV broadcast is presented in [McG99], based on parallel investigation of several feature types. Also a sophisticated feature-based analysis is applied in [Fis95] in order to classify video programs in different genres. Another class of approaches [DeM98], [Han97c], [Han00], [Pfe96] concentrates on *video abstraction*, i.e. compact representation of long video sequences by extracting and organizing a number of its most representative frames and segments.

Even the feature-based content-analysis techniques belonging to the current technological state-of-the-art and developed with the objective of search-space reduction can be used to build efficient tools for multimedia information retrieval, since they provide the user with reliable directions for browsing efficiently through a digital library and lead them quickly to the information of interest. The MPEG-7 standardization platform [ISO97] addresses ways to define standard sets of descriptors for multimedia information based on features which should provide further directions for the development of feature-based content-analysis algorithms. And with new solutions, the performance of information retrieval systems can only improve, leading to a further reduction of the search space and user's involvement during the search procedure. The material presented in Chapters 14 to 17, which is briefly outlined in the next two sections of this chapter, is a further contribution to this positive development.

13.2 Scope of Part III

Part III of this book concentrates on a feature-based analysis of the *visual content* of images and video, enabling an easier image and video retrieval from large-scale multimedia digital libraries.

The scheme in Figure 13.2 presents a series of video processing/analysis steps which provide an organizational structure allowing efficient reviewing of the global video content (e.g. story flow of a movie, topic series of a news program, etc.) and a fast access to and retrieval of any arbitrary part of a video (e.g. an arbitrary movie episode, a news report on a certain topic, a highlight of a sport program, etc.). The scheme depicts a generally known video-analysis procedure which first breaks up a video into temporally homogeneous segments called *video shots*, then condenses these segments into a set of characteristic frames called *key frames* and finally performs a high-level analysis of a video content. This high-level analysis basically includes determining "semantic" relationships among shots (e.g. their grouping into news reports, movie episodes, etc.) using temporal characteristics of shots and suitable features of their key frames. As indicated in the scheme, beside of being used for high-level video analysis key frames also directly participate in forming the organizatorial video-content structure described above. There, they provide visual keys to different aspects of a video content. A large number of algorithms was presented in recent literature for all three mentioned processing/analysis steps, aiming at a robust and high-quality performance with as much automation as possible. We contribute to these efforts in this book and dedicate each of the Chapters 14 to 16 to one of the processing/analysis steps in Figure 13.2.

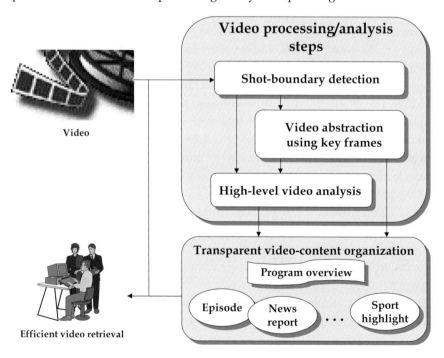

Figure 13.2: *A video-content analysis scheme*

Then we consider the fact that the prevailing amount of information reaching the digital libraries and being stored there will be in a *compressed* form. This is because large and fast advances in the compression area are gladly employed to maximally utilize the available storage space in digital libraries, but also to increase the information-transmission rate and density. Consequently, compressed images and video need to be expected as inputs into feature-based content-analysis algorithms, which, however, must not influence the efficiency of these algorithms compared to the case where they operate on uncompressed data. The most important issue related to this efficiency is the possibility to easily reach all the necessary features in a compressed image/video. We address this issue in Chapter 17 for the case of image compression.

13.3 Overview of Part III

Dividing a video sequence into *shots* is the first step towards video-content analysis and content-based video browsing and retrieval. A video shot is defined as a series of interrelated consecutive frames, taken contiguously by a single camera and representing a continuous action in time or space [Bor93]. As such, shots are considered to be the primitives for higher-level content analysis, indexing and classification, discussed in later chapters of this book. Chapter 14 presents a statistical framework for shot-boundary detection based on minimization of the average detection-error probability. We model the required statistical functions using a robust metric for visual content discontinuities (based on motion compensation) and take into account knowledge about the shot-length distribution and visual discontinuity patterns at shot boundaries. Major advantages of the proposed framework are its robust and sequence independent detection performance, as well as its capacity to detect different types of shot boundaries simultaneously.

Abstracting a video by extracting a number of characteristic or *key frames* is useful for different applications in video libraries. The form and the size of the key-frame abstract needs, however, to be adapted to the structure of the video material, as well as to the targeted application. Chapter 15 presents two methods for extracting key frames, aiming at different applications in video-retrieval systems. The first method is characterized by the possibility to control the total number of key frames extracted for the entire sequence. While this number does not exceed the prespecified maximum, key frames are spread along a video such that the quality of capturing all relevant variations of its visual content is maximized and that a storyboard of a video is provided. The objective of the second approach to key-frame extraction presented in Chapter 15 is to minimize the size of the key-frame abstract while providing all the necessary aspects of the visual content of a video. This algorithm is designed to produce a set of key frames which capture the content of a video in a similar way as when key frames are extracted manually based on human cognition.

As already mentioned in Section 13.1, information retrieval from digital libraries by formulating queries on the highest semantic level is not realistic in view of the current technological state-of-the art. However, examples were also shown where certain semantic components can be recognized in the stored information and be used to organize the information in such a way that the overall large search space is reduced as far as possible. Using, for instance, the algorithm from [Vai98] for city-versus-landscape image classification, the time for finding an image showing the New York skyline can be considerably reduced since only relevant images, i.e. those showing cities, are submitted to the user. Although he/she still needs to browse through the city image collection and must search for the particular image of interest (New York), the number of images he/she needs to check is much smaller than the entire image library.

In Chapter 16 we first present an idea how to translate the above image-search example to the case of video retrieval, and especially retrieval of *movies*, which is a very important program category in video storage systems. We assume that a typical movie can be represented as a series of high-level semantic contexts called *episodes*, which correspond to different classes in an image database. If a movie is segmented into episodes, a search for different movie segments showing specific faces or sceneries can be performed only within the relevant episode, which reduces the overall search space and, therefore, also the retrieval time. We develop a feature-based algorithm for automatically segmenting movies into *logical story units*, which are the approximates for the actual movie episodes.

Movie segmentation into logical story units is followed by the description of an algorithm for analyzing TV news programs at a high level. The algorithm detects the appearance of anchorperson shots, which can be considered as the first step in recovering the report structure of a news program at a later stage.

Chapter 17 addresses the issue of content accessibility in compressed images and video. This accessibility is analog to the efficiency of regaining the features of content elements being important for a given retrieval task. Since current compression standards, like JPEG or MPEG are not optimized regarding the content accessibility, a high computational load in reaching image and video features combined with large amount of information stored in a database, can negatively influence the efficiency of the interaction with that database.

In order to make the interaction with a database more efficient, it is necessary to develop compression methods which explicitly take into account the content accessibility of images and video, together with the three classical optimization criteria that are (1) minimizing the resulting bit rate, (2) maximizing the quality of the reconstructed image and video and (3) minimizing the computational costs. This challenge can also be formulated as to reduce the computational load in obtaining the features from a compressed image or video. As a concrete step in this direction a novel image compression methodology is presented where a good synergy among the four optimization criteria is reached.

Chapter 14
Statistical Framework for Shot-Boundary Detection

14.1 Introduction

The basis of detecting shot boundaries in video sequences is the fact that frames surrounding a boundary generally display a significant change in their visual contents. The detection process is then the recognition of considerable *discontinuities* in the visual-content flow of a video sequence. The process of shot-boundary detection, having as input two frames k and $k+l$ of a video sequence, is illustrated in Figure 14.1. Here l is the interframe distance with a value $l \geq 1$.

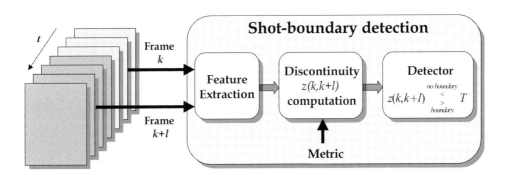

Figure 14.1: *Illustration of the process for detecting a shot boundary between frames k and k+l*

In the first step of the process, *feature extraction* is performed. Within the context of this book, extracted features depict various aspects of the visual content of a video. Then, a *metric* is used to quantify the feature variation from frame k to frame $k+l$. The discontinuity value $z(k,k+l)$ is the magnitude of this variation and serves as an input into the *detector*. There, it is compared against a *threshold T*. If the threshold is exceeded, a shot boundary between frames k and $k+l$ is detected.

To be able to draw reliable conclusions about the presence or absence of a shot boundary between frames k and $k+l$, we need to use the features and metrics for computing the discontinuity values $z(k,k+l)$, that are as discriminating as possible. This means that a clear separation should exist between discontinuity-value ranges for measurements performed *within shots* and *at shot boundaries*. In the following, we will refer to these ranges as $\overline{R}$ and R, respectively. The problem of having unseparated ranges $\overline{R}$ and R is illustrated in Figure 14.2, where some discontinuity values within shot 1 belong to the overlap area. Such values $z(k,k+l)$ make it difficult to decide about the presence or absence of a shot boundary between frames k and $k+l$ without avoiding detection mistakes, i.e. *missed* or *falsely detected* boundaries.

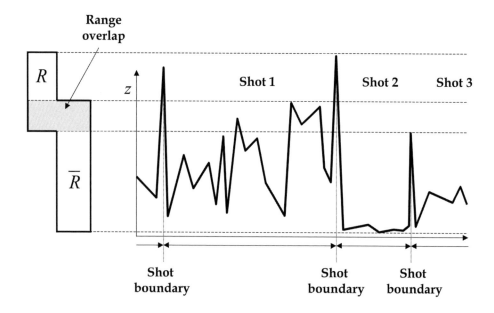

Figure 14.2: *The problem of unseparated ranges $\overline{R}$ and R*

We realistically assume that the visual-content differences between consecutive frames within the same shot are mainly caused by two factors: *object/camera motion* and *lighting changes*. Depending on the magnitude of these factors, the computed discontinuity values within shots vary and sometimes lie in the overlap area, as shown in Figure 14.2. Thus, an effective way to better discriminate between discontinuity values belonging to ranges $\overline{R}$ and R is to use features and metrics that are insensitive to motion and lighting changes. However, this is not the only advantage of using such features and metrics. Since different types of sequences can globally be characterized by their average rates and magnitudes of object/camera motion and lighting changes (e.g. high-action movies vs. stationary dramas), eliminating these distinguishing factors also provides a high level of *consistency* of ranges $\overline{R}$ and R across different sequences. If the ranges $\overline{R}$ and R are consistent, the parameters of the detection system (e.g. the threshold T) can first be optimized on a set of training sequences to maximize the detection reliability, and then the system can be used to detect shot boundaries in an arbitrary sequence without any human supervision, while retaining a high detection reliability.

As will be shown in the following section, *motion compensating* features and metrics can be found, capable of considerably reducing the influence of motion on discontinuity values. However, the influence of strong and abrupt lighting changes, induced by flashes or a camera directed to a light source, cannot be reduced in this way. For instance, one could try working only with chromatic color components, since the common lighting changes can mostly be captured by luminance variations. But this is not an effective solution in extreme cases, where all color components are changed. Strong and abrupt lighting changes can result in a series of high discontinuity values, which can be mistaken for the actual shot boundaries. In the remainder of this chapter we define possible causes for high discontinuity values within shots as *extreme factors*. These factors basically include strong and abrupt lighting changes, as well as some extreme motion cases, which cannot be captured effectively by motion compensating features and metrics.

While the influence of extreme factors on discontinuity values cannot be neutralized by choosing suitable features and metrics, it is possible to neutralize such influences by embedding *additional information* in the shot-boundary detector. For instance, the *temporal patterns* formed by consecutive discontinuity values can be investigated for this purpose. Then, the decision about the presence or absence of a shot boundary between frames k and $k+l$ made by the detector is not only based on the comparison of the computed discontinuity value $z(k,k+l)$ and the threshold T, but also based on the match between the pattern formed by consecutive discontinuity values surrounding $z(k,k+l)$ and a known pattern that is specific for a shot boundary. This is illustrated in Figure 14.3.

Different types of shot boundaries need to be taken into account during the detection process, where each of these types is characterized by its own characteristic temporal pattern. We can distinguish *abrupt boundaries*, which are the most common boundaries and occur between two consecutive frames k and $k+1$, from *gradual transitions*, such as fades, wipes and dissolves, which are spread over several frames.

Beside the information on temporal boundary patterns, the *a priori* information describing global knowledge about the visual-content flow can also be taken into account when detecting shot boundaries. An example of such information is the dependence of the probability for a shot boundary on the shot length. While being almost zero at the beginning of a shot, this probability rises with increasing shot length and converges to "1". In this way, the information on shot lengths is also highly efficient in preventing false detections due to extreme factors.

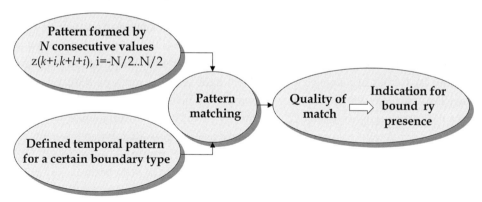

Figure 14.3: *Matching of the temporal pattern formed by N consecutive discontinuity values and a temporal pattern characteristic for a shot boundary. The quality of match between two patterns provides an indication for boundary presence between frames k and k+l*

If we combine the usage of motion compensating features and metrics for computing the discontinuity values with embedding the additional information in the detector to reduce the influence of extreme factors on these values, we are thus likely to obtain highly reliable detection results. The scheme of such a detection procedure is illustrated in Figure 14.4.

Compared to the detector in Figure 14.1, the threshold T does not remain constant but has a new value at each frame k. This is the consequence of the embedded additional information which regulates the detection process by continuously adapting the threshold to the quality of the pattern match for each new series of consecutive discontinuity values and the time elapsed since the last detected shot boundary. The remaining issue is to find the function $T(k)$ providing the optimal detection performance. Statistical detection theory provides means for solving this problem efficiently. Using the statistical properties of discontinuity values and the additional information embedded in the detector, we can compute the threshold function $T(k)$ such, that the average probability for detection mistakes is minimized.

After reviewing existing approaches to shot-boundary detection in Section 14.2, we develop in Section 14.3 a statistical framework for shot-boundary detection as shown in Figure 14.4, which addresses all the issues discussed above. Due to the consistent ranges $\bar{R}$ and R, a high generality of functions and parameters used is provided, so that our framework can operate without human supervision and is suitable for implementation into fully automated video-analysis systems. In Section 14.4 we apply the proposed detection framework to abrupt shot boundaries and evaluate the detection performance. Finally, some conclusions about the material presented in this chapter can be found in Section 14.5.

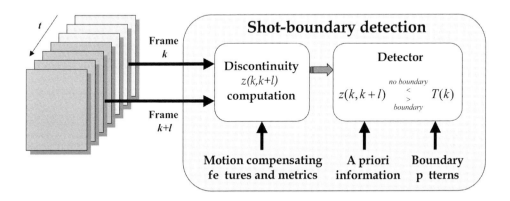

Figure 14.4: *A shot-boundary detector with improved detection performance regarding a reduction of the false-detection rate*

14.2 Previous work on shot-boundary detection

The problem of reliably detecting shot boundaries in a video has been the subject of substantial research over the last decade. In this section we give a concise overview of the relevant literature. The overview concentrates, on the one hand, on the capability of features and metrics to reduce the motion influence on discontinuity values. On the other hand, it investigates existing approaches to shot-boundary detection, involving the threshold specification, treatment of different boundary types and usage of additional information to improve the detection performance.

14.2.1 Discontinuity values from features and metrics

Different methods exist for computing discontinuity values, employing various features related to the visual content of a video. Characteristic examples of features used are pixel values, histograms, edges and motion smoothness. For each selected feature, a number of suitable metrics can be applied. Good comparisons of features and metrics used for shot-boundary detection with respect to the quality of the obtained discontinuity values can be found in overview papers [Fur95], [Aha96], [Bor96] and [Lie99].

The simplest way of measuring the discontinuity between two frames is to compute the mean absolute intensity change for all pixels of a frame [Kik92]. We first define $I(x,y)$ as the intensity of the pixel at coordinates (x,y) and compute the absolute intensity change of that pixel between frames k and $k+l$ as

$$D_{k,k+l}(x,y) = |I_k(x,y) - I_{k+l}(x,y)| \qquad (14.2.1)$$

The values (14.2.1) are then summarized over all pixels of the frame with dimensions X and Y, and averaged to give the discontinuity value, that is

$$z(k, k+l) = \frac{1}{XY} \sum_{x=1}^{X} \sum_{y=1}^{Y} D_{k,k+l}(x,y) \qquad (14.2.2)$$

A modification of this technique is only counting the pixels that change considerably from one frame to another [Ots91]. Here, the absolute change of the intensity $I(x,y)$ is compared with the prespecified threshold T_1, and is only considerable if the measured absolute difference exceeds the threshold, that is

$$D_{k,k+l}(x,y) = \begin{cases} 1 & \text{if } |I_k(x,y) - I_{k+l}(x,y)| > T_1 \\ 0 & \text{else} \end{cases} \qquad (14.2.3)$$

An important problem of the two approaches presented above is the sensitivity of discontinuity values (14.2.2) to camera and object motion. To reduce the motion influence, a modification of the described techniques was presented in [Zha93], where a 3x3 averaging filter was applied to frames before performing the pixel comparison. Much higher motion independence show the approaches based on motion compensation. There, a *block matching* procedure is applied to find for each block $b_i(k)$ in frame k a corresponding block $b_{i,m}(k+l)$ in frame $k+l$, such that it is most similar to the block $b_i(k)$ according to a chosen criterion (difference formula) D, that is:

$$D_{k,k+l}(i) = D\big(b_i(k), b_{i,m}(k+l)\big) = \min_{j=1..N_{Candidates}} D\big(b_i(k), b_{i,j}(k+l)\big) \tag{14.2.4}$$

Here, $N_{Candidates}$ is the number of candidate blocks $b_{i,j}(k+l)$, considered in the procedure to find the best match for a block $b_i(k)$. If k and $k+l$ are neighboring frames of the same shot, the values $D_{k,k+l}(i)$ can generally be assumed low. This is because for a block $b_i(k)$ almost the identical block $b_{i,m}(k+l)$ can be found due to a global constancy of the visual content within a shot. This is not the case if frames k and $k+l$ surround a shot boundary, since, in general, the difference between corresponding blocks in the two frames will be large due to a radical change in visual content across a boundary. Thus, computing the discontinuity value $z(k,k+l)$ as a function of differences $D_{k,k+l}(i)$ is likely to provide a reliable base for detecting shot boundaries.

An example of computing the discontinuity values based on the results of block-matching procedure is given in [Sha95a]. There, a frame k is divided into $N_{Blocks} = 12$ nonoverlapping blocks and differences $D\big(b_i(k), b_{i,j}(k+l)\big)$ are computed by comparing pixel-intensity values within blocks. Then, the obtained differences $D_{k,k+l}(i)$ are sorted and normalized between 0 and 1 (where 0 indicates a perfect match), giving the values $d^s_{k,k+l}(i)$. These values are multiplied with weighting factors c_i and summarized over the entire frame to give the discontinuity values, that is

$$z(k,k+l) = \sum_{i=1}^{N_{Blocks}} c_i d^s_{k,k+l}(i) \tag{14.2.5}$$

A popular alternative to pixel-based approaches is using histograms as features. Consecutive frames within a shot containing similar global visual material will show little difference in their histograms, compared to frames on both sides of a shot boundary. Although it can be argued that frames having completely different visual contents can still have similar histograms, the probability of such a case is small. Since histograms ignore spatial changes within a frame, histogram differences are considerably more insensitive to *object* motion with a constant background than pixel-wise comparisons are. However, a histogram difference remains sensitive to *camera* motion, such as panning, tilting or zooming. If histograms are used as features, the discontinuity value is obtained by bin-wise computing the difference between frame histograms. Both grey-level and color histograms are used in literature, and their differences are computed by a number of metrics. A simple metric is the sum of absolute differences of corresponding bins, with N_{Bins} being the total number of bins, that is

$$z(k,k+l) = \sum_{j=1}^{N_{Bins}} |H_k(j) - H_{k+l}(j)| \qquad (14.2.6)$$

when comparing grey-level histograms and

$$z(k,k+l) = \sum_{j=1}^{N_{Bins}} |H_k^R(j) - H_{k+l}^R(j)| + |H_k^G(j) - H_{k+l}^G(j)| + |H_k^B(j) - H_{k+l}^B(j)| \qquad (14.2.7)$$

if color histograms are compared [Yeo95a]. In (14.2.6), $H_k(j)$ is the *j*-th bin of the grey-value histogram belonging to frame *k*. In (14.2.7), $H_k^R(j)$, $H_k^G(j)$ and $H_k^B(j)$ are the *j*-th bins of histograms of the R-, G- and B-color component of the image *k*. Another popular metric is the so-called χ^2-test, proposed in [Nag92] for grey-level histograms:

$$z(k,k+l) = \sum_{j=1}^{N_{Bins}} \frac{|H_k(j) - H_{k+l}(j)|^2}{H_{k+l}(j)} \qquad (14.2.8)$$

However, according to experimental results reported in [Zha93], the metric (14.2.8) does not only enhance the discontinuities across a shot boundary, but also the effects caused by camera/object motion. Therefore, the overall detection performance of (14.2.8) is not necessarily better than that from (14.2.6), whereas it does require more computational power.

A metric involving histograms in the *HVC* color space [Fur95] (Hue – color type, Value – intensity, luminance, Chroma – saturation, the degree to which color is present) exploits the advantage of the invariance of Hue under different lighting conditions. This is useful in reducing the influence of common (weak) lighting changes on discontinuity values. Such an approach is proposed in [Arm93a], where only histograms of H and C components are used. These one-dimensional histograms are combined into a two-dimensional surface, serving as a feature. Based on this, the discontinuity is computed as

$$z(k,k+l) = \sum_{x=1}^{X}\sum_{y=1}^{Y} \{|\delta_{k,k+l}(x,y)| \times \Delta_{Hue} \times \Delta_{Chroma}\} \qquad (14.2.9)$$

where $\delta_{k,k+l}(x,y)$ is the difference between the bins at coordinates (x,y) in HC-surfaces of frames k and $k+l$, and Δ_{Hue} and Δ_{Chroma} are the resolutions of Hue and Chroma components used to form the two-dimensional histogram surface.

Also the histograms computed block-wise can be used for shot-boundary detection, as shown in [Nag92]. There, both the images k and $k+l$ are divided into 16 blocks, histograms $H_{k,i}$ and $H_{k+l,i}$ are computed for blocks $b_i(k)$ and $b_i(k+l)$ and the χ^2-test is used to compare corresponding block histograms. When computing the discontinuity as a sum of region-histogram differences, 8 largest differences were discarded to efficiently reduce the influence of motion and noise. An alternative to this approach can be found in [Ued91], where first the number of blocks is increased to 48, and then the discontinuity value is computed as the total number of blocks within a frame, for which the block-wise histogram difference exceeds a prespecified threshold T_1, that is

$$z(k,k+l) = \sum_{i=1}^{48} D(b_i(k), b_i(k+l)) \qquad (14.2.10)$$

with

$$D(b_i(k), b_i(k+l)) = \begin{cases} 1 & \text{if } \dfrac{1}{N_{Bins}} \sum_{j=1}^{N_{Bins}} \dfrac{(H_{k,i}(j) - H_{k+l,i}(j))^2}{H_{k,i}(j)} > T_1 \\ 0 & \text{else} \end{cases} \qquad (14.2.11)$$

According to [Ots93], the approach from [Ued91] is much more sensitive to abrupt boundaries than the one proposed in [Nag92]. However, since emphasis is put on blocks, which change most from one frame to another, the approach from [Ued91] also becomes highly sensitive to motion.

Another characteristic feature that proved to be useful in detecting shot boundaries is edges. As described in [Mai95], first the overall motion between frames is computed. Based on the motion information, two frames are registered and the number and position of edges detected in both frames are compared. The total difference is then expressed as the total edge change percentage, i.e. the percentage of edges that enter and exit from one frame to another. Due to registration of frames prior to edge comparison, this feature is robust against motion. However, the computational complexity of computing the discontinuity values is also high. Let p_k be the percentage of edge pixels in frame k, for which the distance to the closest edge pixel in frame $k+l$ is larger than the prespecified threshold T_1. In the same way, let p_{k+l} be the percentage of edge pixels in frame $k+l$, for which the distance to the closest edge pixel in frame k is larger than the prespecified threshold T_1. Then, the discontinuity value between these frames is computed as

$$z(k,k+l) = \max(p_k, p_{k+l}) \tag{14.2.12}$$

Finally, we discuss the computation of the discontinuity value $z(k,k+l)$ using the analysis of the motion field measured between two frames. An example for this is the approach proposed in [Aku92], where the discontinuity value $z(k,k+1)$ between two consecutive frames is computed as the inverse of *motion smoothness*. For this purpose, we first compute all motion vectors $\vec{v}(b_i(k), b_{i,m}(k+1))$ between frames k and $k+1$ and then check if they are significant by comparing their magnitude with a prespecified threshold T_1:

$$w_{i,1}(k) = \begin{cases} 1 & if \ |\vec{v}(b_i(k), b_{i,m}(k+1))| > T_1 \\ 0 & otherwise \end{cases} \tag{14.2.13a}$$

Then, we also take into consideration the frame $k+2$ and check if a motion vector between frames k and $k+1$ significantly differs from the related motion vector measured between frames $k+1$ and $k+2$. This is done by comparing their absolute difference with a prespecified threshold T_2:

$$w_{i,2}(k) = \begin{cases} 1 & if \ |\vec{v}(b_i(k), b_{i,m}(k+1)) - \vec{v}(b_i(k+1), b_{i,m}(k+2))| > T_2 \\ 0 & otherwise \end{cases} \tag{14.2.13b}$$

The sum of values (14.2.13a) for all blocks $b_i(k)$ is the number of significant motion vectors between frames k and $k+1$, and can be understood as a measure for object/camera velocity. Similarly, the sum of values (14.2.13b) is the number of motion vectors between frames k and $k+1$ that are "significantly" different from their corresponding vectors between frames $k+1$ and $k+2$, and can be understood as the measure for motion continuity along three consecutive frames of a sequence. Using these two sums, we can now compute the motion smoothness at frame k as

$$M(k) = \frac{\sum_{i=1}^{N_{Blocks}} w_{i,1}(k)}{\sum_{i=1}^{N_{Blocks}} w_{i,2}(k)} \tag{14.2.14}$$

The more motion vectors change across consecutive frames, the lower is the motion smoothness (14.2.14). Finally, the discontinuity value can be obtained as an inverse of (14.2.14), that is

$$z(k, k+1) = \frac{1}{M(k)} = \frac{\sum_{i=1}^{N_{Blocks}} w_{i,2}(k)}{\sum_{i=1}^{N_{Blocks}} w_{i,1}(k)} \tag{14.2.15}$$

14.2.2 Detection approaches

Threshold specification

The problem of choosing the right threshold for evaluating the computed discontinuity values has not been addressed extensively in literature. Most authors work with heuristically chosen global thresholds [Nag92], [Ots91], [Arm93a]. An alternative is given in [Zha93], where the authors first measure the statistical distribution of discontinuity values within a shot. Then they model the obtained distribution by a Gaussian function with parameters μ and σ, and compute the threshold value as

$$T = \mu + r\sigma \tag{14.2.16}$$

where r is the parameter related to the prespecified tolerated probability for false detections. For instance, when $r=3$, the probability of having falsely detected shot boundaries is 0.1%. The specification of the parameter r can only explicitly control the rate of false detections. The rate of missed detections is implicit and cannot be regulated, since the distribution of discontinuity values measured on boundaries is not taken into account.

However, even if they can be specified in a non-heuristic way, as shown by (14.2.16), the crucial problem related to the global threshold still remains, as illustrated in Figure 14.5. If the prespecified global threshold is too low, many false detections will appear in the shot, where high discontinuity values are caused by extreme factors, as defined in Section 14.1. If the threshold is made higher to avoid falsely detected boundaries, then the high discontinuity value corresponding to the shot boundary close to frame 500 will not be detected.

A much better alternative is to work with adaptive thresholds, i.e. with thresholds computed locally. The improved detection performance that results from using adaptive threshold function $T(k)$ instead of the global threshold T is also illustrated in Figure 14.5. If the value of the function $T(k)$ is computed at each frame k based on the extra information embedded in the detector (Figure 14.4), high discontinuity values computed within shots can be distinguished from those computed at shot boundaries. Three detection approaches applying adaptive thresholds can be found in recent literature.

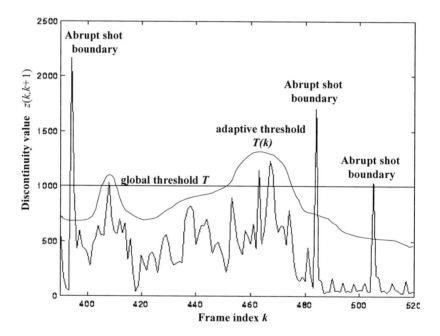

Figure 14.5: *Improved detection performance when using an adaptive threshold function $T(k)$ instead of a global threshold T.*

A method for detecting abrupt shot boundaries using an adaptive threshold is presented in [Yeo95a]. There, the values $T(k)$ are computed using the information about the temporal pattern that is characteristic for abrupt boundaries. The authors compute the discontinuity values with the interframe distance $l=1$. As shown in Figure 14.6, the N last computed consecutive discontinuity values are considered, forming a sliding window. The presence of a shot boundary is checked at each window position, in the middle of the window, according to the following criterion:

$$\text{if } z(k,k+1) = \max_{i=-\frac{N}{2},...,\frac{N}{2}} \left(\forall z(k+i, k+1+i) \right) \wedge z(k,k+1) \geq \alpha \, z_{sm} \quad (14.2.17)$$
$$\Rightarrow \text{ abrupt shot boundary}$$

In other words, an abrupt shot boundary is detected between frames k and $k+1$ if the discontinuity value $z(k,k+1)$ is the window maximum and α times larger than the second largest discontinuity value z_{sm} within the window. The parameter α can be understood as the *shape parameter* of the boundary pattern. This pattern is characterized by an isolated sharp peak in a series of discontinuity values. Applying (14.2.17) to such a series at each position of a sliding window is nothing else than matching the ideal pattern shape and the actual behavior of discontinuity values found within the window. The major weakness of this approach is the heuristically chosen and fixed parameter α. Because α is fixed, the detection procedure is too coarse and too inflexible, and because it is chosen heuristically, one cannot make statements about the scope of its validity.

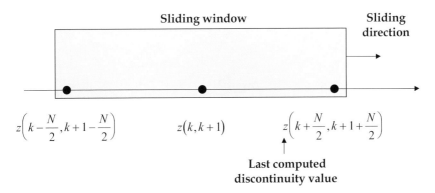

Figure 14.6: *Illustration of a sliding window approach from [Yeo95a]*

In order to make the threshold specification in [Yeo95a] less heuristic, a detection approach was proposed in [Han97a] and [Han97b], which combines the sliding window methodology with the Gaussian distribution of discontinuity values proposed in [Zha93]. Instead of choosing the form parameter α heuristically, this parameter is determined indirectly, based on the prespecified tolerable probability for falsely detected boundaries. Zhang et al. observe in [Zha93] that the discontinuity values (there obtained by comparing color code histograms) can be regarded as a realization of an uncorrelated Gaussian process if no shot change or motion is present. This observation is extended in [Han97b] to any other temporal segment with a *uniform* content development, independent of the present amount of action. Within a single shot, the series of discontinuity values can then be modeled either as a single uncorrelated Gaussian process or as a temporal concatenation of multiple uncorrelated Gaussian processes. Shots themselves are separated by individual large-valued outliers, or peaks. Based on this a statistical model for the discontinuity values is defined that has the following properties:

- Each discontinuity value measured along a sequence can be assigned one state of a two-state model: the state "S" when it is within a Gaussian shot segment, and the state "D" when it is computed at shot boundaries. A state "S" can be followed by another state "S" or by a state "D". State "D" is always followed by state "S";

- Each state "S" has three parameters, determining the process that generates the discontinuity value $z(k,k+1)$ in that state, namely: the duration of the state L, the mean and the variance of the corresponding Gaussian process;

- State "D" has duration 1.

Figure 14.7 shows the defined statistical model of a fictive series of discontinuity values, with each Gaussian segment "S" represented by its mean value. The detection procedure is activated only if the discontinuity value in the middle of the sliding window is the window maximum. As shown in Figure 14.8a, it is assumed that the series of discontinuity values captured by the window and lying at each side of the window maximum can be described by one and the same Gaussian probability density function. We define these functions as $p_{left}(z,k)$ and $p_{right}(z,k)$. The new threshold value $T(k)$, illustrated in Figure 14.8b together with the defined Gaussian distributions, is computed as the solution of the following integral equation:

$$P = \frac{1}{2} \int_{T(k)}^{\infty} \left(p_{left}(z,k) + p_{right}(z,k) \right) dz \qquad (14.2.18)$$

Here, P is the given tolerable probability for falsely detected boundaries. As in [Zha93], the rate of missed detections cannot be regulated, since the distribution of discontinuity values measured on boundaries is not taken into account. Note that the form parameter α is "hidden" in the computed threshold value $T(k)$.

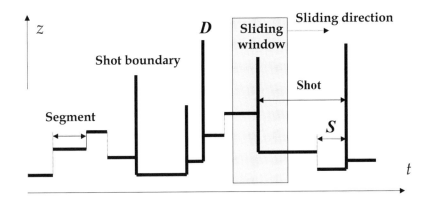

Figure 14.7: *Temporal segment structure of the series of consecutive discontinuity values computed along a sequence*

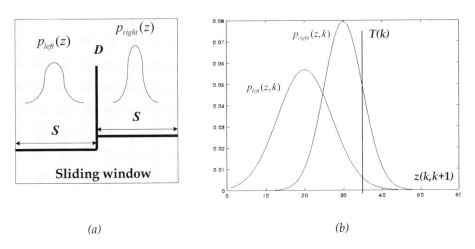

Figure 14.8: *Moment situation within a sliding window. (a) A "D" state in the middle of the window surrounded by unbroken segments of "S" states, each of them described by one and the same Gaussian distribution. (b) The threshold $T(k)$ together with Gaussian probability density functions of discontinuity values on both sides of the window maximum*

Different types of shot boundaries

One way in which the additional information embedded in the detector can influence the process of shot-boundary detection much more effectively is using the *statistical detection theory*. One of the first applications of the statistical detection theory to signal analysis can be traced back to the work of Curran [Cur65]. A characteristic example of recent works in this area can be found in [Vas98]. There, the proposed statistical method for detecting abrupt shot boundaries includes the *a priori* information based on shot-length distributions, which can be assumed consistent for a wide range of sequences. However, this *a priori* information is the only type of information embedded in the detector, and is, by itself not sufficient to prevent false detections caused by extreme factors. A more robust statistical framework for shot-boundary detection is presented in Section 14.3 of this chapter.

Different boundary types were considered in most of the approaches presented in recent literature, although the emphasis was mostly put on the detection of abrupt boundaries. This preference can be explained by the fact that there is no strictly defined behavior for discontinuity values around and within gradual transitions. While the abrupt boundaries are always represented by an isolated high discontinuity value, the behavior of these values around and within a gradual transition is not unique, not even for one and the same type of transition. In the following we will present some recent approaches to detecting non-abrupt boundaries.

One of the first attempts for detecting non-abrupt boundaries can be found in [Zha93], where a so-called twin-comparison approach is described. The method requires two thresholds, a higher one, T_h, for detecting abrupt boundaries, and a lower one, T_l, for detecting gradual transitions. First the threshold T_h is used to detect high discontinuity values corresponding to abrupt boundaries, and then the threshold T_l is applied to the rest of the discontinuity values. If a discontinuity value is higher than T_l, it is considered to be the start of a gradual transition. At that point, the summation of consecutive discontinuity values starts and goes on until the cumulative sum exceeds the threshold T_h. Then, the end of the gradual transition is set at the last discontinuity value included in the sum.

In [Ham94], a model-driven approach to shot-boundary detection can be found. There, different types of shot boundaries are considered to be editing effects, and are modeled based on the video production process. Especially for dissolves and fades, different chromatic scaling models are defined. Based on these models feature detectors are designed and used in a feature-based classification approach to segment the video. The described approach takes into account all types of shot boundaries defined by the models.

One further method for detecting gradual transitions can be found in [Men95], which investigates the temporal behavior of the variance of the frame pixels. Since within a dissolve different visual material is mixed, it can be assumed that frames within a dissolve loose their sharpness. This can be observed in the temporal behavior of the frame variance, which starts to decrease at the beginning of the transition, reaches its minimum in the middle of the transition and then starts to increase again. A characteristic parabolic pattern of variance behavior is reported. The detection of the transition is then reduced to detecting the parabolic curve pattern in a series of measured variances. In order to be recognized as a dissolve, the potential pattern has to have a width and the depth that exceeds the prespecified thresholds.

In [Son98], a chromatic video edit model for gradual transitions is built based on the assumption that discontinuity values belonging to such a transition form a pattern consisting of two piece-wise linear functions of time; one decreasing and one increasing. Such linearity does not apply outside the transition area. Therefore, the authors search for close-to-linear segments in the series of discontinuity values by investigating the first and the second derivative of the slope in time. A close-to-linear segment is found if the second derivative is less than a prespecified percentage of the first derivative.

Although each of the described models is reported to perform well in most cases, strong assumptions are made about the behavior of discontinuity values within a transition. Furthermore, several (threshold) parameters need to be set heuristically. The fact that patterns which are formed by consecutive discontinuity values and correspond to a gradual transition can strongly vary over different sequences still makes the detection of gradual transitions an open research issue [Lie99].

14.3 A robust statistical framework for shot-boundary detection

In this section we develop the statistical framework for shot-boundary detection, which is in accordance to the scheme in Figure 14.4. In contrast to detection methodologies we discussed earlier, our statistical framework includes all aspects discussed until now relevant for maximum detection performance:

- In order to provide a high level of discrimination between ranges $\overline{R}$ and R, we compute the discontinuity values using motion compensating features and metrics.

- We use both the information on temporal boundary patterns and on shot-length distributions in the detector to compute the adaptive threshold $T(k)$. Here we apply the sliding window methodology and compute the threshold value at each window position.

- We apply statistical detection theory to build a robust boundary-detection framework. This theory provides means to effectively embed the extra information from the previous item and compute the threshold value $T(k)$ using the criterion that the average probability for detection mistakes must be minimized.

In terms of the statistical detection theory, shot-boundary detection can be formulated as the problem of deciding between the two hypotheses:

- S - boundary present between frames k and $k+1$
- $\overline{S}$ - no boundary present between frames k and $k+1$

In order to take into account the information about temporal boundary patterns, we consider the N last computed consecutive discontinuity values together, in this way forming a sliding window. We define the vector $\underline{z}(k)$ as

$$\underline{z}(k) = \left\langle z(k-i, k+1-i), \quad i = -\frac{N}{2}, \ldots, \frac{N}{2} \right\rangle \tag{14.3.1}$$

We also define the *likelihood functions* $p(\underline{z}|S)$ and $p(\underline{z}|\overline{S})$, which indicate at which degree an arbitrary series of discontinuity values $\underline{z}(k)$, defined by (14.3.1), belongs to series not containing any shot boundary and those containing a shot boundary, respectively, that is

$$p(\underline{z}(k)|S) = p\left(z\left(k-\frac{N}{2}, k+1-\frac{N}{2}\right), \ldots, z\left(k+\frac{N}{2}, k+1+\frac{N}{2}\right) \Big| S \right)$$

and $\hspace{10cm}$ (14.3.2)

STATISTICAL FRAMEWORK FOR SHOT-BOUNDARY DETECTION

$$p(\underline{z}(k)|\overline{S}) = p\left(z\left(k-\frac{N}{2}, k+1-\frac{N}{2}\right),...,z\left(k+\frac{N}{2}, k+1+\frac{N}{2}\right)\bigg|\overline{S}\right)$$

In terms of statistical detection theory, the defined likelihood functions can be considered analogous to previously used ranges of discontinuity values $\overline{R}$ and R. Consequently, the requirements for a good discrimination between ranges can now be transferred to the likelihood functions $p(\underline{z}|S)$ and $p(\underline{z}|\overline{S})$. Further, we define the *a priori probability function* $P(S,k)$, which defines the probability that there is a boundary between frames k and $k+l$ based on the number of frames elapsed since the last detected shot boundary. As the criterion for deriving the rule for deciding between the two hypotheses, we choose minimizing the average probability for detection mistakes, given as

$$P_e(k) = (1 - P(S,k)) \int_{Z_S} p(\underline{z}(k)|\overline{S}) d\underline{z}(k) + P(S,k) \int_{Z_{\overline{S}}} p(\underline{z}(k)|S) d\underline{z}(k) \qquad (14.3.3)$$

Minimization of (14.3.3) provides the following decision rule at the frame k:

$$\frac{p(\underline{z}(k)|S)}{p(\underline{z}(k)|\overline{S})} \underset{S}{\overset{\overline{S}}{\lessgtr}} \frac{1 - P(S,k)}{P(S,k)} \qquad (14.3.4)$$

which can be transformed into

$$f\left(z\left(k-\frac{N}{2}, k+1-\frac{N}{2}\right),...,z\left(k+\frac{N}{2}, k+1+\frac{N}{2}\right)\right) \underset{S}{\overset{\overline{S}}{\lessgtr}} T(k) \qquad (14.3.5)$$

We call $\underline{Z}_S$ and $\underline{Z}_{\overline{S}}$ the *discontinuity-value domains* belonging to the two hypotheses. The domain $\underline{Z}_{\overline{S}}$ contains all vectors $\underline{z}(k)$, for which the hypothesis $\overline{S}$ is chosen in (14.3.5), and vice versa. However, the N-dimensional likelihood functions (14.3.2) are difficult to compute. Therefore, we simplify the shot-boundary detector (14.3.4) in several respects, under the condition that the detection performance is not degraded:

- We keep the sliding-window concept, but use only the scalar likelihood functions $p(z|S)$ and $p(z|\overline{S})$ evaluated for the discontinuity value $z(k,k+l)$ lying in the middle of the window.

- Instead of capturing the dependencies between elements of the vector $\underline{z}(k)$ via their mutual likelihood functions $p(\underline{z}|S)$ and $p(\underline{z}|\overline{S})$, we pursue the following procedure. We first investigate the temporal pattern belonging to a certain boundary type. Each of these patterns is characterized by specific relationships among discontinuity values. A typical example is an isolated peak of an abrupt shot boundary, which can be fully captured by finding the ratio between the maximal and the second largest value in a discontinuity value series. The higher this ratio, the more probable is the presence of an abrupt shot boundary at the place of the maximal discontinuity value. The ratio between the maximal and the second largest discontinuity value can now be defined as *pattern-matching indication (PMI)*, i.e. an indication that the pattern formed by consecutive discontinuity values is similar to the one that is characteristic for a certain boundary type, and therefore also as an indication of having a boundary of a certain type between frames k and $k+l$. Thus the PMI can be defined for any arbitrary type of shot boundary by the following generalized function:

$$\psi(k,k+l) = F\left(z\left(k-\frac{N}{2}, k+l-\frac{N}{2}\right), \ldots, z\left(k+\frac{N}{2}, k+l+\frac{N}{2}\right)\right) \qquad (14.3.6)$$

- At last, we define the conditional probability function $P_{Patt}(\psi(k,k+l)|S)$, which is the probability of having a shot boundary between frames k and $k+l$, based on matching of temporal patterns. It is computed at each window position, and serves as the modifier for the *a priori* probability $P(S,k)$. The lower the indication $\psi(k,k+l)$, the less likely is the presence of a shot boundary between frames and the lower are the values of $P_{Patt}(\psi(k,k+l)|S)$. In such cases the *a priori* probability is modified downwards. This modification becomes crucial if the *a priori* probability and the likelihood functions are in favor of the hypothesis S, whereby $\overline{S}$ is the proper hypothesis. In this way, boundaries detected falsely due to extreme factors can be eliminated. On the other hand, large values $\psi(k,k+l)$ indicate a similarity between the pattern formed by the elements of the vector $\underline{z}(k)$ and the pattern of a shot boundary. In such cases the probability that high discontinuity values are caused by extreme factors is small and the correction of the *a priori* probability by $P_{Patt}(\psi(k,k+l)|S)$ is not necessary.

On the basis of the simplifications described above, the general vector detection rule (14.3.4) has been now reduced to the scalar rule (14.3.7):

$$\frac{p(z(k,k+l)|S)}{p(z(k,k+l)|\overline{S})} \overset{\overline{S}}{\underset{S}{\gtrless}} \frac{1-P(S,k)P_{patt}(\psi(k,k+l)|S)}{P(S,k)P_{patt}(\psi(k,k+l)|S)} \qquad (14.3.7)$$

Since a different function (14.3.6) is required for each boundary type, we cannot use one generalized detector (14.3.4) for detecting all shot boundaries, but need separate scalar detectors (14.3.7) operating in parallel, each being used for one specific type of shot boundary.

In Section 14.4 we develop the detector (14.3.7) for *abrupt* shot boundaries. We start with the computation of discontinuity values based on suitable features and metrics. This is followed by the definition of the *a priori* probability function $P(S,k)$ and by finding the scalar likelihood functions $p(z|S)$ and $p(z|\bar{S})$. At last, PMI function $\psi(k, k+l)$ and the conditional probability function $P_{Patt}(\psi(k, k+l)|S)$ are defined.

14.4 Detector for abrupt shot boundaries

Abrupt shot boundaries take place between two consecutive frames of a sequence. For this reason it is handy to work with discontinuity values, computed with interframe distance $l=1$.

14.4.1 Features and metrics

In order to maximize the discrimination of likelihood functions $p(z|S)$ and $p(z|\bar{S})$ we compute the discontinuity values by compensating the motion between video frames using a block matching procedure, described in Section 14.2.

Similarly as in [Sha95a], we divide frame k into N_{Blocks} nonoverlapping blocks $b_i(k)$ and search for their corresponding blocks $b_{i,m}(k+1)$ in frame $k+1$. The block-matching criterion used here is the comparison of average luminance values of blocks $b_i(k)$ and $b_{i,m}(k+1)$, that is

$$D\big(b_i(k), b_{i,j}(k+1)\big) = \big|Y_{ave}\big(b_i(k)\big) - Y_{ave}\big(b_{i,j}(k+1)\big)\big| \tag{14.3.8}$$

After the corresponding blocks $b_{i,m}(k+1)$ have been found using the formula (14.2.4), we obtain the discontinuity value $z(k,k+1)$ by summarizing the differences between blocks $b_i(k)$ and $b_{i,m}(k+1)$ in view of block-wise average values of all three color components Y_{ave}, U_{ave} and V_{ave}, that is

$$z(k, k+1) = \frac{1}{N_{Blocks}} \sum_{i=1}^{N_{Blocks}} D\big(b_i(k), b_{i,m}(k+1)\big) \tag{14.3.9}$$

with

$$D(b_i(k), b_{i,m}(k+1)) = |Y_{ave}(b_i(k)) - Y_{ave}(b_{i,m}(k+1))| + \\ |U_{ave}(b_i(k)) - U_{ave}(b_{i,m}(k+1))| + \\ |V_{ave}(b_i(k)) - V_{ave}(b_{i,m}(k+1))|$$
(14.3.10)

14.4.2 A priori probability function

Studies by Salt [Sal73] and Coll [Col76], involving statistical measurements of shot lengths for a large number of motion pictures, have shown that the distribution of shot lengths for all the films considered matches the Poisson function well [Pap84]. Therefore, we integrate the Poisson function to obtain the *a priori* probability for a shot boundary between frames k and $k+1$, that is

$$P(S, k) = \sum_{w=0}^{\lambda(k)} \frac{\mu^w}{w!} e^{-\mu}$$
(14.3.11)

The parameter μ represents the average shot length of a video sequence, w is the frame counter, which is reset each time a shot boundary is detected, and $\lambda(k)$ is the current shot length at the frame k. Although in [Col76] and [Sal73] the Poisson function was obtained for motion pictures, we assume that this conclusion can be extended further to all other types of video programs. However, to compensate for possible variations in program characteristics, we adapt the parameter μ to different program types (movies, documentaries, music video clips, etc.) and sub-types (e.g. an action movie vs. drama). The adjustment of the parameter μ is easy and can be performed automatically, if the program type is known at the input into the video analysis system. In our experiments we kept μ constant at the value 70.

14.4.3 Scalar likelihood functions

We now perform a parametric estimation of scalar likelihood functions $p(z|S)$ and $p(z|\bar{S})$, to be used in the detection rule (14.3.6). In order to get an idea about the most suitable analytical functions used for such estimation, the normalized distributions of discontinuity values $z(k,k+1)$ computed within shots and at shot boundaries are obtained first, using several representative test sequences.

A normalized distribution of discontinuity values computed within shots is shown in Figure 14.9a. The shape of the distribution indicates that a good analytic estimate for this distribution can be found in the family of functions given as

$$p(z|\overline{S}) = h_1 z^{h_2} e^{-h_3 z} \qquad (14.3.12)$$

The most suitable parameter combination (h_1, h_2, h_3) is then found experimentally, such that the rate of detection mistakes for the test sequences is minimized. The optimal parameter triplet is found as (1.3, 4, -2). The corresponding analytical function, serving as parametric estimate of the likelihood function $p(z|\overline{S})$, is also shown in Figure 14.9a.

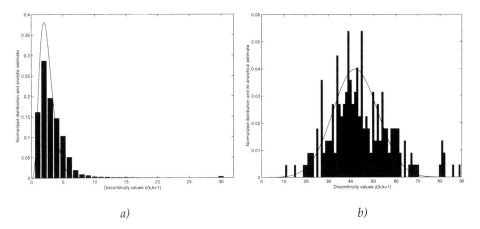

a)   b)

Figure 14.9: (a) The normalized distribution of values z(k,k+1) computed within shots (discrete bins) and its analytic estimate (continuous curve), (b) normalized distribution of values z(k,k+1) computed at shot boundaries (discrete bins) and its analytic estimate (continuous curve)

The analog procedure is applied to obtain the parametric estimate of the likelihood function $p(z|S)$. Figure 14.9b shows the normalized distribution of discontinuity values z(k,k+1), computed at shot boundaries, for which the same set of test sequences as above is used. Judging by the form of the distribution, a Gaussian function

$$p(z|S) = \frac{1}{\sigma\sqrt{2\pi}} e^{-\frac{1}{2}\left(\frac{z-\mu}{\sigma}\right)^2} \qquad (14.3.13)$$

can be taken as a good analytic estimate of it. Again we found the optimal values for the pair of parameters (μ,σ) by experimentally minimizing the rate of detection mistakes for the set of test sequences. This pair of values was obtained as (42, 10), resulting in the Gaussian function presented in Figure 14.9b.

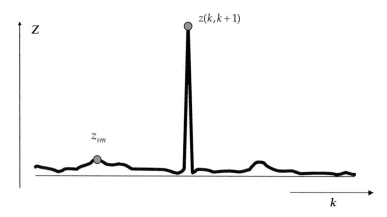

Figure 14.10: *Abrupt boundary pattern with characteristic parameters*

14.4.4 PMI and the conditional probability functions

Based on the discussion in the previous sections, we can state that the presence of an isolated sharp peak belonging to an abrupt shot boundary in the middle of the sliding window can efficiently be described by the ratio of the discontinuity value $z(k,k+1)$ in the middle of the window and the second largest discontinuity value z_{sm} within that window. A typical peak of an abrupt shot boundary with values $z(k,k+1)$ and z_{sm} is illustrated in Figure 14.10. The corresponding PMI function to be used in the detector (14.3.7) is now given as

$$\psi(k,k+1) = \frac{z(k,k+1)}{z_{sm}} \qquad (14.3.14)$$

The value of the PMI function (14.3.14) serves as the argument of the conditional probability function $P_{patt}(\psi(k,k+1)|S)$, defined as

$$P_{patt}(\psi(k,k+1)|S) = \frac{1}{2}\left(1 + erf\left(\frac{\psi(k,k+1)-d}{\sigma_{erf}}\right)\right) \qquad (14.3.15)$$

with

$$erf(x) = \frac{2}{\pi} \int_0^x e^{-t^2} dt \qquad (14.3.16)$$

The parameters d and σ_{erf} are the "delay" from the origin and the spreading factor determining the steepness of the middle curve segment, respectively. The optimal parameter combination (d, σ_{erf}) is found experimentally such that the detection performance for the test sequences is optimized. The resulting optimal pair of parameters was found as (13, 5). The conditional probability (14.3.15) is illustrated in Figure 14.11.

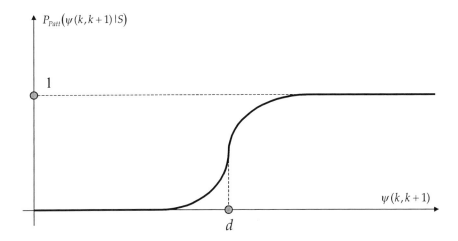

Figure 14.11: *The conditional probability function $P_{patt}(\psi(k, k+1)|S)$*

14.4.5 Experimental validation

Achieving a high detection performance was an important issue when we developed the statistical detection framework. To test the performance of the detector (14.3.7) for abrupt boundaries, we used 5 sequences that belong to 2 different categories of programs, movies and documentaries, and that were not previously employed for training the detection procedure. The results presented in Table 14.1 illustrate a high detection rate and no falsely detected boundaries. Furthermore, the obtained good results remain consistent over all sequences.

Test material	Length in frames	Total	Detected boundaries	Falsely detected boundaries
Documentary 1	700	3	3	0
Documentary 2	800	5	5	0
Documentary 3	900	6	6	0
Movie 1	10590	90	90	0
Movie 2	17400	95	94	0
Total	30390	199	198	0

Table 14.1: *Detection results for abrupt shot boundaries*

14.5 Conclusions

Most existing approaches for shot boundary detection are based on explicitly given thresholds or relevant threshold parameters, which directly determine the detection performance. Due to such a direct mutual dependence, the detection performance is highly sensitive to specified parameter values. For instance, a threshold set to 2.3 will interpret a discontinuity value 2.31 as a shot boundary and a value 2.29 as a regular value within a shot. Beside the sensitivity, the problem of specifying such a precise threshold remains. And, consequently, the scope of the validity of such a precise threshold is highly questionable.

Manual parameter specification clearly cannot be avoided in any of the detection approaches. However, the influence of these parameters on the detection performance can be diminished and the detection can be made more robust if the parameters are used at lower levels of the detection framework, so only for the purpose of globally defining the framework components. Each component then provides the detector with nothing more than an indication of the presence of a boundary based on a specific criterion. The decision making about the presence of a shot boundary is then left solely to the detector, where all the indications coming from different sources are evaluated and combined. In this way, the importance of a single manually specified parameter is not as great as when that parameter is directly a threshold, and can therefore be assumed valid in a considerably broader scope of sequences. In the statistical detection framework presented in this chapter, this is the case with parameter sets (h_1, h_2, h_3) and (μ, σ), which are used to define the likelihood functions (14.3.12) and (14.3.13), as well as with parameters d and σ_{erf} used to formulate the conditional probability function (14.3.15).

The only parameter which needs to be adjusted depending on the type of sequence is μ, which is used in the formula (14.3.11) to define the *a priori* probability. However, setting the value for μ is easy, since it determines the average shot length characteristic for a certain program type. For instance, the μ value for movies can be set to a value within a range 80-100, and for music TV clips to 30-40. The adjustment of the μ value can be performed fully automatically if the program type information is available in the shot-boundary detection system. An example is a video analysis system, as illustrated in Figure 13.2, which operates directly on DVB streams. Here, each transmitted program compliant to DVB standard also contains a header, which – among other data – contains the program type (movie, documentary, music TV clip, etc.). Therefore, μ can be set easily by means of a simple look-up table.

Since the parameters used in our framework can either be assumed generally valid or be adjusted automatically, no human supervision is required during the detection procedure. At the same time, since the parameters are optimized for a general case, similar high detection performance can be expected for any input sequence. Both of these aspects make the developed framework suitable for an implementation in a fully automated sequence analysis system. The facts that the detection method presented in this chapter can operate on a wide range of video sequences without human supervision, and keep the constant high detection quality for each of them, are the major advantages the proposed detection framework has over the methods from recent literature.

Chapter 15

Automatically Abstracting Video using Key Frames

15.1 Introduction

A structured collection of selected video frames, or *key frames*, is a compact representation of a video sequence and is useful for various applications on a video. For instance, it can provide a quick overview of the video-database content, enable access to shots, episodes and entire programs in video-browsing and retrieval systems and be used for making a commercial for a video. Furthermore, a video index may be constructed based on visual features of key frames, and queries by example may be directed at key frames using image-retrieval techniques [Zha97a]. Also the higher-level video processing and analysis steps involving comparisons of shots can benefit from visual features captured in key frames [Yeu95a], [Yeu97], [Han99b]. To enable these applications, key frames can be extracted in various fashions, such as

- **Extracting the *most memorable* video frames**: It is in human nature to remember some most memorable segments of a video, e.g. a zoom of an actor in a funny pose, a slow camera pan along a beautiful landscape or an impressive action scene. A number of key frames can be extracted to represent each of these segments.

- **Summarizing the visual content of a video**: The visual content of a video can be "compressed" by first collecting fragments showing all of its relevant elements, such as landscapes, objects, persons, situations, etc., and by then searching for a limited number of frames to represent each of these elements. An alternative summarizing approach is to investigate the story flow of a video sequence and to represent each successive logical segment (event, episode) by suitable frames. Then, by concatenating these frames chronologically, a *storyboard* can be obtained giving a compact video overview [Pen94a].

Key frames can be extracted manually or automatically. Both possibilities are illustrated in Figure 15.1. If key frames are extracted manually, they comply with human cognition, that is, human understanding of a video content and human perception of representativeness and technical quality of a frame. For instance, each key frame can be extracted based on the role the persons and objects captured therein play in the context of the target application. From several candidate frames, the one being most representative (e.g. taken under the best camera angle) is chosen. Furthermore, it is expected that no blurred or "dark" frames are extracted, or those with coding artifacts, interlacing effects, etc.

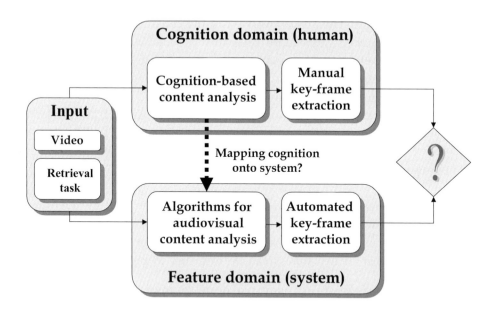

Figure 15.1: *Manual vs. automated key-frame extraction*

In order to develop feature-based algorithms for automatically extracting key frames that have the same quality as those extracted manually, we must map the extraction criteria complying with human cognition onto the *machine criteria*. However, such mapping is highly problematic, not only technically (Chapter 13), but also due to the missing *ground truth* for the key-frame extraction.

If several users manually extract key frames from one and the same video and for the same target application, it can realistically be assumed that each of the obtained sets will be unique, concerning both the total number of frames contained therein and the specific frames extracted. One reason for this is the subjectivity of human perception of a video content. Especially when choosing the most memorable video segments and extracting the corresponding key frames are concerned, the dispersion among extraction results obtained by different users will be high [Paa97]. However, even if there is a consensus among users about which segments should be represented in the visual abstract, again different key-frame sets can be expected. A trivial example is a stationary shot showing an anchorperson in a news program. Such a shot can equally well be represented by any of its frames.

Based on the discussion above we conclude that automatically extracting key frames for the purpose of capturing the most memorable moments of a video sequence is a difficult problem, mainly due to the subjectivity of the definition what is *memorable*. Compared to this, the role of subjectivity in extracting key frames for making a visual summary of a video is significantly smaller. This can be explained by the fact that such a summary ideally contains *all* relevant visual-content elements (faces, objects, landscapes, situations, etc.) and not a subjective selection of these elements. In this way, we understand the key-frame based video summary as a unity of all possible subjective key-frame selections. This makes the extraction of "summarizing" key frames easier to automate. The only aspect which remains subjective and therefore difficult to take into account by automation is to choose a representative frame out of several equally acceptable candidate frames. However, as illustrated in the example that involves a stationary anchorperson shot, selecting any of the candidate frames does not considerably influence the quality of the resulting key-frame set. For this reason, instead of considering the possibility of selecting any frame out of equally acceptable candidates as a problem for automation, we hold that it is an additional degree of freedom in the automation of the key-frame extraction procedure.

We now define the objective of this chapter so as to provide methods for automatically extracting key frames which summarize the visual content of a video. Since the complex extraction criteria related to human cognition are difficult to map onto the system level, we circumvent this mapping by applying a practical extraction methodology which is based on reducing the visual-content redundancy among video frames. In the following, we define and discuss three different groups of key-frame extraction techniques belonging to this methodology: *sequential extraction in a local context* (SELC), *sequential extraction in a global context* (SEGC) and *non-sequential extraction* (NSE).

A typical video can be seen as a concatenation of frame series, each characterized by a high visual-content redundancy. These frame series can be entire video shots or shot segments. Then, the redundancy of the visual content found in such series can be reduced by representing each of them by one key frame. Taking again as an example a stationary shot showing an anchorperson in a news program, the frames of such a shot are almost identical and can be compressed to a single frame. Applied to the entire video sequence, key frames can then be seen as its (non)equally distributed sample frames [Pen94a]. This we call *sequential extraction in a local context* (SELC).

Since a SELC technique extracts key frames only in the local context, similar key frames may be extracted from different (remote) sequence fragments, which results in a redundancy within the obtained key-frame set. This indicates that by using some alternative techniques one can further reduce the number of extracted key frames while still keeping all the relevant visual information of a sequence. One of possibilities is to modify SELC approaches by taking into account all previously extracted key frames each time a new frame is considered. Then, a new key frame is extracted only if it is considerably different from all other already extracted key frames. We call such a technique *sequential extraction in a global context* (SEGC). Another alternative is a *non-sequential extraction* (NSE), where all frames of a sequence are taken and grouped together, based on the similarity of their visual content. The key-frame set is then obtained by collecting representatives of each of the groups.

If concatenated, the key frames obtained by means of a SELC technique represent a "red line" through the story of a video and closely provide a *storyboard*. However, for some applications involving video content, having a storyboard of that video is not required. This is the case with key-frame based video queries in standard image retrieval tools. In such applications, the redundancy among key frames makes the query database too large, slows down the interaction process and puts larger demands on storage space for keeping the key frames than actually necessary. In these cases, SEGC or NSE techniques are more suitable. While SELC and SEGC techniques allow for on-the-fly (on-line) key-frame extraction and are computationally less expensive than the NSE techniques, the NSE techniques consider the key-frame extraction as a postprocessing step and mostly involve complex clustering procedures. However, a higher complexity of NSE techniques is compensated by the fact that they are more sophisticated and, therefore, provide a higher representativity of key frames while keeping the number of key frames minimal.

After a review of existing approaches to automated key-frame extraction in Section 15.2, we present in Sections 15.3 and 15.4 two novel extraction methods. The first method belongs to the SELC group of approaches and aims at providing a good video summary, also including its storyboard, while keeping the total number of extracted key frames for the entire sequence close to the prespecified maximum. This controllability is, on the one hand, an important practical issue, regarding the available storage space and the interaction speed with a video database, but, one the other hand, it also means an additional constraint that needs to be taken into account during the key-frame extraction procedure. In contrast to the method in Section 15.3, the major objective of the method presented in Section 15.4 is minimizing the redundancy among video frames and providing a set of key frames which is similar to the one based on human cognition for a given video sequence. We can explain this objective with the example of a simple dialog sequence, where stationary shots of each of the two characters participating in a dialog are alternated. Since a user would summarize such a sequence by taking only two frames, one for each of the characters, this should be obtained automatically as well. The approach in Section 15.4 belongs to the NSE group; it is based on cluster validity analysis and is designed to work without any human supervision. Conclusions to this chapter can be found in Section 15.5.

15.2 Previous work on key-frame extraction

A number of methods for automating the key-frame extraction procedure can be found in recent literature. As will be shown in this section, some of the methods are based on the criterion of reducing the visual-content redundancy among consecutive frames, as defined above. However, some characteristic key-frame extraction methods based on other criteria will be described as well.

A first attempt to automate key-frame extraction was done by choosing as a key frame the frame appearing after each detected shot boundary [Sha95b]. However, while one key frame is sufficient for stationary shots, in dynamic sequences it does not provide an acceptable representation of the visual content. Therefore, methods were needed to extract key frames that are in agreement with the visual-content variations along a video sequence. One of the first key-frame extraction approaches developed in view of this objective is presented in [Zha95a], with all details given in [Zha97b]. Key frames are extracted in a SELC fashion separately for each shot. The first frame of a shot is always chosen as a key frame. Then, similar methodology is applied as for detecting shot boundaries. The discontinuity value $z(F_{last}, k)$ is computed between the current frame k of a sequence and the last extracted key frame F_{last} using color histograms as spatial features (Chapter 14). If this discontinuity value exceeds a given threshold T, the current frame is selected as a new key frame, that is

Step 1: $F_{last} = 1$
Step 2: $\forall k \in [2, S]$ *if* $z(F_{last}, k) > T \Rightarrow F_{last} = k$ (15.2.1)

Here, S is the number of frames within a shot. The extraction procedure (15.2.1) is then adapted by means of the information on dominant or global motion resulting from camera operations and large moving objects, according to a set of rules. For a zooming-like shot, at least two frames will be extracted, at the beginning and at the end of a zoom. The first frame represents a global and the other one a more detailed view of a scene. In case of panning, tilting and tracking, the number of frames to be selected depends on the rate of visual-content variation: ideally, the visual content covered by each key frame has little overlap, or each frame should capture different object activities. Usually frames that have less than 30% overlap in their visual content are selected as key frames. A key-frame extraction method similar to (15.2.1) can also be found in [Yeu95a]. There, however, the motion information is not used.

Another SELC extraction approach is proposed in [Gun98], where the authors first compute the discontinuity value between the current frame k and the N previous frames. This is done by comparing the color histogram of the frame k and the average color histogram of the previous N frames, that is

$$z(k, \{k-1, .., k-N\}) = \sum_{j=k-1}^{k-N} \sum_{e=Y,U,V} \left| H_k^e(i) - \frac{1}{N} \sum_{j=k-1}^{k-N} H_j^e(i) \right| \quad (15.2.2)$$

If the discontinuity value (15.2.2) exceeds the prespecified threshold T, the current frame k is extracted as a new key frame F_{last}, i.e.

if $z(k, \{k-1, .., k-N\}) > T \Rightarrow F_{last} = k$ (15.2.3)

A possible problem with the extraction methods described above is that the first frame of a shot is always chosen as a key frame, as well as those frames lying in shot segments with varying visual content. As discussed in [Gre97], when a frame is chosen that is close to the beginning or end of a shot, it is possible that that frame is part of a dissolve effect at the shot boundary, which strongly reduces its representative quality. The same can be said for frames belonging to shot segments of great camera or object motion (e.g. strong panning or a zoomed object moving close to the camera and hiding most of the frame surface). Such frames may be blurred, and thus in some cases not suitable for extraction. A solution to this problem can be found in [DeM98], where the authors first represent a video sequence as a curve in a high-dimensional feature space. The 13-dimensional feature space is formed by the time coordinate and 3 coordinates of the largest "blobs" (image regions), where 4 intervals (bins) are used for each luminance and chrominance channel. Then the authors simplify the curve using the multidimensional curve-splitting algorithm. The result is, basically, a linearized curve, characterized by "perceptually significant" points, which are connected by straight lines. A key-frame set of a sequence is finally obtained by collecting frames found at perceptually significant points. With a splitting condition that checks the dimensionality of the curve segment that is split, the curve can be recursively simplified at different levels of detail, that is with different densities of perceptually significant points. The final level of detail depends on the prespecified threshold, which evaluates the distance between the curve and its linear approximation. We consider the main problem of this approach to be evaluating the applicability of obtained key frames, as it is not clear which level and objective of video representation is aimed at. For instance, it is unlikely that the objective of the approach is to provide a good video summary, since there is no proof that extracted key frames lying at "perceptually significant points" capture all important aspects of a video. On the other hand, the connection between perceptually significant points and most memorable key frames according to user's cognition is not clear either.

An example of NSE key-frame extraction approaches can be found in [Zhu98]. There, all frames in a video shot are classified into M clusters, where this final number of clusters is determined by a prespecified threshold T. A new frame is assigned to an existing cluster if it is similar enough to the centroid of that cluster. The similarity between the current frame k and a cluster centroid c is computed as the intersection of two-dimensional HS histograms of the HSV color space (H - Hue, S - Saturation, V - Value). If the computed similarity is lower than the prespecified threshold T, a new cluster is formed around the current frame k. In addition, only those clusters that are larger than the average cluster size in a shot are considered as key clusters, and the frame closest to the centroid of a key cluster is extracted as a key frame.

Extraction of key frames in all approaches discussed above is based on threshold specification. The thresholds used in [Zha95a], [DeM98] and [Zhu98] are heuristic, while the authors in [Gun98] work with a threshold they obtained by means of the technique of Otsu [Sah88]. By adjusting the threshold, the total number of extracted key frames can be regulated. However, such regulation can be performed only in a global sense, meaning that a lower threshold will lead to more key frames, and vice versa. An exact or at least an approximate control of the total number of extracted key frames is not possible. First, it is difficult to relate a certain threshold value to the number of extracted key frames. Second, one and the same threshold value can lead to a different number of extracted key frames in different sequences. A practical solution for this problem is to make the threshold more meaningful and to relate it directly to the extraction performance. An example is the threshold specification in form of the maximum tolerable number of key frames for a given sequence. An NSE approach using this sort of thresholds can be found in [Sun97]. There, two thresholds need to be prespecified: r, controlling which frames will be included in the set and N, being the maximum tolerable number of key frames for a sequence. Key frame extraction is performed by means of an iterative partitional-clustering procedure. In the first iteration step, a video sequence is divided into consecutive clusters of the same length L. The difference is computed between the first and the last frame in each cluster. If the difference exceeds the threshold r, all frames of a cluster are taken as key frames. Otherwise, only the first and the last frame of the cluster are taken as key frames. If the total number of extracted frames is equal to or smaller than the tolerable maximum N, the extraction procedure is stopped. If not, a new sequence is composed out of all extracted frames and the same extraction procedure is applied. The biggest disadvantage of this method is the difficulty of specifying the threshold r, since it is not possible to relate the quality of the obtained key-frame set to any specific r value.

If the total number of extracted key frames is regulated by a threshold, the qualities of the resulting key-frame set and of the set obtained for the same sequence but based on human cognition are not necessarily comparable. For instance, if the threshold is too low, too many key frames are extracted and characterized by a high redundancy of their visual contents. As a result of a threshold set too high, the key-frame set might be too sparse. Especially if the rate of visual-content change allows for only one optimal set of key frames for the best video representation, finding the threshold value providing such a key-frame set is very difficult.

Authors in [Avr98] and [Wol96] aim at avoiding this problem and propose threshold-free methods for extracting key frames. In [Avr98], the temporal behavior of a suitable feature vector is followed along a sequence of frames; a key frame is extracted at each place of the curve where the magnitude of its second derivative reaches the local maximum. A similar approach is presented in [Wol96], where local minima of motion are found. First, the optical flow is computed for each frame and then a simple motion metric is used to evaluate the changes in the optical flow along the sequence. Key frames are then found at places where the metric as a function of time has its local minima. However, although the first prerequisite for finding good key frames was fulfilled by eliminating threshold dependence of the extraction procedure, the two described methods have the same disadvantage as the method proposed in [DeM98], namely an unclear applicability of the resulting key frames.

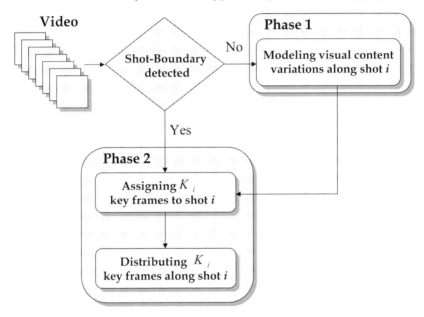

Figure 15.2: *Scheme of the key-frame extraction approach with controlled number of key frames*

15.3 Extracting key frames by approximating the curve of visual-content variations

In the key-frame extraction method presented in this section we aim at providing a good video summary while keeping the number of extracted key frames close to the prespecified maximum. This SELC method can be considered as an alternative to the approach from [Sun97]. However, it has the advantage that the number of thresholds is reduced to one; it is the maximum allowed number of key frames N for the entire sequence.

As illustrated in Figure 15.2, key frames are extracted for each shot of a sequence separately. This is done in two major phases. The first phase starts at the beginning of a shot i and lasts until the boundary to the shot $i+1$ is detected. During this time, the variation of the visual content is modeled along a shot i. The result of this phase is twofold. First, a curve is obtained which models the visual-content variations along shot i. Second, the total magnitude C_i of visual-content variations along a shot i is available at the moment the boundary between shots i and $i+1$ is detected. The second phase starts at the moment the boundary to the shot $i+1$ is detected, and consists of two consecutive steps. In the first step, a fraction K_i of the prespecified N key frames is assigned to shot i, proportional to the computed value C_i, and such that the sum of key frames assigned to all shots of a sequence does not exceed the prespecified maximum N. The number N can be adjusted if we know *a priori* the type of the program to be processed. In the second step, a threshold-free procedure is applied to find optimal positions for the assigned number of key frames along a shot i. Such an optimal distribution is obtained iteratively, by means of a suitable numerical algorithm. In the following subsections, we will describe both extraction phases and all of their steps in more detail.

15.3.1 Modeling visual content variations along a shot

In order to model the variations of the visual content along a shot i, we must consider *relevant* content variations, i.e. those that make the extraction of a new key frame necessary. For instance, object motion by a constant background is not as relevant for key-frame extraction as, for instance, camera panning, tilting and tracking. This is because the object motion alone does not result in a drastic change of the visual content, and does not need to be captured by several key frames. Opposed to this, a camera motion constantly introduces new visual material, which needs to be represented by more than one key frame. For efficiently capturing camera operations while excluding the sensitivity of key-frame extraction to object motion, we compute the discontinuity values $z(k,k+1)$ using color histograms and according to (2.2.7), but here in the YUV color space and for $l=1$:

$$z(k,k+1) = \sum_{j=1}^{N_{Bins}} \left(|H_k^Y(j) - H_{k+1}^Y(j)| + |H_k^U(j) - H_{k+1}^U(j)| + |H_k^V(j) - H_{k+1}^V(j)| \right) \quad (15.3.1)$$

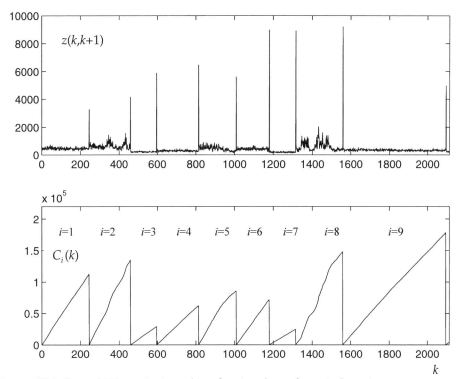

Figure 15.3: *(upper) Discontinuity values for nine shots of a typical movie, (lower) Functions (15.3.2) modeling the visual content variations*

Accumulating the discontinuity values (15.3.1) along a shot and taking the current cumulative value at each frame k results in the function $C_i(k)$, which we consider as the model for visual-content variations along a shot i:

$$C_i(k) = \sum_{j=f_{1,i}}^{k-1} z(j, j+1) \qquad (15.3.2)$$

The frame $f_{1,i}$ is the first frame of the shot i and the summation process (15.3.2) is reset at the shot boundary. Since $z(k,k+1)$ can only have non-negative values, $C_i(k)$ is a non-decreasing function. It has a close-to-linear behavior in shot segments with a uniform rate of visual content variations (e.g. a stationary segment or a constant camera motion) and changes in steepness wherever changes in the variation rate occur (e.g. camera motion after a stationary segment). Figure 15.3a shows the discontinuity values computed for nine shots of a typical movie sequence and Figure 15.3b the behavior of the corresponding functions $C_i(k)$. When the end of a shot is reached, we obtain with (15.3.3) the total magnitude of visual content variations along the shot i, with S_i being the number of frames in that shot:

$$C_i = C_i(S_i) = \sum_{j=f_{1,i}}^{S_i-1} z(j, j+1) \tag{15.3.3}$$

15.3.2 Distributing N key frames over the sequence

After the total magnitude of visual content variations for the shot i has been obtained by means of (15.3.3), the total prespecified number N of key frames is distributed along all shots of a sequence proportional to values C_i. The higher C_i, the more diverse visual content is assumed in the shot i, which then requires more key frames in order to be represented well. As N_{shots} is the number of shots in the entire sequence, we assign K_i key frames to shot i according to the following ratio:

$$K_i = \frac{C_i}{\sum_{j=1}^{N_{shots}} C_j} N \tag{15.3.4}$$

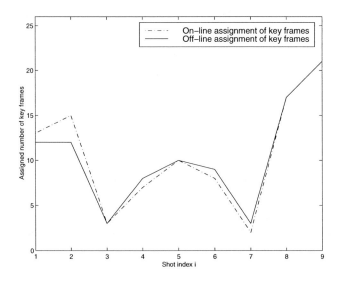

Figure 15.4: *Key-frame assignments according to the procedures (15.3.4) and (15.3.5)*

Equation (15.3.4) assumes that the values C_i are known for all shots of a sequence, so the denominator in (15.3.4) can now be computed. Since, in practice, on-line key-frame extraction is more appealing, we adapt the assignment rule (15.3.4) so that a suitable number of key frames can be assigned to a shot i immediately after the boundary between shots i and $i+1$ has been detected. The adapted assignment rule is given as follows:

$$K_i = \frac{C_i}{\sum_{j=1}^{N_{shots}} C_j} N = C_i \frac{N}{S} \frac{S}{\sum_{j=1}^{N_{shots}} C_j} \approx \text{int}\left(C_i \frac{N}{S} \frac{\sum_{j=1}^{i} S_j}{\sum_{j=1}^{i} C_j} \right) \quad (15.3.5)$$

Here, S is the total sequence length and S_j is the length of the shot j. Compared to the off-line assignment (15.3.4), the rule (15.3.5) uses only the information available at the moment when K_i is computed. Since the total cumulative variations of the visual content along the entire sequence (denominator in (15.3.4)) is not known, we can only summarize until the shot i. This disadvantage is, however, compensated by taking into consideration also the time parameter, e.g. shot lengths. Thereby we assume that the ratio between the total sequence length and the values C_i for all shots of a sequence can be well approximated by the ratio between these two quantities, where both are only taken up to the current shot i.

Assignment results obtained using (15.3.4) and (15.3.5) may differ in the beginning, that is, for a low shot index i. However, with increasing i we expect the value (15.3.5) to converge towards the value (15.3.4). In order to show this, we chose to distribute an unusually large number of $N=100$ key frames along nine shots of the sequence illustrated in Figure 15.3a. Assignment results using both methods are presented in Figure 15.4.

15.3.3 Distributing key frames within a shot

In the final step, K_i assigned key frames need to be located within a shot. For the sake of notation and derivation, in the following we will consider k in (15.3.2) to be a continuous variable, although a practical implementation will use a discretized version. If we would interpolate $C_i(k)$ for non-integer values of k from neighboring values for integer k, i.e. $C_i(\lfloor k \rfloor)$ and $C_i(\lceil k \rceil)$, then $C_i(k)$ becomes a non-decreasing function. We will assume this property in the sequel. The underlying theory used here for distributing key frames along a shot is that K_i key frames should be distributed along a shot such that the visual content is summarized in the best possible way. Since the function $C_i(k)$ represents the variations of the visual content along the shot, it also provides the information about the amount of redundancy present in each of the shot segments. The steeper the function, the less redundancy is to be found among consecutive frames in that segment, and vice versa.

Consequently, properly distributing key frames along a shot is equivalent to finding a suitable way of representing the function $C_i(k)$ by K_i (non-) equidistant samples, where the sample density is dependent on function steepness and where each sample is a key frame representing a series of consecutive frames around it. A key frame F_j, $j=1,...,K_i$, lies in the middle of the interval (t_{j-1}, t_j) and represents all frames in that interval. We approximate the function $C_i(k)$ along this interval by its value at frame $k = F_j$, that is, by $C_i(F_j)$. By doing this for each key frame, a *step curve* is obtained, which closely approximates the function $C_i(k)$ along a shot i. Maximizing the quality of such an approximation is now equivalent to properly placing the horizontal line segments and defining their optimal lengths, which is, again, equivalent to properly positioning the key frames F_j in the middle of these segments. To achieve such optimal positioning, we choose to minimize the following L_1 error function:

$$g(F_1,...,F_{K_i},t_1,...,t_{K_{i-1}}) = \sum_{j=1}^{K_i} \int_{t_{j-1}}^{t_j} |C_i(k) - C_i(F_j)| \, dk \qquad (15.3.6)$$

Note that t_0 and t_{K_i} are the (known) temporal starting and endpoints of the shot i. Figure 15.5 illustrates the meaning of (15.3.6). It shows the function $C_i(k)$ and how the key frames are distributed such that the area between this function and the approximating rectangles, defined by F_j and t_j, is minimized. The minimization of (15.3.6) is carried out in two steps. First, if we assume that the breakpoints t_{j-1} and t_j are given, then the partial integral

$$g(F_j) = \int_{t_{j-1}}^{t_j} |C_i(k) - C_i(F_j)| \, dk \qquad (15.3.7)$$

is minimized by taking as key frame the center of the interval considered:

$$F_j = \frac{t_{j-1} + t_j}{2} \quad \text{for} \quad j = 1, 2, ..., K_i \qquad (15.3.8)$$

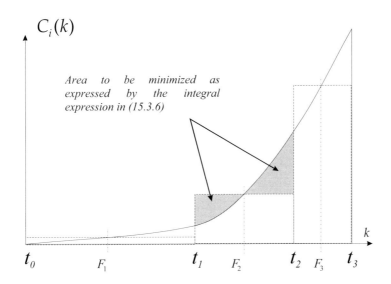

Figure 15.5: *Illustration of the function $C_i(k)$, the distribution of key frames F_j and breakpoints t_j*

Note that this result is independent of the actual cumulative action function on this interval as long as $C_i(k)$ is a non-decreasing function. After substituting (15.3.8) into (15.3.6), we can minimize the resulting expression with respect to the breakpoints t_j. The resulting solution is given by the following set of K_i equations:

$$C_i(t_j) = \tfrac{1}{2}\bigl(C_i(F_j) + C_i(F_{j+1})\bigr) \quad \text{for } j = 1, \ldots, K_i \tag{15.3.9}$$

The interpretation of this set of equations is that the breakpoint t_j is chosen such that the value of the function $C_i(k)$ at that breakpoint is the average of the $C_i(k)$ values at the key frames preceding and following that breakpoint. Together, (15.3.8) and (15.3.9) form the solution of the desired key-frame distribution according to the criterion (15.3.6). To solve the key- frame positions from (15.3.8) and (15.3.9), one can employ a recursive search algorithm. To this end, we rewrite (15.3.9) as follows:

$$C_i(F_{j+1}) = 2C_i(t_j) - C_i(F_j) \quad \text{for } j = 1, \ldots, K_i \tag{15.3.10}$$

If we start with assuming a breakpoint t_1, then we can compute key frame F_1 using (15.3.8). From (15.3.10) we can then compute breakpoint t_2 (substitute $j=1$ in (15.3.10)). Subsequently, from t_2 we can compute F_2 using (15.3.8), from which

t_3 follows (substitute $j=2$ in (15.3.10)). In this way we can recursively compute for the assumed value of t_1 the value of t_{k_K}, which should be identical to the given length of the i-th shot. Depending on the mismatch between the computed and actual value, the position of the breakpoint t_1 can be adjusted. Note that this recursive search procedure is very close to the one often used for designing scalar quantizers [Max60].

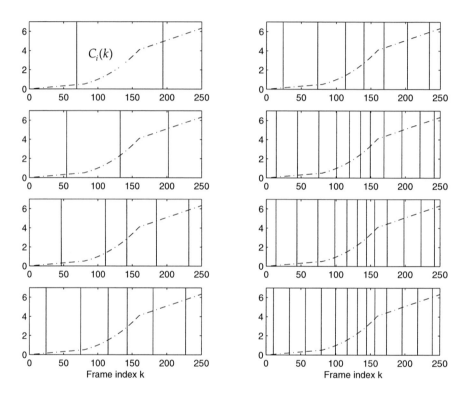

Figure 15.6: *Distribution of different number of key frames along a fictive video shot with a variable rate of visual-content variations.*

15.3.4 Experimental validation

The major issue in the key-frame extraction approach presented in this section is related to distributing a given number of key frames along a shot, such that the best possible summarization of the visual content of a shot is obtained. We therefore concentrate here on testing the optimization process (15.3.6) in the controlled situation. Visual content variations along an arbitrary shot i are modeled by two artificially produced functions $C_i(k)$. The form of the first function is given in the diagrams of Figure 15.6 and indicates that there is a constant low rate of visual content variation in the beginning of the shot, followed by an exponentially increasing variation rate, while the shot ends with a segment having the constant variation rate, but one that is higher than in the first shot segment. The exponential form of the second function, shown in diagrams of Figure 15.7, indicates a steadily increasing rate of visual-content change, for instance, in the case of an accelerated camera panning, tilting or tracking.

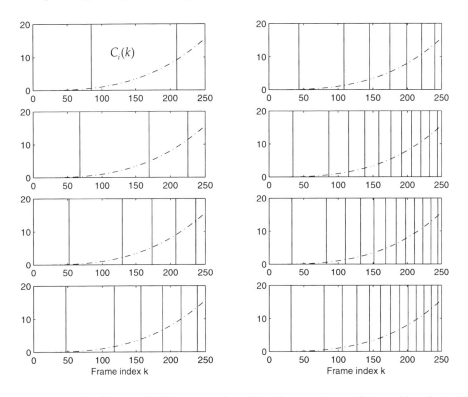

Figure 15.7: *Distribution of different number of key frames along a fictive video shot with exponential visual-content variations.*

Independent of the number K_i of assigned key frames, the varying key-frame density along shot i should follow the visual-content variations modeled by the function $C_i(k)$. Furthermore, in shot segments with a constant variation rate, key frames are distributed homogeneously and all shot segments need to be represented. The last requirement should prevent that all or a large majority of key frames are concentrated on one small shot segment, while the rest of the shot's visual material is not captured by key frames.

We show in Figures 15.6 and 15.7 the results of distributing 2, 3, 5, 6, 7, 11, 12 and 13 key frames along a shot i for both modeling functions $C_i(k)$. In all cases, key frames – represented by vertical lines - were distributed as expected. On the one hand, it can be seen how in each case the concentration of key frames follows the dynamic of the function $C_i(k)$. On the other hand, from the fact that key frames are always distributed along the entire shot, it can be concluded that the visual content of all shot segments is well captured in each of the key-frame sets.

15.4 Key-frame extraction based on cluster-validity analysis

The objective of the key-frame extraction method presented in this section is to minimize the redundancy among video frames and provide a set of key frames for a given video sequence, which is similar to the one based on human cognition. While in the method from Section 15.3 key frames are extracted separately for each shot, the extraction procedure described here can be applied to a sequence containing an arbitrary number of shots. Furthermore, the method presented in this section does not require any human supervision or parameter (threshold) specification. This makes the extraction procedure very user friendly and it supplies the user with a stable quality of obtained key frames for any arbitrary sequence.

The visual-content redundancy is reduced here by applying a partitional clustering [Jai88] to all video frames. The underlying idea is that all frames with the same or similar visual content will be clustered together. Each cluster can be represented by one characteristic frame, which then becomes a key frame of a sequence, capturing all the visual material of that cluster. Since frames in different clusters contain different visual material, the redundancy among obtained key frames is low. At the same time, all variations of the visual material along a sequence is captured in its key-frame set.

Consequently, the problem of finding the optimal number of key frames for a given sequence is reduced to finding the optimal number of clusters in which the frames of a video can be classified based on their visual content. The main difficulty here is that the optimal number of clusters needs to be determined automatically. To solve this, we apply known tools and methods of cluster validity analysis and tailor them to our specific needs.

As illustrated in Figure 15.8, the extraction approach in this section consists of three major phases. First, we apply N times a partitional clustering to all frames of a video sequence. The prespecified number of clusters starts at 1 and is increased by 1 each time the clustering is applied. In this way N different clustering possibilities for a video sequence are obtained. In the second step, the system automatically finds the optimal combination(s) of clusters by applying the cluster-validity analysis. Here, we also take into account the number of shots in a sequence. In the final step, after the optimal number of clusters is found, each of the clusters is represented by one characteristic frame, which then becomes a new key frame of a video sequence.

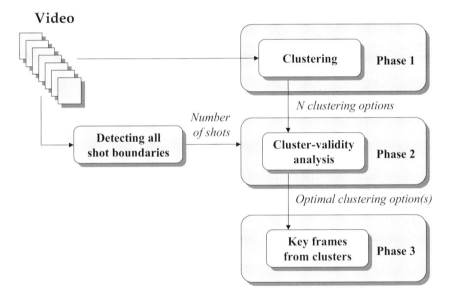

Figure 15.8: *Key-frame extraction scheme based on cluster validity analysis*

15.4.1 Clustering

The clustering process is performed on all video frames. For this purpose, each frame k of a video sequence is represented by a D-dimensional feature vector $\vec{\phi}(k)$, consisting of features $\varphi_v(k)$. The feature vector can be composed using texture, color, shape information, or any combination of those. Similarly as in the previous section, we wish to efficiently capture with key frames the changes introduced in the visual material, by e.g. camera panning, while the key frames must remain relatively insensitive to object motion. Therefore, we have chosen a D-dimensional feature vector, consisting of the concatenated $D/3$-bin color histograms for each of the component of the YUV color space. Furthermore, since $\vec{\phi}(k)$ is easily computable, we also compensate in this way for an increased computational complexity of the overall extraction approach due to the extensive cluster validity analysis, but still achieve an acceptable frame content representation. The feature vector used in this chapter is given as

$$\vec{\phi}(k) = \langle \varphi_v(k) | v = 1,..,D \rangle =$$
$$= \left\langle H_k^Y(1),..H_k^Y\left(\frac{D}{3}\right), H_k^U(1),..H_k^U\left(\frac{D}{3}\right), H_k^V(1),..H_k^V\left(\frac{D}{3}\right) \right\rangle \qquad (15.4.1)$$

By taking into account the *curse of dimensionality* [Jai82], we made the parameter D dependent on the sequence length. Now we compute it as $S/5$ [Jai82], whereby S is the number of frames to be clustered, and in this case also the number of frames in the sequence.

Since the actual cluster structure of the sequence is not known *a priori*, we first classify all frames of a sequence into 1 to N clusters. Thereby, the number N is chosen as the maximum allowed number of clusters within a sequence by taking into account the sequence length. Since each cluster corresponds to one key frame, the number N is equivalent to the maximum allowed number of key frames used in the previous section; here we use the same notation. Although N can be understood as a threshold parameter, its influence on the key-frame extraction result is minimal. This is because here we choose N much higher than the largest number of clusters to be expected for a given sequence. The longer the sequence, the higher is the potential number of clusters for classifying its video material. We found the variation of N with the number of sequence frames S defined by the function (15.4.2) suitable for the wide range of sequences tested:

$$N = N(S) = 10 + \text{int}\left(\frac{S}{25}\right) \qquad (15.4.2)$$

When we defined (15.4.2), we took into account that enough alternative options should be offered to the cluster validity analysis to obtain reliable results and that the number of options should increase with sequence length. On the other hand, the value N needs to be kept in limits, since the "noisy" clustering options become more probable with an increasing number of clusters and can negatively influence the cluster validity analysis.

After the clustering phase we perform a cluster-validity analysis to determine which of the obtained N different clustering options, i.e. which number of clusters, is the optimal one for the given sequence. In the following we will explain this procedure in full detail.

15.4.2 Cluster-validity analysis

For each clustering option characterized by n clusters ($1 \leq n \leq N$), we find the centroids c_i ($1 \leq i \leq n$) of the clusters by applying the standard k-means clustering algorithm to feature vectors (15.4.1) for all frames in the sequence. In order to find the optimal number of clusters for the given data set, we compute the *cluster separation measure* $\rho(n)$ for each clustering option according to [Dav79] as follows:

$$\rho(n) = \frac{1}{n} \sum_{i=1}^{n} \max_{1 \leq j \leq n \wedge i \neq j} \left(\frac{\xi_i + \xi_j}{\mu_{ij}} \right), \quad n \geq 2 \tag{15.4.3}$$

with the following parameters:

$$\xi_i = \left(\frac{1}{E_i} \sum_{v=1}^{E_i} \left| \vec{\phi}(v | v \in i) - \vec{\phi}(c_i) \right|^{\eta_1} \right)^{\frac{1}{\eta_1}}, \quad \mu_{ij} = \left(\sum_{v=1}^{D} \left| \varphi_v(c_i) - \varphi_v(c_j) \right|^{\eta_2} \right)^{\frac{1}{\eta_2}} \tag{15.4.4}$$

The better all of the n clusters are separated from each other, the lower is $\rho(n)$ and the more likely is that the clustering option with n clusters is the optimal one for the given video material. The value ξ_i is called *dispersion* of the cluster i, while μ_{ij} is the Minkowski metric [Fri67] of the centroids characterizing the clusters i and j. For different parameters η_1 and η_2, different metrics are obtained [Dav79]. Consequently, the choice of these parameters has also a certain influence on the cluster-validity investigation. We found that the parameter setting $\eta_1=1$ and $\eta_2=2$ gave the best performance for our purpose. E_i is the number of elements in the cluster i. Note that the $\rho(n)$ values can only be computed for $2 \leq n \leq N$ due to the fact that the denominator in (15.4.3) must be nonzero. We now take all $\rho(n)$ values measured for one and the same sequence and for $2 \leq n \leq N$, and normalize them by their global maximum. Three different cases are possible for the normalized $\rho(n)$ curve, as illustrated in Figure 15.9a-c.

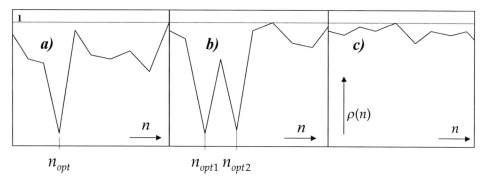

Figure 15.9: *Illustration of three possible cases for the normalized $\rho(n)$ curve.*

Case 1: The normalized $\rho(n)$ curve is characterized by a pronounced global minimum at $n=n_{opt}$, as shown in Figure 15.9a. This can be interpreted as the existence of n_{opt} clear natural clusters in the video material with $n_{opt}>1$. In this case, we assume a set of n_{opt} clusters to be the optimal cluster structure for the given video sequence.

Case 2: The normalized $\rho(n)$ curve has s distinct low values. This means that it is possible to classify the given video material into s different numbers of clusters with a similar quality of content representation. An example of this is illustrated in Figure 15.9b for $s=2$ with options containing n_{opt1} or n_{opt2} clusters.

Case 3: All values of the normalized $\rho(n)$ curve are high and remain in the same range (around 1), as illustrated in Figure 15.9c. This case can be interpreted twofold: either there is no clear cluster structure within the given video material (e.g. an action clip with high motion) or the video sequence is stationary and it can be treated as one single cluster. In the remainder of this chapter we will consider a sequence as *stationary* if there is no or only non-significant camera or object motion (e.g. a zoom of a person talking, characterized by head and face motion). In general, if $\rho(n)$ curve is obtained as shown in Figure 15.9c, the decision about the optimal cluster structure is made depending on the detected number of shots in that sequence.

As a result of the above, the problem of finding the optimal cluster structure for any video sequence given by the normalized $\rho(n)$ values for $2 \leq n \leq N$ is reduced to recognizing the most suitable of the three above cases. To be able recognize this, we first must sort all the normalized values $\rho(n)$, $2 \leq n \leq N$, in the ascending order, resulting in a sorted set $\rho_{sorted}(m), 1 \leq m \leq N-1$. Then, we introduce the reliability measure $r(m)$, $1 \leq m \leq N-2$, defined as:

$$r(m) = \frac{\rho_{sorted}(m)}{\rho_{sorted}(m+1)} \qquad (15.4.5)$$

Finally, we search for the value of the index m for which the function $r(m)$ has its minimum. Two possible results of the minimization procedure are given by the expressions

$$\min_{1 \leq m \leq N-2}(r(m)) = r(1) \qquad (15.4.6a)$$

$$\min_{1 \leq m \leq N-2}(r(m)) = r(s), \quad s \neq 1 \qquad (15.4.6b)$$

We will interpret these results for two different types of sequences, namely sequences containing several video shots and sequences corresponding to single video shots.

Sequences containing several video shots

We first analyze the situation involving sequences which contain more than one video shot. If there is a pronounced global minimum of the $\rho(n)$ curve at $n = n_{opt}$, as shown in Figure 15.9a, the reliability vector $r(m)$ has its global minimum at $m=1$. Therefore, the validity of (15.4.6a) is equivalent to the defined *Case 1*. Then, the optimal number of clusters is chosen as

$$n_{opt} = \min_{2 \leq n \leq N}(\rho(n)) \qquad (15.4.7)$$

If the equation (15.4.6b) is valid, the scope of possible options is constrained either to *Case 2* or to *Case 3*, where *Case 3* can be considered less probable for the following two reasons: On one hand, the probability that there is a highly stationary content across several consecutive shots is low. On the other hand, enough distinction of the visual material belonging to different shots of the sequence can be expected, so that – if not only one –also several equally acceptable clustering options can be allowed. Therefore, we relate the validity of (15.4.6b) in case of complex sequences to the defined *Case 2*. That is, all cluster sets belonging to $\rho_{sorted}(i)$, $1 \leq i \leq s$, are taken as possible solutions for grouping the frames of a sequence.

Single video shots

The probability that one finds a natural cluster structure containing more than one cluster in sequences consisting of only one video shot is generally much smaller than finding one in sequences containing several shots. This is because changes of the visual content within a shot are continuous, mostly characterized by a camera/object motion without dominant stationary segments. For this reason, a large majority of $p(n)$ curves obtained for single video shots can be expected to correspond to the model in Figure 15.9c. Hence, it is crucial that a reliable distinction can be made between stationary shots and non-stationary shots of which the natural cluster structure is unclear, as this is the basis of obtaining a suitable abstract structure for single video shots.

If n_{opt} clusters are suggested by (15.4.7) for a given shot, and if that shot is stationary, the average intra-cluster dispersion $\bar{\varepsilon}_{n_{opt}}$ computed over all n_{opt} clusters should be similar to the dispersion ε_{one} computed for one cluster containing all frames of that shot. Otherwise, the dispersion ε_{one} can be assumed to be considerably larger than $\bar{\varepsilon}_{n_{opt}}$. In view of this analysis, we define the decision rule (15.4.8) to distinguish stationary shots from the non-stationary ones. For this purpose we first use (15.4.7) to find n_{opt} clusters for a given shot and compute the dispersion $\bar{\varepsilon}_{n_{opt}}$. Then we also compute the dispersion ε_{one} and compare both with $\bar{\varepsilon}_{ref}$, which can be understood as the reference for the stationarity and is obtained by averaging dispersions measured for a large number of different stationary shots.

$$|\varepsilon_{one} - \bar{\varepsilon}_{ref}| \underset{stationary}{\overset{not\ stationary}{\gtrless}} |\bar{\varepsilon}_{n_{opt}} - \bar{\varepsilon}_{ref}| \qquad (15.4.8)$$

If the shot is stationary, it is represented by only one cluster, including all frames of a shot. With non-stationary shots we proceed by checking the evaluations (15.4.6a-b). If the equation (15.4.6a) is valid, n_{opt} is chosen as the optimal number of clusters, indicating that clearly distinguishable natural clusters exist within the shot. If (15.4.6b) is valid, we can either assume that there are several clustering options for the given shot, or that no natural cluster structure can be recognized by the algorithm. The first possibility is relatively improbable because the range of content variations within a shot of an average length is limited. Therefore, the validity of (15.4.6b) for a single shot is related to an unclear cluster structure, which is difficult to represent. On the one hand, one single cluster is too coarse, since variations of the visual content are present. On the other hand, choosing too many clusters would lead to an over-representation of the shot. For these cases we found the smallest number of clusters proposed by (15.4.6b) as a good solution for this problem. Thus, from s clustering options suggested by (15.4.6b), we choose n_{min} clusters, defined by (15.4.9), to represent a single video shot with an unclear cluster structure:

$$n_{min} = \min_{1 \leq i \leq s}(n_i) \tag{15.4.9}$$

15.4.3 Key frames from clusters

Once a suitable cluster structure is found for the given video sequence, one representative frame is chosen from each of the clusters and taken as a key frame of the sequence. As being usual in the clustering theory, we choose for this purpose the cluster elements being closest to cluster centroids. We find the key frame F_i of the cluster i by minimizing the Euclidean distance between feature vectors (15.4.1) of all cluster elements k and the cluster centroid c_i, that is

$$F_i \Leftarrow \min_{1 \leq k \leq E_i} \sqrt{\sum_{v=1}^{D} |\varphi_v(k) - \varphi_v(c_i)|^2} \tag{15.4.10}$$

15.4.4 Experimental validation

In order to test the video-abstraction method presented in this section, we concentrate here first on the evaluation of the proposed procedure for cluster-validity analysis, since both the key-frame sets and the preview sequences of a video abstract are directly dependent on the number and quality of obtained clusters.

We first tested the algorithm performance on sequences consisting of single video shots. For this purpose, we used 76 shots of a typical Hollywood-made movie and characterized them manually regarding the variations in their visual contents. The value of the parameter $\overline{\varepsilon}_{ref}$ from (15.4.8) was obtained experimentally as 0.0228, for which we used a number of stationary shots of different lengths and containing different visual material, and can therefore be assumed generally valid. As illustrated in Table 15.1, each of the shots belonging to the test set is assigned a description of how its content varies in time. From this description, the most suitable number of clusters for grouping all the frames of a shot is derived and used as a ground truth. For instance, a stationary shot should get assigned 1 cluster, and a shot with Q distinct stationary segments should get assigned Q clusters. For 66 shots (87%) of the test set, their frames were clustered in the same way as given by the ground truth.

Shot 2:	Frames 42-286	stationary with minor object motion (1 cluster)
...		
Shot 10:	Frames 1582-1751	slight zoom (1 or 2 clusters)
...		
Shot 24:	Frames 4197-4358	two stationary camera positions (2 clusters)
...		
Shot 29:	Frames 5439-5776	three stationary camera positions (3 clusters)
...		
Shot 45:	Frames 7218-7330	slow camera panning (1 or 2 clusters)
...		
Shot 51:	Frames 8614-8784	stationary camera, followed by a strong zoom (2 clusters)
...		

Table 15.1: *A fragment of the test set for evaluating the performance of the cluster-validity analysis algorithm for single shots*

In order to test the performance of the cluster-validity analysis algorithm for sequences containing several shots, we established a controlled test environment involving a set of sequences with a clearly defined structure in terms of the possibilities for clustering their frames. For each of these sequences we estimated the suitable number of clusters for organizing their visual content and used this estimation as the ground truth. An indication of the algorithm performance can be found in Table 15.2 for the following test sequences used:

- **Sequence 1:** A dialog between two movie characters. Due to two fixed camera positions, two clearly defined clusters are expected, one for each of the characters.
- **Sequence 2:** Three movie characters in discussion, with the camera showing each of them separately and all together. Four clear clusters are expected.
- **Sequence 3:** Two major camera positions to be captured by two clear clusters.
- **Sequence 4:** A long sequence covering different visual material in a series. Five clear clusters are expected for sequence representation

Although for the fourth sequence a clear cluster structure containing 5 clusters was expected, the algorithm suggested two possible clustering options. However, this was still acceptable, since the 5 clusters found corresponded to the expected ones and the option with 6 clusters contained the same clusters and an additional one, capturing a segment with object motion.

Test sequences	Expected number of clusters	Expected cluster structure	Obtained number of clusters	Obtained cluster structure
Sequence 1	2	Clear	2	Clear
Sequence 2	4	Clear	4	Clear
Sequence 3	2	Clear	2	Clear
Sequence 4	5	Clear	5,6	Unclear

Table 15.2: *Algorithm performance for some video sequences containing more than one video shot*

Based on the results of cluster-validity analysis, key-frame sets and preview sequences were formed. For each of the obtained clusters, a key frame was extracted using (15.4.10). Besides of the fact that in each case the obtained cluster combination corresponded to the one given by the ground truth, we also found the resulting key-frame set providing a good representation of the video content. This implies that frames nearest to cluster centroids are suitable to be used as key frames, and that the cluster-validity analysis is here the crucial step in making the video abstract.

15.5 Conclusions

After discussing the possibilities for automation of the key-frame extraction in the first section of this chapter, we presented in Sections 15.3 and 15.4 two methods by which key frames are automatically extracted for making a summary of a video's visual content. Both methods were developed such that the human intervention in dimensioning the extraction process is either limited to easily specified parameters or not necessary at all. In the method from Section 15.3, the maximal number N of key frames for the entire sequence is prespecified, while the approach from Section 15.4 is capable of functioning without human supervision. There, the value of the reference dispersion for stationary shots $\overline{\varepsilon}_{ref}$ found in subsection 15.4.4 can be used for measurements on a wide range of different sequences. Compared to the majority of key-frame extraction methods from recent literature, such a transparent parameter dependence makes the two approaches described in this chapter highly user-friendly.

Regarding the achieved visual-content representation, we first discuss in more detail the approach from Section 15.3. Two conclusions related to the ability of the method to summarize the visual content of a video can be drawn from the experimentally obtained key-frame distribution in Figures 15.6 and 15.7. First, the "sampling interval" between consecutive key frames is clearly dependent on the rate of visual content variation, i.e. on the steepness of the function $C_i(k)$. The higher the variation rate, the more key frames are used to capture the appearing new visual material. This indicates that all relevant elements of the visual content appearing in a shot will be represented in the resulting key-frame set. Second, although the sampling of the function $C_i(k)$ is generally not equidistant, key frames are always distributed such that the entire visual material of a shot is captured. This is opposed to an alternative where e.g. all K_i frames concentrate only on one shot segment. However, if the total number of key frames or any other threshold parameter is a constraint, it is difficult to prevent the cases of redundant key frames or to prevent ending up with too few key frames for a good sequence representation. Clear practical advantages of this method are the possibility of extracting key frames on-the-fly and of obtaining a good video summary and storyboard of a video, while keeping the amount of extracted information limited and closed to the prespecified one.

By using the extraction method presented in Section 15.4, one can obtain a very compact set of key frames for an arbitrary sequence, the quality of which is similar to a key frame set based on human cognition. Each frame selected using (15.4.10) can be assumed to have a high technical quality, since it corresponds to a cluster centroid, which is by definition the cluster element most similar to all other elements of that cluster. For that reason, having an "outlier" as a key frame, lying e.g. in a high-motion, in a blurred, dissolve or fade segment, is not as probable as having as a key frame a frame lying in a stationary, minimum-motion and maximum-clarity sequence segment. Although this method can be applied to a video segment of an arbitrary length, the segments of interest in this chapter are rather constrained to specific events, like for instance a dialog discussed before. The reason for this constraint is that long video segments are mostly characterized by an enormous variety in their visual contents, which is difficult to classify in a number of distinct clusters and, consequently, to represent by a limited number of key frames.

Chapter 16
High-Level Video Content Analysis

16.1 Introduction

Segmenting a video into shots, as discussed in Chapter 14, can be considered an elementary or a *low-level* video-analysis step. The reason for such a characterization is that this process, as well as the obtained results, do not depend on the actual content of the segmented video. In this chapter we concentrate on automatically analyzing a video at a higher level, at which *semantic video segments* can be distinguished.

As illustrated in Figure 16.1, the semantic video segments can be the reports in news programs, episodes in movies, highlights of sport events, topic segments of documentary programs, etc., and are concatenations of interrelated consecutive video shots. This indicates that the objective of high-level video content analysis can be formulated as finding *subsets* of all shot boundaries detected along a video, such that the series of consecutive shots, captured by shot boundaries belonging to these subsets, correspond to the semantic video segments of interest.

Autonomous systems able to analyze a video at a high (semantic) level can effectively be used to facilitate the user interaction with large volumes of video material stored in emerging digital video archives (libraries). Figure 16.2 illustrates how the results of high-level video analysis are used to organize the incoming or already stored video material, in order to provide access to semantic video segments of interest. The target applications of interest can be formulated as, e.g., search requests for all news reports on Bosnia, a movie episode containing the Alpine landscape or a favorite action scene, the "match point" of a tennis game, etc.

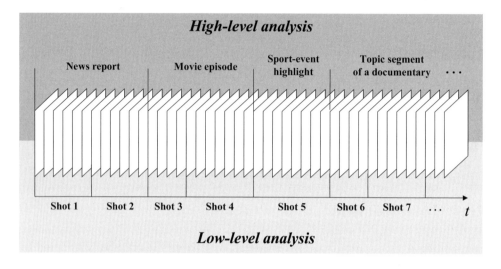

Figure 16.1: *Illustration of two different video-analysis levels*

Similarly as in the case of key-frame extraction, the possibilities for automation of high-level video analysis are not unlimited. The first problem, as already discussed in Chapter 13, is that embedding the human ability of understanding the content of a video into an autonomous system is technically not feasible. A technically feasible solution to this problem is to find ways of relating the video semantics to some specific temporal behavior of suitable low-level features. There are numerous examples, which can indicate the possibilities for developing such methods. Some of them are described in sections 16.2, 16.3 and 16.4 of this chapter, such as detecting TV commercials in various programs, recovering the semantic structure of a news program or detecting the episodes in movies. However, since low-level features are powerless in some cases, for instance, when extracting video segments where a specific actor or the "Alpine landscape" appears, realistic objectives need to be set when choosing the target applications. Thus, instead of attempting to develop algorithms capable of finding the movie episode where "Alpine landscape" appears, alternative algorithms are aimed at, which first find all episode boundaries of a movie, represent them by a number of key frames and then submit the entire episode structure together with the key frames to a browsing tool. There, a user can easily get an overview of the movie content by looking at episode representation, recognize the "Alpine landscape" in one of the key frames and quickly retrieve the corresponding episode. As it will be shown in sections 16.2 and 16.3, detecting episode boundaries in a movie is possible by analyzing only the temporal consistency of low-level visual features of a movie.

The second problem concerns the missing ground truth for the results of high-level video analysis. These are the cases of, for instance, extracting the highlighting or most memorable video segments: due to a highly subjective human perception of the video content in such cases, the dispersion among the results obtained by a subjective analysis of one and the same sequence by several users will be high [Paa97]. However, in many other analysis cases, the problem of missing ground truth is not present, for instance, when detecting TV commercials in an arbitrary video or segmenting news programs into reports. Therefore, only the latter cases are considered in this chapter, although some examples of extracting semantic segments with a "questionable" ground truth will be described in Section 16.2.

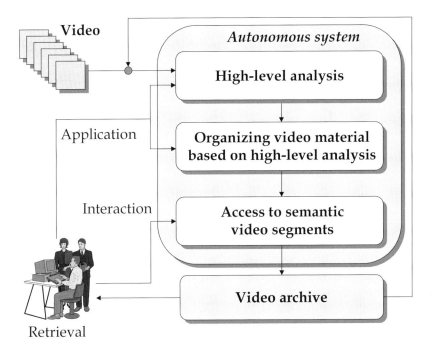

Figure 16.2: *High-level video analysis and related operations embedded into an autonomous system*

An important issue which needs to be taken into account when automating the high-level video analysis is that no generally applicable analysis methods exist. One reason for this is that the semantic video segments of interest vary for different program types, user environments and applications. Furthermore, the characteristics of such segments, in terms of low-level features being most suitable for their detection, vary over different programs. Therefore, rather a specific analysis methodology can be developed for each particular program type. We set as the objective of this chapter to develop methods for automated high-level analysis of two specific video types: *movies* and *news* programs. In the remainder of this section, we will explain our motivation for choosing these two particular program types, and give an overview of methods for their analysis developed in Section 16.3 and Section 16.4 of this chapter.

With respect to the discussion in Chapter 14, we witness a strong development of home video libraries and expect that the digital storage of video material at home will soon overtake the current analog video cassette recording systems [Oka93], [dWi92], [dWi93], [Yan93], and that the volumes of stored data in home video archives will rapidly grow in time. Stored in these archives we find programs of various types, such as movies, news, documentaries, TV shows, sport broadcasts, etc. In view of their large popularity with private users, it can realistically be assumed that now and in the future, movies belong to the most frequently stored programs, covering the highest percentage of the stored data volumes in home video archives. Although the major user interest regarding a movie is simply to watch it, some other applications involving movie content may be desirable to users as well. Such applications include, for instance, retrieving and watching of selected movie scenes, searching for a shot where an actor appears in a funny pose and watching a short preview of a movie. Although they are still new for a common user, these applications can be expected to become more and more popular with the emerging and quickly developing technology [SMA]. The most important objective in this development is to provide methods and tools for automatically analyzing movie content and providing the user with semantic video segments of interest, with minimal user involvement in the analysis process. This is understandable, since users at home want to be entertained; they do not want to be burdened with programming or adjusting their video equipment, especially not if this burden exceeds the level reached by some current VCRs that can be programmed in various ways, but are already too complicated for an average user.

Regarding the movie analysis in this chapter, we follow the objective of automatically providing semantically meaningful entry points into a movie. These points are ideally the boundaries between consecutive movie *episodes*. We define an episode as a series of consecutive shots unified by the same *chronological time-frame* of the story. Since we can base our episode-boundary detection only on low-level features, it is unlikely that the detected boundaries always correspond to the actual episode boundaries. For this reason, the results of our approach generally do not reveal movie episodes but their approximates, which we define as *Logical Story Units* (*LSUs*). Compared to episodes, which are defined by their semantic contents, LSUs are defined in terms of specific spatio-temporal features which were found to be characteristic for an episode. As it will be explained later, we found the global temporal consistency of the visual content of an episode a powerful means for defining an LSU as an episode approximation.

A news broadcast, which is the other program type considered in this chapter, has been widely recognized as a highly interesting "storing object" in emerging large-scale digital video databases [Boy99], [Che97]. The main reason lies undoubtedly in the information content of news programs, which may be useful for applications in many professional areas (e.g. education, journalism, government) as well as for private needs. One could think of building up large information archives containing all available sorts of informative programs, e.g. news, documentaries, TV-debates, political or social discussions, reportages, etc. In such archives, news is at least as important as all other mentioned program types, since it concisely covers huge amounts of topics related to society, daily politics, sports, business, etc. The importance of news programs may even be larger, since not all daily events get a thorough coverage through e.g. a dedicated documentary. Collecting news over a longer time period from different broadcasters can therefore provide a solid top level for an information collection, whereby other informative programs on certain topics, if any, are linked to relevant news reports and serve as lower-level (more detailed) information sources. If large information archives are to be used efficiently, all the information segments need to be organized, either according to their topics or to any other specific criteria. Here, the issue of automating the news-program analysis and reducing human interaction is a great challenge, and becomes more and more important, if not crucial, as increasing information volumes are stored in video archives. Such tools should be capable of autonomously segmenting a news program into reports, recognizing the report topic or fulfilling of the specified application criteria, and should classify it with all other closely related reports, enabling, in this way, direct execution of search requests, such as "find me a business report in a CNN news program from 2.4.1997", or "give me everything what is available on car races".

In this chapter we concentrate on developing methods for automatically detecting anchorperson shots in an arbitrary news program. Since these shots are directly related to news reports and since, in most cases, they directly determine the report boundaries, their detection can be considered as an important step in automatically recovering the report structure of a news program, and also in reaching the overall topic-based organization structure of a news archive. When developing our method, we made use of a specific visual structure of an anchorperson shot, which can also be found repeatedly in different segments of a news broadcast.

Before we present a method for detecting LSU boundaries in movies in Section 16.3 and a method for detecting anchorperson shots in news programs in Section 16.4, in Section 16.2 we give a brief overview of some of the methods reported in recent literature, which indicate the current possibilities in using low-level features in extracting semantic aspects out of different types of video. Conclusions belonging to this chapter can be found in Section 16.5.

16.2 Related work

16.2.1 Detecting different temporal events in a video

We start this section by discussing the method of capturing and characterizing a video by temporal events, such as dialogues, actions and story units [Yeu97]. The method consists of two major steps. In the first step, the *semantic labeling* of all video shots in a video is performed by applying *time-constrained clustering*. There, shots of a video are clustered based on their visual similarity and mutual temporal locality. In other words, two visually similar shots are not clustered together if they are too far from each other. G_i is the i-th cluster, the shots x, y and w are elements, d the distance between shots in terms of their visual similarity, T the maximum allowed temporal distance between two shots within the same cluster and δ the maximum allowed visual dissimilarity between two shots. The clustering procedure can be defined as follows:

- $\max_{w \in C_i} d(x, w) \leq \delta \quad, \forall x \in G_i$ (16.2.1a)

- $\max_{y \in C_i} d_t(x, y) \leq T \quad, \forall x \in G_i$ (16.2.1b)

- $d(x, w) > \delta \text{ or } d_t(x, w) > T, \; \forall x \in G_i, \; \forall w \in G_j, \; j \neq i$ (16.2.1c)

Assuming that all shots of a sequence are clustered in N clusters G_i, all shots within one cluster get assigned a *label*. Then, by replacing each shot of a video by its corresponding label, we can represent the entire video as a series of labels, that is, as

$$ABCADGHBACDKHDBAC\ldots \qquad (16.2.2)$$

In the second step, the label sequence (16.2.2) is investigated for different prespecified patterns appearing therein and corresponding to dialogs, actions, etc. For instance, dialog patterns are found as interchanging labels such as

$$A\underbrace{BAX}_{\text{Dialog}}Y Z\underbrace{ABABAB}_{\text{Dialog}}C\underbrace{DEFEDE}_{\text{Dialog}}GHI\ldots \qquad (16.2.3)$$

Or, as another example, action patterns are characterized by a series of shots with contrasting visual contents, expressed by no or only a minimal repetition of shot labels, that is

$$ABCDEFBGHI\ldots \qquad (16.2.4)$$

16.2.2 Detecting scene boundaries in a movie

In [Ken98], the authors consider a movie as a series of consecutive *scenes* and propose an approach for finding probable boundaries of scenes. The approach is based on investigating the *coherence* measured along a series of consecutive shots and representing the consistence of the visual material contained therein. We first introduce the *recall* between shots s_m and s_n as $\mathrm{SRecall}(s_m, s_n)$ being proportional to the function $Sim(s_m, s_n)$ describing their visual similarity and the function $TR(s_m, s_n)$ taking into account their lengths and their relative temporal positions within a video, that is

$$\mathrm{SRecall}(s_m, s_n) = Sim(s_m, s_n)\, TR(s_m, s_n) \qquad (16.2.5)$$

Then we define the total recall of all the shots older than the boundary by all the shots newer than the boundary as

$$\mathrm{Recall}(s_i, s_{i+1}) = \sum_{m<i}\sum_{n>i+1} \mathrm{SRecall}(s_m, s_n) \qquad (16.2.6)$$

The coherence at the boundary between shots s_i and s_{i+1} is now computed as the total recall $\mathrm{Recall}(s_i, s_{i+1})$ normalized by the maximum potential recall $\mathrm{Ideal}(s_i, s_{i+1})$ possible at that boundary, that is

$$\text{Coh}(s_i, s_{i+1}) = \frac{\text{Recall}(s_i, s_{i+1})}{\text{Ideal}(s_i, s_{i+1})} \qquad (16.2.7)$$

The maximum potential recall $\text{Ideal}(s_i, s_{i+1})$ is computed similarly as $\text{Recall}(s_i, s_{i+1})$, except that $\text{Sim}(s_m, s_n)$ in (16.2.5) is fixed at its maximum value of 1.

The significant local minima of the coherence curve measured along a sequence indicate the potential scene boundaries. A methodology for high-level video segmentation based on similar principles as the one from [Ken98] was published in [Han99b] and [Han99c] and is explained in detail in Section 16.3.

16.2.3 Extracting the most characteristic movie segments

As discussed in the introduction to this chapter, it is very difficult to develop methods which automate the detection of semantic content elements for which no clear ground truth is defined. Therefore, in literature not many approaches can be found dealing with this problem. In [Pfe96] the *most characteristic* movie segments are extracted for the purpose of automatically producing a movie *trailer* (a short summary). Movie segments to be included in such a trailer are selected by investigating the specific visual and audio features and by taking those segments which are characterized by *high motion* (action), *basic color composition* similar to average color composition of a whole movie, *dialog-like audio track*, and *high contrast*. It is claimed that this method yields good quality movie abstracts, since "all important places of action are extracted" [Pfe96].

16.2.4 Automated recognition of video genres

The method for detecting video types (genres) presented in [Fis95] is a good example of an attempt to obtain some conclusions related to an extremely high abstraction level of a video by simply investigating its low-level features. The proposed approach consists of three steps. In the first step, the syntactic properties of a digital video, such as color statistics, shot-boundaries, motion vectors, simple object segmentation and audio-statistics, are analyzed. The results of the analysis are used in the second step to derive *video-style attributes*, such as shot lengths, camera panning and zooming, types of shot boundaries (abrupt ones vs. dissolves, fades, etc.), object motion and speech vs. music, which are considered to be the distinguishing properties for video genres. In the final step, an "educated guess" is made about the genre to which the video belongs, based on a mapping of the extracted style attributes with those corresponding to different prespecified genres.

Experiments were reported using a number of sequences which were to be classified in one of the following genres: news, car races, tennis, commercials and animated cartoon. It is interesting to see in which way the style attributes were related to a particular genre. For instance, for a news program, the appearance of interchanging low- vs. high-motion video segments is investigated. There, low-motion segments correspond to anchorperson shots, which are separated by high-motion report segments. Also, a distinction is made between the anchorperson and some other "talking head" through the requirement that the periodically appearing low-motion segments need to be visually similar. This is done by computing and block-wise comparing the histograms of three subsequent low-motion segments. On the other hand, tennis is a good example of how audio can be used for detecting a video genre. As reported in [Fis95], a tennis game has a highly pronounced structure of the audio stream, characterized by interchanging "bouncing-ball" and speaker phases.

16.2.5 News-program analysis

We now move to high-level analysis of news programs. Due to their defined "container" structure, these programs are popular targets for developing content-analysis algorithms. The guiding objective when developing such algorithms is that these must automatically recognize the report structure of news programs and reach a topic-based organization of the news material on the system level with maximally reduced human interaction. While some of the proposed methods address this objective directly, many of them concentrate only on certain semantic aspects of a news program, which can be used at some later stages to reach the above objective. Examples are given in [Fur95], [Ari96], where the detection of anchorperson shots within a news program is performed.

Anchorperson shot detection using temporal shot characteristics

The approach to anchorperson-shot detection, presented in [Fur95], consists of three major steps. In the first step, potential anchorperson shots are found based on the fact that these shots are more or less stationary, compared to other shots within a news program. So, a shot is considered a candidate anchorperson shot if the following two expressions are valid, with μ and σ^2 being the mean and variance of discontinuity values $z(k,k+1)$, measured along a shot:

$$\mu < T_1$$
$$\sigma^2 < T_2$$
(16.2.8)

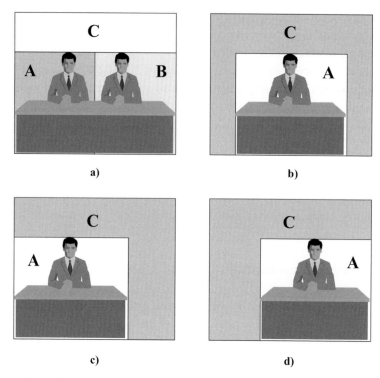

Figure 16.3: *Spatial structure models of four different types of anchorperson shots*

The second step is performed by taking a candidate anchorperson shot and analyzing the temporal changes in regions *A, B* and *C*, indicated on four characteristic types of anchorperson shots in Figure 16.3. Changes between consecutive frames along these shots are expected in frame regions where the speakers are, that is, in regions *A* and *B*. Opposed to this, no motion should be registered in regions *C*. These conditions can mathematically be formulated as follows:

$$\begin{aligned} \mu_A > T_3 > 0 & & \sigma^2_A > T_4 > 0 \\ \mu_B > T_3 > 0 & \quad \text{and} \quad & \sigma^2_B > T_4 > 0 \\ \mu_C \approx 0 & & \sigma^2_C \approx 0 \end{aligned} \qquad (16.2.9)$$

If a candidate anchorperson shot fulfills the conditions (16.2.9) for any combinations of regions, as indicated in Figure 16.3, it can be considered as an anchorperson shot.

Obviously, this procedure can also provide the information on the anchorperson-shot type (e.g. one of the four types from Figure 16.3). In view of this, after the first anchorperson shot is found, it is used to find all anchorperson shots of the same type among the remaining candidates. This is done by computing the average frame of the detected anchorperson (*model image*) and by comparing it to average frames of candidate shots. The second step of the procedure is repeated until model images of all anchorperson-shot types appearing in a news broadcast are computed and all anchorperson shots have been detected.

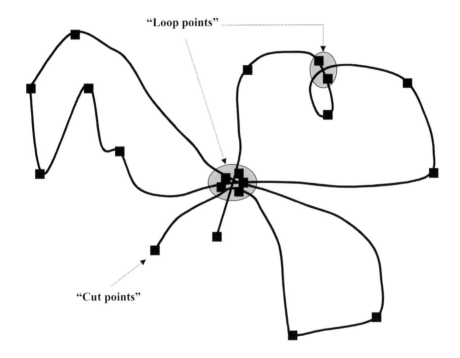

Figure 16.4: *"Cut points" and "loop points"*

Anchorperson-shot detection using planar graphs

The method for anchorperson-shot detection proposed in [Ari96] uses the results of shot boundary detection to detect the appearance of an anchorperson. The first frame of each detected shot is extracted and considered as a "cut point". Then, a planar graph is formed with the cut points as nodes and the shots as edges connecting each two nodes. Under the assumption that each news report starts and ends with the same anchorperson shot, a loop structure can be assumed within the obtained graph, where each loop corresponds to a report starting from one node (anchorperson) and ending in the same node. This is illustrated in Figure 16.4. Then, in order to detect all starting frames of anchorperson shots, nodes forming the loop points need to be detected. For this purpose, a threshold is defined and all nodes with distances smaller than the threshold are considered as starting frames of anchorperson shots.

Recovering news-program structure by combining different media

An attempt to automatically recover the entire semantic structure of a news broadcast can be found in [Hua99b]. The proposed approach for high-level news analysis is based on utilizing cues from different media and has the objective of recovering semantic segments from broadcast news at different levels of abstraction. The authors observe a hierarchy of a typical news program, which consists of four semantic levels. At the lowest level, a news program can be split into news material and commercials. Then, within the news material, anchorperson shots can be separated from shots taken outside the studio. Here, the anchorperson shot usually introduces and summarizes a report, which is followed by detailed reporting from a site. At the next level, anchorperson shots and related shots from different sites can be merged into reports.

In order to recover the first hierarchy level, news is separated from commercials by registering the changes in the audio-waveform, which are mainly caused by the background music in the commercials. In the second step, the news material is classified into segments corresponding to anchorperson shots and the rest using text-independent speaker recognition techniques. These techniques make it possible to distinguish an anchorperson segment from background speech coming from other sources (non-anchorperson shots) as well as from various audio segments (e.g. music in commercials). This step is meant to use the detected anchor's identity to hypothesize about a set of report boundaries that consequently partition the continuous text into adjacent blocks of text, each corresponding to a single report. In further steps this helps in obtaining higher levels of hierarchy by grouping the text blocks into reports.

16.2.6 Methods for analyzing sports programs

Instead of movies and news, some authors considered sports programs when they developed high-level video-analysis methods. The analysis approach presented in

[Sau97] uses spatio-temporal features to classify the video material of a basketball sequence in segments such as wide-angle and close-up views, fast breaks, steals, potential scores, number of ball possessions and possession times. For instance, shots are classified as wide-angles and close-ups, by an investigation of their motion intensity. While wide-angle shots are taken from a distance and are relatively stationary, close-up shots are highly dynamic, since the camera only shows a small portion of a scene and usually follows an object. The term "fast break" is defined as a "fast" movement of the ball from one end of the court to the other. In order to detect fast breaks, one accumulates the magnitude of the motion vectors along a sequence in such a way that the accumulation is reset to zero each time the motion changes direction. If the camera follows the ball during a fast break, a long and persistent pan is registered in these segments. Therefore, the search for fast breaks is actually the search for extremely long segments in the accumulation curve between two reset points. By exploring specific camera motion and lengths of corresponding video segments, one can also characterize steals and ball-possessions.

Also, as referred to in [Sau97], a system is developed in [Gon95], that can automatically parse TV soccer broadcasts. There, the standard layout of a soccer field was used to classify the video material into nine different categories, such as "around the left penalty line" or "near the top right corner".

16.3 Automatically segmenting movies into Logical Story Units

As already discussed in Section 16.1, we here present an approach for high-level movie analysis which was developed with the objective to provide semantically meaningful entry points into a movie. Although we envision such entry points as boundaries between consecutive movie episodes, detecting episode boundaries with great precision is difficult if only spatio-temporal features are used. Approximates of movie episodes captured by the boundaries detected using our approach are defined here as Logical Story Units. We start this section by justifying the episode boundaries as the meaningful semantic entry points into a movie. Then, we choose appropriate low-level features and define the LSU-boundary procedure such that the detected boundaries are as close to the actual episode boundaries as possible.

16.3.1 Hierarchical model of a movie structure

We first define a hierarchical model of a movie structure, which consists of three hierarchy levels, namely

- Shots
- Events
- Episodes

While shots are elementary "technical" temporal units of a video in general, we define an *event* as the smallest semantic segment of a movie. Such an event can be a dialog, an action scene or, generally, any series of shots unified by *location* or *dramatic incident*. However, an event does not need to be an unbroken series of consecutive shots; it can also alternate with another event. This is often used in the process of movie generation to represent several events taking place in parallel. Several alternating events are, all together, a good example of the highest semantic segment, which we define in this chapter as an *episode*. There, all events are unified by the same *chronological time frame* of the story and form a rounded context, which is in a certain sense separated from the neighboring contexts.

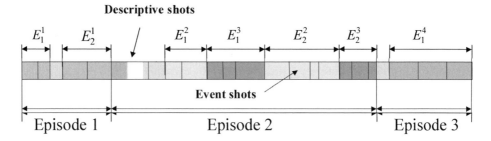

Figure 16.5: *Episodes 1 and 3 cover only one event and have a simple structure. Episode 2 covers two events, presented by their alternating fragments.*

An episode does not need to be related to several events; it can also concentrate on a single event. Since no shot within a movie is isolated but semantically it always belongs to a certain part of the story, each shot can be said to belong to one or to another episode. This implies that a movie can be understood as a *concatenation of episodes*. The hierarchical model of the movie structure, involving shots, events and episodes, is illustrated in Figure 16.5. There, we denote the fragment i of the event j by E_i^j. The model shows how an episode is built up around one movie event or around several of them taking place in parallel. Thereby a shot can either be a part of an event or it can serve for its "description" by, e.g., showing the scenery where the next or the current event takes place, showing a "story telling" narrator in typical retrospective movies, etc. In view of such a distinction, we further refer to shots of a movie as either *event shots* or *descriptive shots*.

Based on the above definitions, it can be said that if a movie is segmented into episodes, each boundary between two consecutive episodes provides an entry point into a new global segment of a story, having a rounded context and therefore being suitable for retrieval separately from the rest of the movie.

16.3.2 Definition of LSU

We now define the procedure of detecting the LSU boundaries such that they closely approximate the actual episode boundaries. In order to do this, we first analyze the characteristics of an episode and investigate the possibilities to efficiently capture them using suitable features.

It can realistically be assumed that an event is related to a specific location (scenery) and to certain movie characters. In other words, every now and then within an event similar *visual content elements* (scenery, background, people, faces, dresses, specific patterns, etc.) appear, and some of them even appear repeatedly. Since an episode is built around events, the same can be assumed for an episode as well; it is either related to only one event or to several of them alternating in time:

> **Assumption:** *An episode can generally be characterized by a global temporal consistency of its visual content, that is, by good matches of its visual-content elements found anywhere within a certain limited time interval.*

According to this assumption, approximate episode boundaries can be found by investigating the temporal behavior of visual low-level features. In this sense, we define the LSU as follows:

> *An LSU is a series of temporally contiguous shots which is characterized by overlapping links that connect shots with similar visual content elements.*

Since the definition of an LSU is based only on an assumption about the episode characteristics, which is not always fulfilled, the LSU boundaries do not exactly correspond to the episode boundaries in some cases. We will now explain some of the most characteristic problematic cases in view of the LSU definition and the movie-structure model in Figure 16.5.

For this purpose, we first investigate a series of shots a to j, as illustrated in Figure 16.6. Let the boundary between episodes m and $m+1$ lie between shots e and f.

We now assume that the shot *e*, although belonging to the episode *m*, has a different visual content than the rest of the shots in that episode. This can be the case if, e.g., *e* is a descriptive shot, which generally differs from event shots. Consequently, the content consistency could be followed by overlapping links in the *LSU(m)* up to shot *d*, so that the LSU boundary is found between shots *d* and *e*. If the shot *e* contains enough visual elements also appearing in the episode *m+1* so that a link can be established, *e* is assumed to be the first shot of the *LSU(m+1)* instead of shot *f*. This results in a *displaced* episode boundary, as shown in Figure 16.6. However, if no content-consistency link can be established between shot *e* and any of the shots from the episode *m+1*, another LSU boundary is found between shots *e* and *f*. Suppose that *f* is a descriptive shot of the episode *m+1*, containing a different visual content than the rest of the shots in that episode, so again no content-consistency link can be established. Another LSU boundary is found between shots *f* and *g*. If the linking procedure can now be started from shot *g*, it is considered to be the first shot of the new *LSU(m+1)*. In this case, not a precise *LSU boundary* is found but one that is spread around the actual episode boundary, where all places where the actual episode boundary can be defined are taken into consideration. Consequently, the shots *e* and *f* are not included in the LSUs, as shown in Figure 16.6.

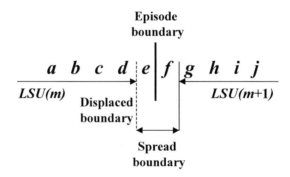

Figure 16.6: *Possible differences between an LSU and an episode boundary.*

We now proceed to define the LSU analytically, using the illustration of the LSU definition in Figure 16.7. The basis of the definition of an LSU given above is that a visual dissimilarity between two video shots can be measured. For now we assume that the dissimilarity $D(k,k+l)$ between the shots k and $k+l$ is quantitatively available. Then, three different cases can be distinguished, depending on the relation of the current shot k and the m-th LSU.

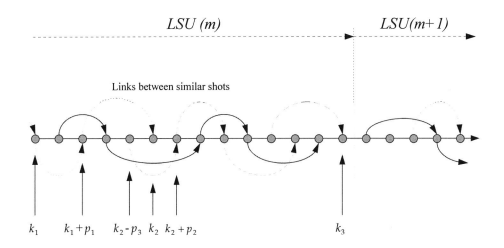

Figure 16.7: *Illustration of LSUs characterized by overlapping links connecting similar shots*

Case 1: Visual content elements from shot k_1 reappear (approximately) in shot $k_1 + p_1$. Then, shots k_1 and $k_1 + p_1$ form a linked pair, illustrated in Figure 16.7 by the arrow. Since shots k_1 and $k_1 + p_1$ belong to the same $LSU(m)$, consequently all intermediate shots also belong to $LSU(m)$:

$$[k_1, k_1 + p_1] \in LSU(m) \quad \text{if} \quad p_1 \Leftarrow \min_{l=1,\ldots,c} D(k_1, k_1 + l) < T(k_1). \quad (16.3.1)$$

Here c is the number of subsequent shots (look-ahead distance) with which the current shot is compared to check the visual dissimilarity. The threshold function $T(k)$ specifies the maximum dissimilarity allowed within a single LSU. Since the visual content is usually time-variant, the function $T(k)$ also varies with the shot under consideration.

Case 2: There are no subsequent shots with sufficient similarity to shot k_2, i.e. the inequality in (16.3.1) is not satisfied. However, one or more shots preceding shot k_2 link with shot(s) following shot k_2 (see Figure 16.7). Then, we enclose the current shot by a pair of shots that belongs to $LSU(m)$, i.e.

$$[k_2 - p_3, k_2 + p_2] \in LSU(m)$$
$$\text{if } (p_3, p_2 > 0) \Leftarrow \min_{i=1,\ldots,r} \min_{l=-i+1,\ldots,c} D(k_2 - i, k_2 + l) < T(k_2). \quad (16.3.2)$$

Here r is the number of shots to be considered preceding the current shot k_2 (look-back distance).

Case 3: If for the current shot k_3 neither (16.3.1) nor (16.3.2) is fulfilled, and if shot k_3 links with one of the previous shots, then shot k_3 is the last shot of *LSU(m)*. This can also be seen in Figure 16.7.

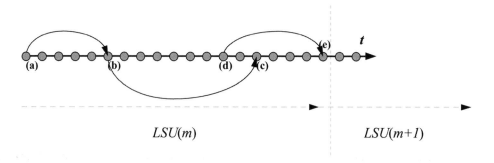

Figure 16.8: *Illustration of the LSU boundary-detection procedure. The shots indicated by (a) and (b) can be linked and are by definition part of LSU(m). Shot (c) is implicitly declared part of LSU(m) since the shot (d) preceding (c) is linked to a future shot (e). Shot (e) is at the boundary of LSU(m) since it cannot be linked to future shots, nor can any of its r predecessors.*

16.3.3 Novel approach to LSU boundary detection

The objective is to detect the boundaries between LSU's, given the definition of an LSU and the concept of linking shots described by Cases 1-3 from the previous section. In principle one can check equations (16.3.1) and (16.3.2) for all shots in the video sequence. This, however, is computationally intensive and also unnecessary. According to (16.3.1), if the current shot k is linked to shot $k+p$ (link between shots (a) and (b) in Figure 16.8), all intermediate shots automatically belong to the same LSU, so they need not to be checked. Only if no link can be found for shot k (shot (c) in Figure 16.8), it is necessary to check whether at least one of r shots preceding the current shot k can be linked with a shot $k+p$ (for $p>0$, as stated in (16.3.2)). If such a link is found (link between shots (d) and (e) in Figure 16.8), the procedure can continue at shot $k+p$; otherwise shot k is at the boundary of *LSU(m)* (shot (e) in Figure 16.8). The procedure then continues with shot $k+1$ for *LSU(m+1)*.

In order to determine whether a link can be established between two shots, we need the threshold function $T(k)$. We compute this threshold recursively from already detected shots that belong to the current LSU. For this purpose we define the *content inconsistency value* $u(k)$ of shot k as the minimum of $D(k,n)$ found in (16.3.1) (or in (16.3.2) if (16.3.1) does not hold), that is

$$u(k) = \begin{cases} D(k_1, k_1 + p_1) & \text{if (4.3.1) holds} \\ D(k_2 - p_3, k_2 + p_2) & \text{if (4.3.2) holds} \end{cases} \tag{16.3.3}$$

Then the threshold function $T(k)$ we propose is:

$$T(k) = \frac{\alpha}{N_k + 1}\left(\sum_{i=1}^{N_k} u(k-i) + u_0\right) \tag{16.3.4}$$

Here α is a fixed parameter whose value is not critical between 1.3 and 2.0. The parameter N_k denotes the number of links in the current LSU that have led to the current shot k, while the summation in (16.3.4) comprises the shots defining these links. Essentially the threshold $T(k)$ adapts itself to the content inconsistencies found so far in the LSU. It also uses as a bias the last content inconsistency value u_0 of the previous LSU for which (16.3.1) or (16.3.2) is valid.

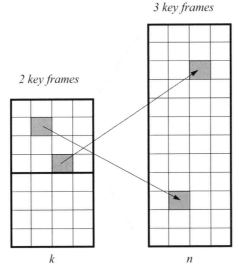

Figure 16.9: *Comparison of shot k with shot n by matching HxW blocks from each key frame of shot image k with shot image n. Shot k had 2 key frames and shot n had 3 key frames.*

16.3.4 Inter-shot dissimilarity measure

The LSU detection algorithm and the computation of the threshold function require the use of a suitable dissimilarity function $D(k,n)$. We assume that the video sequence is segmented into shots, and that each detected shot is represented by one or multiple key frames so that its visual information is captured in the best possible way.

For each shot, all key frames are merged in one large variable-size image, called the *shot image*, which is then divided into blocks of HxW pixels. Each block is now a simple representation of one visual-content element of the shot. Since we cannot expect an exact shot-to-shot match in most cases, and because the influence of those details of a shot's visual content which are not interesting for an LSU as a whole should be as small as possible, we choose to use only those features that describe the blocks *globally*. In view of this we only use the average color in the $L*u*v*$ uniform color space as a block feature.

For each pair of shots (k,n), with $k<n$, we would now like to find the mapping between the blocks b_k and b_n, each being an HxW block from the shot image k and n, respectively, such that

- each block b_k in a key frame of shot image k has a unique correspondence to a block b_n in shot image n. If a block b_n has already been assigned to a block b_k of a key frame belonging to shot image k, no other block of that key frame may use it. All blocks b_n are only available when a new key frame of shot k is to be matched. Figure 16.9 illustrates this in more detail.

- the average distance in the $L*u*v*$ color space between corresponding blocks of the two shot images is minimized:

$$\min_{\text{all possible block combinations}} \sum_{\text{all blocks } b} d(b_k, b_n) \qquad (16.3.5)$$

where

$$d(b_k, b_n) = \sqrt{\left(L^*(b_k) - L^*(b_n)\right)^2 + \left(u^*(b_k) - u^*(b_n)\right)^2 + \left(v^*(b_k) - v^*(b_n)\right)^2} \qquad (16.3.6)$$

and where all possible block combinations are given by the first item. Unfortunately this is a problem of high combinatorial complexity. We therefore use a suboptimal approach to optimize (16.3.5). The blocks b_k of a key frame of shot k are matched in the unconstrained way in shot image n, starting with the top-left block in that key frame, and subsequently scanning in the line-fashioned way to its bottom-right block. If a block b_n has been assigned to a block b_k, it is no longer available for assignment until the end of the scanning path. For each block b_k the obtained match yields a minimal distance value, $d_1(b_k)$. This procedure is repeated for the same key frame in the opposite scanning fashion, i.e. from bottom-right to top-left, yielding a difference mapping for the blocks b_k and a new minimal distance value for each block, denoted by $d_2(b_k)$. On the basis of these two different mappings for a key frame of shot k and corresponding minimal distance values $d_1(b_k)$ and $d_2(b_k)$ per block, the final correspondence and actual minimal distance $d_m(b_k)$ per block is constructed as follows:

- $d_m(b_k) = d_1(b_k)$, if $d_1(b_k) = d_2(b_k)$ (16.3.7a)

- $d_m(b_k) = d_1(b_k)$, if $d_1(b_k) < d_2(b_k)$ and $d_1(b_k)$ is the lowest distance value measured for the assigned block in the shot image n (one block in shot image n can be assigned to two different blocks in a key frame of shot k: one time in each scanning direction) (16.3.7b)
  $d_m(b_k) = \infty$, otherwise. (16.3.7c)

- $d_m(b_k) = d_2(b_k)$, if $d_2(b_k) < d_1(b_k)$ and $d_2(b_k)$ is the lowest distance value measured for the assigned block in the shot image n (16.3.7d)
  $d_m(b_k) = \infty$, otherwise. (16.3.7e)

where ∞ stands for a fairly large value, indicating that no objective best match for a block b_k could be found. The entire procedure is repeated for all key frames of a shot k, leading to one value $d_m(b_k)$ for each block of a shot image k. Finally the average of the distances $d_m(b_k)$ of the B best-matching blocks (those with lowest $d_m(b_k)$ values) in the shot image k is computed as the final inter-shot dissimilarity value:

$$D(k,n) = \frac{1}{B} \sum_{B \text{ best matching blocks}} d_m(b_k) \qquad (16.3.8)$$

The reason for taking only the B best-matching blocks is that two shots should be compared only on a global level. In this way, we allow for inevitable changes within the LSU, which, however, does not degrade the global continuity of its visual content.

16.3.5 Experimental validation

We illustrate the performance of the proposed LSU boundary-detection approach with the example of two full-length movies which belong to quite different categories in view of their dynamics and the variety of their contents. The objective of the evaluation is to compare the obtained LSU boundaries with the actual episode boundaries and to investigate the consistency of results for both different types of movies.

Establishing the ground truth

In order to evaluate the performance of our segmentation procedure, we need reference episode boundaries, serving as a ground truth. Generally, such reference boundaries can be obtained if the information about the movie generation process is available, i.e. the movie script. Since such information was not available for our tests, the first step in the evaluation procedure was to obtain a set of reference boundaries which (closely) correspond to the ground truth. This was done by a number of test subjects, who manually segmented both movies in units which they believed to be episodes. The obtained segmentation results differed mainly in the number of episode boundaries that were detected; this was especially noticeable in the complex movie segments and can be explained by the fact that each subject perceived that episode to be constructed differently. On the basis of manual segmentation results, we defined two different classes of episode boundaries

- *Probable boundaries* – registered by all test subjects
- *Potential boundaries* – registered by some of the test subjects

In total, 19 probable and 17 potential boundaries were detected for the first movie and 26 probable and 16 potential boundaries for the second one. Since the *probable* boundaries were those all test subjects had selected, we considered them to be fundamental, and relevant for evaluating our detection method. This is not the case with *potential* boundaries, and they are, therefore, not considered in the boundary set belonging to the ground truth.

Parametrizing the LSU-boundary detection procedure

After establishing the ground truth, we had our algorithm perform the automatic segmentation of the movies for different values of parameters B and α. Thereby, we limited the range of the parameter α only to [1.4-1.5], while B, here expressed as a percentage of the total number of blocks in a shot, varied in the range 40-70%. We learned that taking less than 30% of the blocks makes the inter-shot comparison too coarse. On the other hand, more than 70% makes the comparison too detailed. Although both parameters determine the sensitivity of the detection procedure and, consequently, also the number and positions of detected boundaries, parameter B is more interesting since it defines the limits of inter-shot comparison, concerning both the amount of detail taken into account and how "global" this comparison should be. On the other hand, we left the parameters c and r, defined in (16.3.1) and (16.3.2), constant at values $c=8$ and $r=3$, since the segmentation results were fairly insensitive to the setting of these parameters. We represented each shot by two subsampled key frames, taken from the first and last shot segment. Dimensions of key frames were 88x72 and 80x64, and the parameters H and W determining the size of the blocks to compute (16.3.8) were chosen correspondingly, as 8.

Evaluation

We now evaluate the performance of the detection algorithm for each parameter pair (B, α). In view of the possible tolerable displacements between an LSU and the corresponding episode boundary (Figure 16.6), we consider here an automatically obtained LSU boundary as properly detected if it was close enough to the one detected manually. For this purpose we set the maximum tolerable distance to 4 shots. Any other automatically detected boundary was considered to be *false*. Also, if no LSU boundary was detected within 4 shots of the actual episode boundary, it was considered *missing*.

In order to quantitatively estimate the quality of the automated boundary detection for a certain parameter combination (B, α), we used the following expression:

$$Q = \frac{\text{Properly detected probable boundaries}}{1 + \text{Falsely detected boundaries}} \qquad (16.3.9)$$

The parameter Q denotes the quality of the boundary detection, depending on the number of properly detected LSU boundaries and the number of falsely detected ones for a given parameter combination. As it will be shown by the obtained experimental results, the quality parameter Q is rather sensitive to the number of falsely detected boundaries. This was also the main intention when we defined the function (16.3.9), since the objective of the detection procedure, presented in this section, is to provide semantically meaningful entry points into a movie. Such points can only be found at properly detected boundaries, while the number of falsely detected ones needs to be kept low. After computing the quality parameter Q for each parameter combination (B, α) belonging to ranges defined above, we sorted all values of Q and ranked them in descending order. The parameter combination having the largest Q gets the rank "1". Parameter combinations having the same value of Q are assigned the same rank.

B, α	MOVIE 1				MOVIE 2				Overall quality ranking
	Detected probable bounds	Detected potential bounds	Falsely detected bounds	Quality ranking	Detected probable bounds	Detected potential bounds	Falsely detected bounds	Quality ranking	
40, 1.4	11/19	2/17	0	(2)	18/26	6/16	4	(6)	(4)
40, 1.5	9/19	1/17	0	(3)	18/26	5/16	3	(4)	(3)
50, 1.4	12/19	3/17	0	(1)	19/26	4/16	2	(3)	**(1)**
50, 1.5	11/19	1/17	0	(2)	18/26	4/16	3	(4)	(2)
60, 1.4	14/19	4/17	1	(4)	19/26	4/16	2	(3)	(3)
60, 1.5	12/19	4/17	1	(6)	19/26	5/16	1	(2)	(4)
70, 1.4	14/19	6/17	2	(7)	21/26	4/16	1	(1)	(5)
70, 1.5	13/19	4/17	1	(5)	20/26	7/16	4	(5)	(6)

Table 16.1: *LSU boundary-detection results for different parameter settings. Bold numbers indicate the parameter combination providing the optimal overall detection performance. Combinations with the same Q values have been assigned the same ranking.*

The first column of Table 16.1 shows all parameter combinations (B, α) used in the experiments. The other columns show for each of the movies the number of probable and potential boundaries that were detected, the number of false alarms and the ranking for each parameter combination according to the computed detection quality Q. In the final step, ranks of all pairs (B, α) obtained for both movies have been added up and the obtained results have been sorted in ascending order. The parameter combination with the lowest sum of two ranks was assigned the overall rank "1" and considered as the optimal combination for both movies.

As shown by the overall ranking list in the last column of the table, the best performance for both movies is obtained when 50% of blocks are considered for computing the overall inter-shot difference value and when the threshold multiplication factor α is 1.4. It can also be observed that the quality of a parameter combination decreases the more it differs from the optimal parameter set. This is mainly due to the influence of parameter B: if less blocks are taken into account when (16.3.8) is computed, the inter-shot comparison becomes too global, resulting in an unacceptably low number of detected boundaries. On the other hand, the large number of blocks considered in (16.3.8) can make the boundary detection too sensitive, resulting in an increased number of falsely detected boundaries.

For the chosen optimal parameter combination $B=50\%$ and $\alpha=1.4$, the average percentage of detected *probable* boundaries is 69%, with only 5% of false detections. This is compatible with the requirement that, while as many boundaries as possible are properly detected, the number of falsely detected boundaries should be kept low, since they do not correspond to semantically meaningful entry points into the movie. However, absolutely seen, the obtained total percentage of 69% of properly detected boundaries for the optimal parameter combination is low. This is mainly the consequence of insufficient changes of visual features at certain episode boundaries or, in other words, of having two consecutive episodes each containing mutually similar visual content.

Table 16.1 also shows that the efficiency of the algorithm concerning the detection of probable and potential boundaries is not the same. The higher percentage of probable boundaries that were detected can be explained by the fact that those boundaries were characterized by a radical change of the scenery, which could easily be recognized by the algorithm. On the other hand, most of the potential boundaries were marked by some of the users in highly complex parts of the movies, where clearly distinguishing different episodes was a difficult task. Since our assumption about the temporal consistency of the visual content within an episode, i.e. its change at an episode boundary, was often not fulfilled in such complex movie segments, no good detection performance could be expected there.

16.4 Detecting anchorperson shots in news programs

A typical news report consists of one or several consecutive segments, each of them containing one or several concatenated video shots and belonging to one of the following categories:

- An anchorperson shot
- A news shot series (e.g. a series of shots taken by a reporter on a site, outside the studio)

Although the commercial segments can also be found in many news broadcasts, we do not consider them here, since they can easily be detected and separated from the actual news program by using any of the approaches proposed in recent literature (e.g. [Liu98]). In order to recover the next semantic level of a news-program structure, we must first classify the entire news material into one of the above two categories. Such classification is required since the beginning and the end of an anchorperson shot represents a potential report boundary which cannot be determined otherwise, e.g. by just analyzing the audio track of a news broadcast. After the classification is completed, the reports can be formed by merging related anchorperson shots and news shot series. A method to recover the report structure in this way can be found in [Hua99b] (explained in Section 16.2).

In this section, we concentrate on the problem of automatically detecting *anchorperson shots* in an arbitrary news program and propose a new approach for performing this operation. Compared to already existing anchorperson-shot detection methods described in Section 16.2, we believe our method can yield an increase in detection robustness, mainly due to the minimized usage of different thresholding parameters and, at the same time, maximal exploitation of inherent properties of the news program structure, related to anchorperson shots.

16.4.1 Assumptions and definitions

We base our anchorperson shot detection approach on the assumption that an anchorperson shot is the only type of video shots in a news program that has multiple matches of most of its visual content along the entire news program. Other (news) shots may match well only in their closest neighborhood (e.g. within a single report) where they can eventually find enough similar visual features. Such an assumption is realistic due to specific visual characteristics of anchorperson shots and their regular appearance along a news sequence. We also assume that the first anchorperson shot k_{ap} in a news program containing S video shots certainly appears within the interval $[1, N]$, where $N<S$ is assumed to be around 5 shots. In order to make the detection as robust as possible, we took into account different types of anchorperson shots, including non-stationary ones. We introduce now the following definition:

> *Anchorperson shots are visually characterized by studio background and by one or two news readers sitting at the desk, appearing separately or together, also with some possible variations of a camera angle and the magnitude of a zoom. These shots can be static or dynamic (containing some camera operations like zooming or panning). They all generally contain a certain (high) percentage of the same or similar visual features.*

During the detection procedure we compare video shots based on their *key frames*. Hereby, we assume that, prior to the anchorperson shot detection procedure, a news sequence has already been segmented into video shots, and that each shot is represented by a visual abstract consisting of a limited number of key frames. The proposed anchorperson shot detection approach consists of two steps:

- A threshold-free procedure of finding the sequence-specific template for anchorperson shots,
- Using the template to detect all anchorperson shots in a sequence by applying adaptive thresholding.

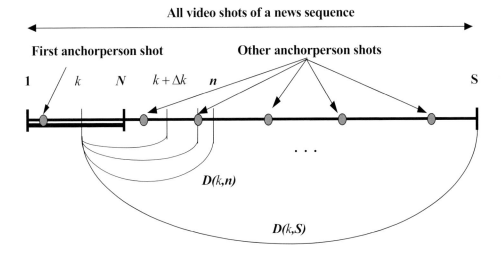

Figure 16.10: *Obtaining a dissimilarity values set for the shot k*

16.4.2 Finding a template

Based on assumptions made above, we start the procedure for finding the anchorperson- shot template by matching each shot $k \in [1, N]$, $N<S$, with all other news shots $n \in [k + \Delta k, S]$, as shown in Figure 16.10. In this way, a set of dissimilarity values $\{D(k, k + \Delta k), ... , D(k,S)\}$ is obtained for each shot k. The dissimilarity measure used here to compute values $D(k,n)$ compares two shots on basis of their abstracts (key frames) and is the same as the one used in the previous section. The "security" interval $[k, k + \Delta k]$ serves to avoid a possible good match of a news shot

with its surrounding shots and, consequently, to separate the shot k_{ap} even stronger from the rest. For each shot $k \in [1, N]$ we now take the P best matches (lowest values) from the set of dissimilarities and average them to compute the overall matching value. The shot with the lowest overall matching value is assumed to be an anchorperson shot, and is used as the template for finding all other anchorperson shots of a news sequence. With $k_{min,i,j}$ being the j-th of the P shots, between which and the shot k_i the lowest dissimilarity D is measured, we find the shot k_{ap} using the following expression:

$$k_{ap} \Leftarrow \min_i \sum_{j=1}^{P} D(k_i, k_{min,i,j}) \qquad (16.4.1)$$

16.4.3 Template matching

After the template has been found, again the inter-shot dissimilarity metric $D(k,n)$ is used on all shots of a sequence to test which are anchorperson shots. Low dissimilarity values will be obtained when the template is matched with another anchorperson shot. For each shot k of a sequence we now define its similarity with the template shot as

$$s(k) = \frac{1}{D(temp, k)} \qquad (16.4.2)$$

whereby $D(temp,k)$ is the dissimilarity between the template $temp$ and the shot k. In order to perform the detection of anchorperson shots automatically, we use the similar adaptive threshold $T(k)$ as in the previous chapter, defined here as the function of the similarity (16.4.2):

$$T(k) = \frac{\alpha}{N_k + 1} \left(\sum_{i=1}^{N_k} s(k-i) + s_0 \right) \qquad (16.4.3)$$

Here α is again a fixed threshold parameter, as in the previous section. The parameter N_k denotes the number of shots until k and since the last detected anchorperson shot. It also uses the similarity value s_0 computed before the last detected anchorperson shot as a bias. For each shot k, a value $s(k)$ is available as well as the threshold value $T(k)$. An anchorperson shot is detected when $s(k) > T(k)$.

16.4.4 Experimental validation

We now illustrate the performance of the developed algorithm on the example of two news sequences produced by different broadcasting companies and having the following global characteristics:

- *Sequence 1*: 12 minutes long, 5 anchorperson shots, one news reader, first appearance in the first sequence shot,
- *Sequence 2*: 25 minutes long, 17 anchorperson shots, two news readers, first appearance in the third sequence shot.

We represented each video shot by two subsampled key frames with sizes 165x144 for *Sequence 1* and 180x144 for *Sequence 2*. The parameter setting for both sequences was $N=5$, $P=3$, $\Delta k =25$ and $\alpha =3.1$. For computing the inter-shot differences (16.3.8) we chose the dimensions of the blocks in shot images $H=W=8$ and found 70% of all blocks in a shot image to be a good value for B. With this parameter setting we will now evaluate each of the two steps separately.

	Relative distance $\delta(\psi,\lambda)$	Total number of anchorperson shots	Detected anchorperson shots	False detections
Sequence 1	73 %	5	5	0
Sequence 2	17 %	17	17	1

Table 16.2: *Reliability evaluation of the template finding procedure and AP detection results*

On both sequences we applied the template-finding procedure and managed to find the proper template for each of them. Figure 16.11 shows the matching results of two template-candidates along the *Sequence 2*. We then measured the relative distance

$$\delta(\psi,\lambda) = 100\left(\frac{\psi}{\lambda}-1\right)\% \qquad (16.4.4)$$

between the chosen minimum overall matching value λ corresponding to the template, and the second smallest matching value ψ corresponding to the major other competitor shot for template selection. The larger the relative distance, the more reliable is the found template. Table 16.2 shows in its second column these relative distances for both sequences. The lower relative distance in the second sequence is most probably the result of the particular sequence structure, which shows an introduction for the coming reports after the first anchorperson shot. This introduction contains very similar visual information as the shots in the later parts of that sequence, which partially violates the assumptions made at the beginning of this section.

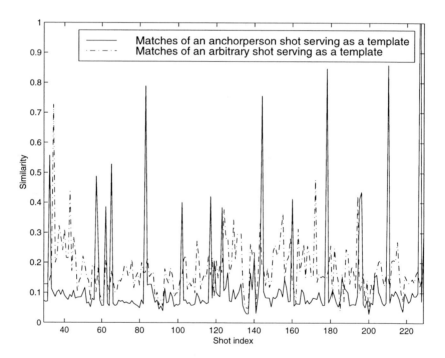

Figure 16.11: *Results of the matching procedure for two different templates $k \in [1, N]$ and shots $[k + \Delta k, S]$*

We then matched the found templates along the corresponding sequences to detect all anchorperson shots. The results of the template-matching procedure are given in the third and fourth column of Table 16.2 in terms of missed and false detections. Only one shot of *Sequence 2* was falsely interpreted as the anchorperson shot. This shot featured an interview between the news reader and a reporter outside the studio. Both the news reader and the reporter were positioned within their "windows" and the background of the screen in terms of its color composition fully corresponded to the studio background found in regular anchorperson shots. Similar color compositions were, thus, the most probable reason for a falsely interpretation of this shot as a regular anchorperson shot.

An idea about the robustness of the method presented in this section can be obtained by analyzing the types of anchorperson shots detected by each of the templates, and the visual content of a template itself. The first sequence contained three different variations of an anchorperson shot with one news reader. In some cases, the news reader was on the left side, zoomed in or zoomed out, with a news icon in the top right corner. In one of the shots, the news reader was in the middle of the screen and no news icon was present. This shot was also chosen as the template for *Sequence 1*.

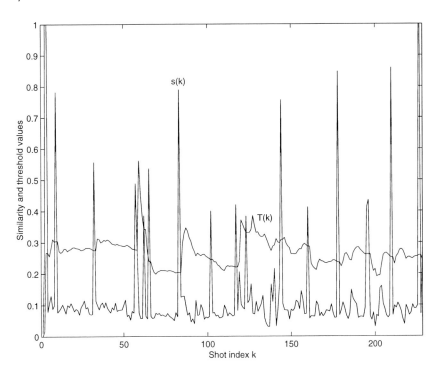

Figure 16.12: *Detection diagram for Sequence 2 and* $\alpha = 3.1$

All 17 anchorperson shots from *Sequence 2* were distributed as follows: 13 of them show the first (7 shots) or the second (6 shots) news reader on the right side of the screen with a news icon on the opposite side, 2 of them show the first news reader in the middle of the screen and no news icon, and 2 of them show both news readers together from two different camera positions. Two anchorperson shots were highly dynamic and characterized by a strong zoom from the studio to a news reader. As the template, one of the shots showing the first news reader on the right side of the screen was chosen.

Reliability of the detection process can be evaluated by analyzing the heights of the detection peaks in $s(k)$ curves. One such curve, corresponding to the second sequence, is shown in Figure 16.12 together with the adaptive threshold $T(k)$.

16.5 Conclusions

As already mentioned in the introduction to this chapter, the need for tools capable of automatically managing large amounts of information will steadily become larger with increasing volumes of video contents stored in emerging video archives. A high level of sophistication is required by such tools, since video material needs to be analyzed at the semantic level. The examples described in Section 16.2, as well as the methods for high-level analysis of movies and news in Sections 16.3 and 16.4, respectively, have shown a high potential of the low-level feature space in recovering the semantic information. This potential needs to be further exploited in the future.

In Section 16.3 an approach was presented for automatically segmenting movies into units which closely approximate actual movie episodes. The segmentation is based on an investigation of the visual content of a movie sequence and its temporal variations, as well as on the assumption that the visual content within a movie episode is temporally consistent. Consequently, an LSU is defined on the basis of overlapping links, which connect shots with similar visual content. We determine whether a link between two shots exists or not by applying an adaptive threshold function to shot dissimilarities. Based on the assumptions and definitions made in Section 16.3, the number of missed episode boundaries for a particular movie primarily depends on the degree with which an episode boundary corresponds to a large discontinuity in the global visual content flow. Similarly, the number of falsely detected boundaries is directly related to the global temporal consistency of the visual content within an episode.

Regarding the results in Table 16.1, it can be seen that, although the percentage of the detected LSU boundaries is relatively low, the large majority of all detected boundaries indeed provide meaningful entry points into a movie. This is because the percentage of non-meaningful entries (falsely detected boundaries) is low. Since this corresponds to the objective of the approach, the results obtained for the optimal parameter combination can be considered good. A strong improvement of the performance, in terms of increasing the percentage of properly detected boundaries, is not possible by using only the visual information. We expect that involving of the audio-track analysis into the proposed procedure will be helpful. Also, the results of applying the algorithm to two movies belonging to quite different movie categories did not differ much, indicating that the detection performance, and therefore also the defined LSU model, are sufficiently consistent for different types of movies. And, finally, as the proposed technique computes the detection threshold recursively, and only looks ahead at a limited number of shots, the entire process, including the shot-change detection, key-frame extraction, and LSU boundary detection, can be carried out in a single pass through a sequence.

Reports in a news program can be considered equivalent to episodes in a movie, since they can also be retrieved separately from the news program due to their rounded context. In this sense, a report boundary is the same type of a meaningful entry point into a news program as the episode boundary is for a movie. However, while episode boundaries can approximately be determined by investigating only the visual content of a movie, this cannot be said for the report boundaries. This is due to the fact that a news report is composed out of "lossy" shots, describing the report topic from different aspects and having generally a totally different visual content. Besides this, also no visual content can be related to a certain topic. An example for this is a report about a soccer match consisting of 4 higher-level segments: an anchorperson shot characterized by a news reader and a studio background, a series of shots from the soccer field, another anchorperson shot and the series of shots showing the press conference.

The furthest we can get by analyzing only the visual content of a news program is detecting the anchorperson shots. This is because anchorperson shots are characterized by a relatively constant visual content along the entire news program. A technique developed for the detection of anchorperson shots was demonstrated in Section 16.4. As shown by experimental results, the detection can be performed with acceptable reliability under the given assumptions. The most important assumption is that no shot of a news sequence other than anchorperson shot can be used to find P good matches along the entire sequence. And indeed, a definite probability for failure of this condition can be the major reason for lowering the algorithm's robustness in a general case, which can be observed on a lower relative distance for the second sequence in Table 16.2. We believe that this problem can be solved by further improving the inter-shot dissimilarity metric so that different types of anchorperson shots are distinguished better from the rest of the sequence, while at the same time it allows for variation among these types.

Chapter 17

Compression Trends: The "Fourth Criterion"

17.1 Introduction

For a long time, a considerable scientific and technical effort has been invested in the development and improvement of high-quality image and video compression with respect to three important "classical" criteria:

- minimization of bit rate,
- minimization of distortion in decompressed images and video
- reaching balanced and low computational costs on both the encoder and decoder side.

It is without a doubt that the excellent performance of the existing compression methodologies has strongly contributed to a fast growth of telecommunication networks, visual-information production, distribution and exchange. However, it becomes obvious that these methods will soon not be able to cope with all aspects being consequent to this growth. The most characteristic example of these aspects are content-based operations in large-scale image and video databases, such as video analysis and abstraction steps explained in Chapters 14 to 16, a query-by-example or a content-based classification. These applications require a high *content accessibility* for all images and videos stored in a database. In view of the analysis in Chapter 13 we define here the image and video content on the cognitive level as a set of content elements like objects, persons and sceneries captured by a camera as well as their spatio-temporal positions and mutual relations in an image or a video clip. Content elements are characterized by features including their color, shape, texture and (mutual) spatio-temporal position coordinates. Then, the content accessibility on the system level becomes analog to the efficiency of regaining the features of content elements which are important for a given content-based operation. This efficiency is low in case the existing compression methodologies are applied to images and videos stored in a database, because most of the features being interesting for content-based operations can be obtained only after performing complex decompression steps. Since query or classification generally involves large number of images or videos which all need to be decompressed first, the cumulative computational load resulting from the decompression alone can considerably increase the total interaction time between the user and a database and so decrease the user friendliness of a database. In order to illustrate this extra computational load we analyze a database of 100.000 JPEG compressed images. If each image needs to be decompressed prior to performing a query-by-example and if we assume that JPEG decompression lasts only for 0.01 second, there are 1000 seconds of extra time in interacting with the database due to decompression alone. Such increase in the interaction time in case of video query or classification is expected to be even higher in view of considerably larger amount of data contained therein compared to single images.

The problem of quickly accessing the content of compressed images and video has been known already for some years. The solutions towards speeding up the interaction with large volumes of compressed images or video were, however, so far mostly proposed in the way not to jeopardize the existing compression standards. In other words, a large majority of attempts to regain the content from compressed images or video were constrained by what the structural properties of e.g. the JPEG or MPEG format allowed. Since JPEG is based on a frequency transformation (DCT – Discrete Cosine Transform), methods were proposed in [Cha95a], [Cha95b] and [Smi94] for extracting some image features for retrieval applications in the frequency domain, that is, directly from the DCT-properties of JPEG-compressed images. Similarly to this, the authors in [Men95] and [Men96] propose techniques for extracting certain content elements directly from MPEG compressed video by exploiting properties of the MPEG video format. So in [Men95], an algorithm is proposed for detecting boundaries between neighboring video shots which requires minimal decoding of an MPEG compressed video and detects abrupt and gradual shot boundaries using DCT DC coefficients and motion vectors. This technique is extended in [Men96], where motion vectors are used for camera operation detection and estimation, such as zoom and pan, as well as for moving object detection and tracking. An indication about the increase in efficiency when performing the operations on images in a DCT-compressed domain directly is given in [Smi93] where the authors propose a series of methods for image manipulation without the need for their decompression. Very good results are reported, such as 50 to 100 times faster processing compared to the case where images need to be decompressed first.

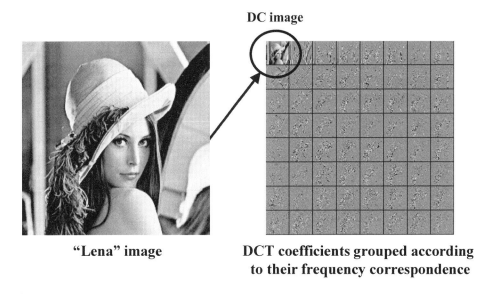

"Lena" image   DCT coefficients grouped according to their frequency correspondence

Figure 17.1: *Subsampled version of the original image, which can be obtained after partial image decompression*

As an alternative to feature extraction approaches described above, a fast access to content-related information in JPEG-compressed images and MPEG-compressed video can be provided by performing a partial decompression. Such decompression involves only the steps preceding the computationally expensive inverse DCT and results in a low-resolution (subsampled) version of an image or a video. As shown in Figure 17.1 on the example of a "Lena" image, the subsampled version obtained from a JPEG-compressed image consists of collected DC coefficients of all DCT image blocks. The method for obtaining a so-called *DC sequence* from MPEG-compressed video was proposed in [Yeo95b]. In that approach, DC images are created for all I frames and the approximations of subsampled P and B frames are obtained by performing a motion compensation in the DCT domain. The proposed approach results in a video sequence consisting of subsampled frames, which proved to be suitable for performing some basic steps of video analysis, such as shot-boundary detection, key-frame extraction and even for some higher-level analysis steps (Chapter 16). However, the possibilities for performing content-based operations on subsampled images and video frames are generally rather limited. This is mainly due to the missing high-frequency components and small dimensions of subsampled image versions (e.g. eight times smaller height and width in case of the JPEG-DC image or the MPEG-DC video frame), which allow for performing content-based operations only in view of some global content aspects of an image or a video, such as color composition and some dominant shapes (objects) or motion.

A possibility to provide an access to the content of images and video without having to analyze and process them first is to provide them a priori with side information containing so-called "content descriptors". These descriptors can be of various types, and are meant to represent certain aspects of the image or video content which may be of interest for a potential content-based operation, such as query or classification. Then, a content-based operation on images or videos from a database can be performed using weakly coded descriptors, without any need for decompressing the images or videos themselves. The development and standardization of suitable content descriptors for audiovisual information is the objective of MPEG-7 [ISO97]. Nevertheless, this alternative for providing a fast content access to compressed images and video has the disadvantage that the descriptors reveal only certain aspects of image content and cannot take into account all possible image features required for an arbitrary query or classification scenario, applied to a database. Thus, while it is highly practical for specific applications, this alternative is not sufficiently general to ensure unconstrained interaction with an image or a video database.

The first move in an entirely new direction regarding the compression of visual information was made in [Pen94b]. There, images in a large thematic database were, instead of compressing them using e.g. JPEG, represented by small sets of perceptually significant coefficients, making in this way direct search on compressed image content possible and introducing a new great challenge for the research community. In [Pic95b], providing a fast access to the content of compressed images and video has been proposed as the additional, *fourth optimization criterion* when developing new compression methods.

Figure 17.2 shows an idea of how to include the fourth criterion into the development of an image or video CODEC[*].

---

[*] CODEC is a common abbreviation for a joint COder-DECoder system

COMPRESSION TRENDS: THE "FOURTH CRITERION"

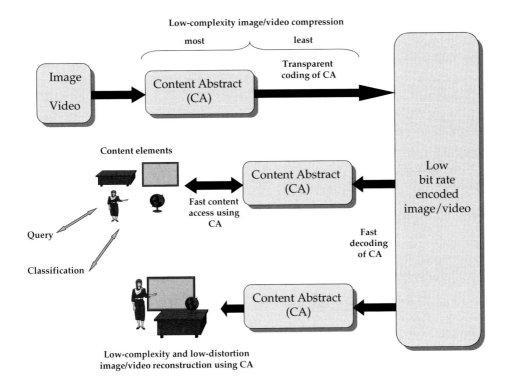

Figure 17.2: *Image/video compression enabling fast content access. Content is easily accessible by quickly decoding transparently encoded CA*

In this scheme, an arbitrary image or video is represented by its *Content Abstract* (CA) for which we set the following requirements:

- CA is considerably compacter than the original representation of an image or video,
- CA is easy to generate from the original image or video,
- It is possible to reconstruct the original image or video from CA quickly and with a low distortion,
- The characteristics of any content element being relevant to given content-based application (e.g. analysis, query or classification) can be regained from CA much more efficiently than if one of the existing compression methodologies (JPEG, MPEG) is applied.

If CA is sufficiently compact, a good compression of the original is guaranteed already by generating the CA. Figure 17.2 also shows the possibility of encoding the generated CA in order to further reduce the bit rate. This encoding needs, however, to be "transparent", i.e. such that only a small computational load is required for decoding the CA.

In the remaining sections of this chapter we present a newly developed image CODEC which by far complies to the scheme in Figure 17.2. The concept on which this CODEC is based is presented in Section 17.2. This is followed by an analysis of CODEC components in Section 17.3 and a performance evaluation in Section 17.4. The conclusions relevant to this chapter are given in Section 17.5.

17.2 A concept of an alternative image CODEC

We approach the development of our CODEC by considering the following issues. First, the definition of an image or video content from the previous section indicates that the content elements are to be searched for mainly in the spatio-temporal domain. This means that any transformation of an image or a video signal into a frequency domain, such as DCT or Wavelet, actually decreases the content accessibility in a general case. Exceptions can only be found by those content elements which are easily identifiable in the frequency domain as well. For this reason, we concentrate here on *spatial-domain* image compression techniques, examples of which are Vector Quantization (VQ) [Ger92] and Fractal Image Compression [Jac92]. Second, it is not realistic to expect that a full-resolution spatial (pixel-level) image content is available in the compressed domain if a good compression ratio needs to be obtained. This is simply because of the fact that the compression is based on reducing the redundancy and the irrelevancy of this content, so that only non-redundant and relevant content components are available in the compressed format. However, in order to obtain an increase in the efficiency of image-database operations we require that non-redundant and relevant image information contained in the compressed format is already usable for performing some of the image-database operations. We also require that the full-resolution spatial image content should quickly be reconstructable from its non-redundant and relevant elements, or in other words, that the complexity of the image-decompression procedure is considerably reduced if compared to transform-based decoders. In the following, we first recapitulate the principles of fractal image compression from Section 8.4.4 and describe vector quantization. Subsequently we choose the most suitable of the two techniques as the base for developing our CODEC.

In case of fractal compression, an image is first partitioned at two different levels: in range blocks of size NxN at the first level, and in domain blocks of size $2Nx2N$ at the second level. A transformed domain block is searched for each range blocks such that the mean square error between the two blocks is minimal. Hereby the following transformations are performed on the domain blocks: they are first subsampled by factor two to get the same dimensions as the range blocks. Then, eight isometries of subsampled domain blocks are found, including the rotated original block and its mirrored versions (mirroring over 0, 90, 180 and 270 degrees). Finally, an adjustment of the scale factor and the luminance offset is performed. Consequently, a fractal compressed image is defined by a set of relations for each range block, the index number and the orientation of the best fitting domain block, the luminance scaling and the luminance offset. Using this description, the decoder can reconstruct the compressed image by taking any initial random image and by calculating the content of each range block from its associated domain block. This reconstruction is repeated iteratively by taking the resulting image as a new initial image until the desired quality of the reconstructed image is reached.

As illustrated in Figure 17.3, compressing an image using VQ is the process of taking an image block of NxN pixels and finding its corresponding (most similar) block in a *code book*. A code book is a collection of representative blocks, constructed on the basis of a number of training images. Each image block is then represented by the code-book address, where the corresponding block is found. Consequently, a VQ-compressed image is simply a concatenation of addresses, collected for all image blocks. If the same code book is available at the receiver side, a VQ-compressed image can easily be decoded by filling in the blocks from a code book in the proper positions in the image, according to the addresses received by the decoder.

Because of the above descriptions of Fractal and VQ image CODECs, we find the CODEC based on Vector Quantization more suitable for our needs. First, it realizes image decompression as a fast "look-up-and-fill" procedure and involves no iterations. Second, VQ-compressed images can be compared and classified based on their block correspondences. This is because these correspondences directly depict the image content with respect to the code book used. Compared to this, a list of geometric and luminance transformations of domain blocks describing the Fractal-compressed images do not provide a clear impression about the image content and, therefore, cannot be used as efficiently for image-database operations as the block correspondences and code book of the VQ. However, not all the characteristics of the basic VQ scheme are suitable for direct usage for the CODEC development in this chapter. Therefore, we adapt the basic VQ scheme in order to better suit the applications addressed here.

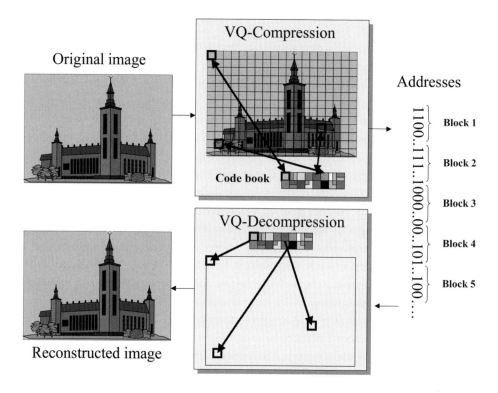

Figure 17.3: *Scheme of an image CODEC based on vector quantization*

The adaptation is related mainly to the highly complex and time-consuming process of code-book generation. This process basically includes a *partitional clustering* of the visual material collected from a set of *training images*. Its high complexity is due to a large amount of data to be clustered and due to the iterative nature of the clustering process. Consequently, the code book is made only once and used to compress and decompress all images in a database. It is also optimized to provide the maximal quality of all reconstructed images from that database. This optimization is performed such that, first, the training images are selected as the most representative for all the images contained in a database. Second, the clustering process is designed to take into account all linear and non-linear dependencies among blocks to be found in training images. Each cluster is then represented by one most representative image block, which then becomes an element of a code book. The described process of code-book generation by the basic VQ implies that the code book can be used effectively only for compressing images that belong to the same categories as those from the training set. This is, however, unpractical for applications in general image databases because of the following reasons. First, images can be very diverse, so that one single code book might not be sufficient for coding all of them with an acceptable quality. Second, if a database is extended by new images belonging to a different class, a time-consuming update of a code book is required. Third, image exchange among different databases (users) is difficult if different code books are used.

To provide a good solution to these problems, we apply in our CODEC a strongly simplified procedure for code-book generation, which – due to a reduced complexity - allows for generating a code book for each individual image. Using image-specific code books not only makes it unnecessary to perform highly complex code-book generation/update and to have one code book for the entire database; it has several other important advantages as well. First, the quality of reconstructed images can only improve since an image is abstracted and later reconstructed using the same (its own) blocks. Second, in contrast to the basic VQ, here the code book needs to be included in the compressed image format. As will be shown in Section 17.4, this makes it even easier to perform various image-database operations without the need for image decompression.

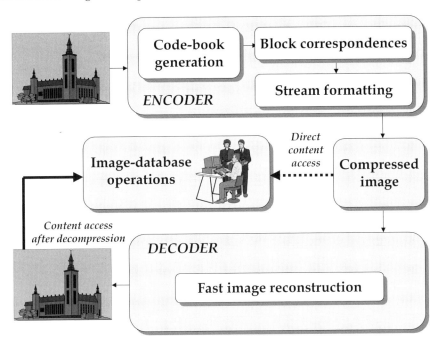

Figure 17.4: *Image CODEC enabling an easy content access in compressed images*

17.3 Image CODEC based on simplified VQ

Figure 17.4 illustrates all components of our new image CODEC in form of a block diagram. The first two steps of the encoder are the processes of making a code book and of finding the correspondences of image blocks with those belonging to a code book. Subsequently, a compressed image stream is formatted, where we only use as many bits as necessary to encode all the addresses in order to minimize the resulting bit rate. Apart from the fact that the code book used is image specific and therefore included into the compressed stream, the decompression process fully complies with the one of the basic VQ. As indicated in the scheme by the full arrow, a low computational complexity on the decoder side is already one possibility for a fast content access. The other possibility indicated by the dashed arrow is related to the direct usage of the image-specific code book and block correspondences for content-based operations.

In view of the scheme in Figure 17.2, the code book obtained for an image forms together with block correspondences the CA of that image. Similarly, the bit-stream formatting with the objective of minimizing the number of bits used for block correspondences can be understood as "transparent" coding of the CA. The issues regarding the compactness of obtained CA, the total computational load related to CA generation and image reconstruction from CA, as well as the content accessibility on the obtained CA will be discussed in detail in Section 17.4. In the following we proceed by defining all major components of the CODEC scheme in Figure 17.4.

17.3.1 Code-book generation

We first define an efficient methodology for generating an image code book. For this purpose, an image is first divided into non-overlapping square pixel blocks b_i with dimensions NxN, and each block is represented by the average color ($L^*u^*v^*$ color space) of all block pixels. We choose to work with relatively small blocks, i.e. with $N=2$ as the code book will have to be included into the compressed image format. The experiments have shown that if a similar quality of reconstructed images is to be achieved, and larger blocks, e.g. $N=4$, are used, the code-book size becomes unacceptably large (up to 20% of an image).

As shown in Figure 17.5, the code-book generation starts by including the first image block b_1 into the code book. Each further block along the arrow is compared with all blocks already in the code book. For this purpose, the Euclidean distance is computed for the three components of the average colors of blocks, that is

$$d(b_k, b_n) = \sqrt{\left(L^*(b_k) - L^*(b_n)\right)^2 + \left(u^*(b_k) - u^*(b_n)\right)^2 + \left(v^*(b_k) - v^*(b_n)\right)^2} \qquad (17.3.1)$$

A block b_i joins the code book if it cannot be matched well with any of already selected blocks, i.e. when the distance (17.3.1) between the block b_i and each block from the code book exceeds the threshold T.

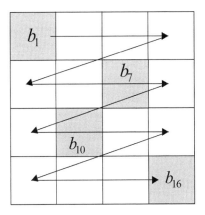

Figure 17.5: *Illustration of the simplified procedure for making a VQ code book. Grey blocks are included in the code book.*

While the code book of the basic VQ is obtained by a sophisticated procedure which optimally represents the visual material of training images, the major objective of the sequential procedure from Figure 17.5 is to *quickly* generate a code book. In order to achieve a code-book quality similar to that of the basic VQ, the described fast procedure for code-book generation requires some *fine tuning*. For this purpose, we make the threshold T locally adaptive, based on the following analysis. Since coding artifacts are particularly visible in smooth image regions (e.g. artifacts like *false contours*), the threshold function needs to be chosen such that these regions are represented by a sufficient number of code-book vectors. On the other hand, the number of blocks extracted from textured regions can be kept low since the coding artifacts are less visible there. This implies that it is convenient to make the threshold value at each block b_i dependent on the amount of texture present in its surroundings, that is

$$T = T(b_i) = f(\text{texture around } b_i) \qquad (17.3.2)$$

By properly choosing the threshold function (17.3.2), a number of code-book blocks can be extracted that is similar to using a fixed value of T. However, the extracted blocks are distributed better over an image, providing at a later stage a higher overall quality of the reconstructed image. In other words, the number of code-book vectors representing textured image regions slightly decreases in favor of those representing smooth regions.

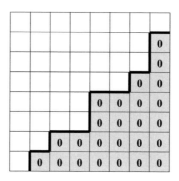

Figure 17.6: *Zero coefficients in a DCT block obtained by applying the quantization*

We now define the threshold function $T(b_i)$ by suitably modeling the variations of the amount of texture over an image. This is done by first dividing the gray-scale version of an image into nonoverlapping blocks B_j with dimensions 8x8 pixels, and by applying the Discrete Cosine Transform (DCT) to each of them. Then, all the elements of the DCT block are quantized according to the following procedure, which is analog to the one from JPEG:

$$\mathrm{ROUND}\left(\frac{DCT(u,v)}{Q(q)*W(u,v)}\right), \text{ with } Q(q) = \begin{cases} \dfrac{50}{q}, & q < 50 \\ \dfrac{100-q}{50}, & q \geq 50 \end{cases} \quad (17.3.3)$$

We call q the *quantization parameter* which can vary in the range $0<q<100$. Q is the gain factor depending on q and $W(u,v)$ is the corresponding element of the JPEG luminance quantization table.

As a consequence of the quantization, a number m_i of DCT coefficients of the block B_j will become zero, which is mainly the case with those corresponding to higher frequencies, as shown in Figure 17.6. The more texture is present in an image area, the stronger are the high-frequency components of the image signal in that area. Then, the DCT coefficients corresponding to these components are also high, and therefore will hardly ever become zero after quantization. This is not the case with smooth regions, where a large number of zero coefficients are present after the quantization. For this reason we relate the number m_i of zero-DCT coefficients to the presence or absence of a texture and formulate the threshold function as follows:

$$\text{if } b_i \in B_j \Rightarrow T(b_i) = p_1 - p_2 \left(e^{\frac{m_j - 64}{p_3}} \right) \tag{17.3.4}$$

Parameters p_1, p_2 and p_3 define the behavior of the threshold function and are to be specified experimentally. Since the parameter q directs the DCT quantization process (17.3.3), the threshold-function behavior can indirectly be adjusted by specifying a value for q. The higher the q, the lower is the gain factor Q, the smaller is the quantization step, the less DCT coefficients are zero and the threshold function (17.3.4) is shifted upwards.

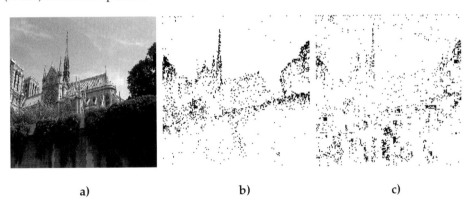

a)   b)   c)

Figure 17.7: *Code book extraction for an image using a constant and a variable threshold: a) original image, b) image blocks included in the code book by using a constant threshold, c) image blocks included in the code book by using the threshold function (17.3.4)*

Figure 17.7a shows an image from our test set, for which we generated two code books of a similar size. The first one is obtained by using a fixed threshold. The positions from which image blocks are taken and included into the code book are indicated as black spots in Figure 17.7b. Then we used the variable threshold to generate another code book, where we adjusted the function (17.3.4) by choosing a suitable q such that a similar number of blocks is extracted, as in Figure 17.7.b. The positions from where image blocks are taken using the threshold (17.3.4) are indicated by black spots in Figure 17.7c. It can be seen that the prevailing majority of blocks in Figure 17.7b are concentrated in high-texture regions, leaving the smooth regions insufficiently represented.

17.3.2 Finding the block correspondences

The step of generating the code book is followed by the search for correspondences between image blocks b_i and blocks c_j of the code book. In this way, each block b_i is represented by an address in the code book, which is embedded into the compressed image format and determines which block c_j is used to approximate block b_i during the image reconstruction in the decoder.

We find that block c_j corresponds to the image block b_i by comparing b_i with all blocks c_j using the distance function (17.3.1) and then by minimizing (17.3.1) for all indices j. As S is the total number of blocks b_i in an image, and M the number of code-book blocks, the procedure of establishing the block correspondences can analytically be formulated as

$$\forall i \in [1, S] \quad b_i \propto c_t \quad \Leftrightarrow \quad d(b_i, c_t) = \min_{1 \leq j \leq M} d(b_i, c_j) \tag{17.3.5}$$

17.3.3 Compressed image format specification

The format of an image, compressed using the CODEC presented in this chapter, is illustrated in Figure 17.8, and consists of the following information:

- File header
- Code book
- List of addresses for block correspondences

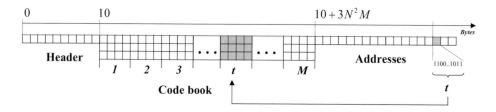

Figure 17.8: *Format of a compressed image file*

We will now define each of these components in detail.

File header

The file header contains 10 Bytes, which represent (in the order of appearance):

- the format identification mark (3 Bytes),
- image width and height (4 Bytes)
- the number M of blocks in a code book (3 Bytes)

Code book

All code-book blocks c_j, $j \in 1..M$ are addressed in the order in which they are extracted. In the compressed bit stream, they are represented by the RGB triplets of all of their pixels, ordered in the 1-dimensional uncompressed array of $3MN^2$ Bytes, that is

$$R^1_{11}G^1_{11}B^1_{11}...R^1_{1N}G^1_{1N}B^1_{1N}...R^1_{NN}G^1_{NN}B^1_{NN}R^2_{11}G^2_{11}B^2_{11}...R^2_{NN}G^2_{NN}B^2_{NN}...R^M_{NN}G^M_{NN}B^M_{NN} \quad (17.3.6)$$

Block correspondences

The code book is followed by the list of addresses for block correspondences. For the total number of S blocks b_i in an image, there are S addresses varying in the range 1..M. In order to reduce the size of this bit stream component, we use only so many bytes as are necessary to represent all addresses of characteristic blocks. For M blocks in a code book, the minimum required number w of bytes is computed as

$$w = \frac{1}{8}(\lfloor \log_2 M \rfloor + 1) \quad (17.3.7)$$

17.4 Performance evaluation

In this section we evaluate the performance of the developed image CODEC. We concentrate in Subsection 17.4.1 on the CODEC performance with respect to the obtained compression factor, the quality of reconstructed images and the overall computational costs. For this purpose we use a test-image set containing 54 different color images with dimensions 320x320 pixels. We experimentally found good parameter values in (17.3.4) as $p_1=5.35$, $p_2=4$, and $p_3=3$, and used them in our experiments. Subsequently, in Subsection 17.4.2 we evaluate the content accessibility for the CODEC we developed.

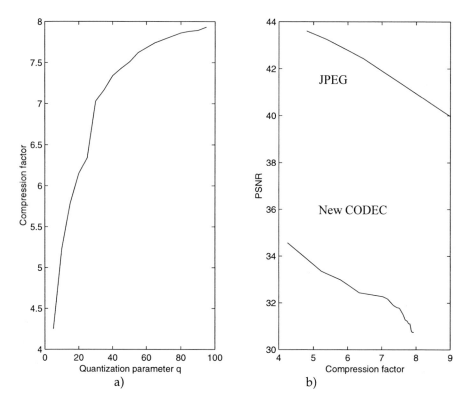

Figure 17.9: *Measurements for one test image: a) Compression factor as function of q, b) average PSNR of the R, G and B color component as a function of the compression factor*

17.4.1 CODEC performance regarding classical criteria

We first investigate a typical range of the compression factor, which is to be obtained using our CODEC. For this purpose we took one image from our test set and compressed it for values of the quantization parameter q in (17.3.3) varying between 5 and 95. The compression factor as a function of the parameter q is displayed in Figure 17.9a. The obtained range for this factor is [4.25, 7.93] for the test image used.

Then, we took the same test image and measured the PSNR (Peak-to-Peak Signal-to-Noise Ratio) for R, G and B color component over the entire range of the compression factor in order to see how the quality of reconstructed images depends on compression efficiency. We averaged the PSNR values of the three color channels at each measurement point and displayed them over corresponding compression factors in Figure 17.9b. A range of the average PSNR values was obtained as [30.74, 34.57]. In order to get a better impression of the above results, we also compressed the same test image using JPEG. We let the JPEG quality factor vary in its entire effective range from 5 to 95 and obtained the compression factors between 9.02 and 94.3 and PSNR between 24.62 and 39.97. Especially for the range of the compression factor obtained for our CODEC, the PSNR varied in the case of JPEG compression between 39.97 and 43.6, as also shown in Figure 17.9b. A comparison of the curves in Figure 17.9b indicates that JPEG performs better in terms of compression efficiency and resulting image distortion.

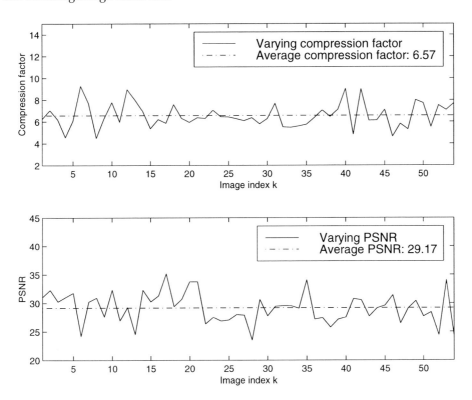

Figure 17.10: *Variations of the compression factor and PSNR for all test images and q=30*

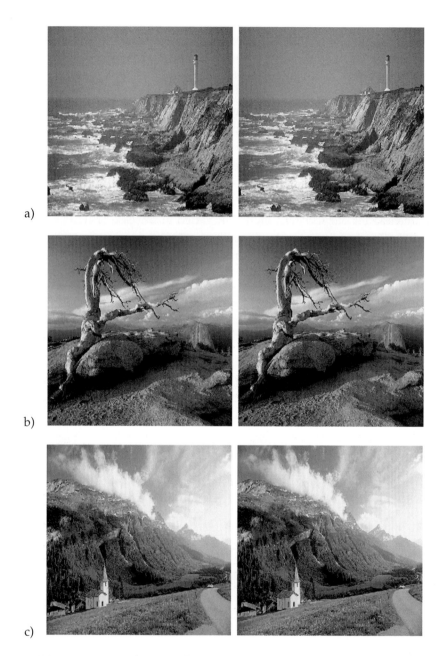

Figure 17.11: *Reconstructed images followed by the originals. The following quantitative data were obtained for q=30: a) code-book size 2%, compression factor 8, b) code-book size 5.3 %, compression factor 6, c) code-book size 4.1%, compression factor 6.4*

Besides compression efficiency and image distortion, the classical criteria also include the overall computational complexity. Compared to the basic VQ, the JPEG CODEC is characterized by well-balanced computational costs on the encoder and decoder side, which makes it more practical. Thus, in order to make the VQ competeable with JPEG regarding the cost balance, a strong reduction of the encoder complexity would be required. This was one of the objectives when developing the methodology for a simplified code-book generation in Subsection 17.3.1. Although we managed to considerably reduce the encoding complexity in our CODEC compared to the basic VQ, this complexity is still relatively large. This is mainly due to small block dimensions, which, as explained before, were chosen such to increase the compression efficiency. For instance, only 4% of image information, contained in the code book of an image with dimensions 320x320 pixels, corresponds to 1024 blocks with dimensions 2x2 pixels. Consequently, for each of the 25600 blocks of that image, 1024 computations of difference values (17.3.1), threshold (17.3.4) and their mutual comparisons are required.

The above comparison of our CODEC and JPEG regarding the classical optimization criteria has shown that JPEG has a better performance. We, however, took this into account with a reference to the fact that a CODEC that performs well regarding the three classical criteria is not necessarily optimal when it comes to the fourth criterion: providing a higher content accessibility [Pic95b].

To complete the evaluation of the CODEC performance regarding the compression efficiency and image distortion, we also investigated the consistency of the CODEC performance regarding these criteria for different images. For this purpose we fixed the quantization parameter q to the value 30 and computed the compression factor and PSNR for all images from our test set. The results are displayed in Figure 17.10. The variations of computed values can be explained by the variability of the code-book size, which depends on image content and is the only variable segment of the compressed image stream. In our measurements, the relative code-book size varied between 1.4% and 9.78%, with an average of 4.6%. Perceptual quality of the reconstructed images for q=30 can be evaluated by comparing the originals and decompressed images in Figures 17.11a-c.

17.4.2 Increase of content accessibility

We address in this section some of the possibilities for easy content access in images which are compressed using the CODEC developed in this chapter. These access possibilities are:

- Easy image or image-region decompression
- Easy access to some spatial image features directly in the compressed domain (dominant and less important colors, image and image-region histograms)

An advantage of having the VQ as the underlying principle of the CODEC presented here is that decompressing an image is a simple "look-up-and-fill" procedure. Namely, a VQ-compressed image can easily be decoded by filling it with blocks from a code book, according to addresses received by the decoder. This is a clear advantage regarding the computational complexity of the decoder, if compared i.e. with the JPEG decompression procedure, characterized by a computationally complex inverse DCT. Further, since the list of addresses in the compressed image format (Figure 17.8) preserves the information about image structure, no decompression of the entire image is required in order to fully reconstruct any of its regions. Such a reconstruction is easily performed by simply choosing the region blocks in the address list, finding their corresponding blocks in the code book and filling the image regions of interest. However, an image, or any of its regions does not need to be reconstructed in order to obtain certain spatial image features; an image-database application involving these features can be performed directly on compressed images.

As a first example, the image-specific code book itself, which is directly accessible in the compressed image format (Figure 17.8), can effectively be used for performing global classification of images or some more general image queries. In our approach, a block is selected in a code book if the average color of its pixels is sufficiently representative for that image. This implies that if average colors are computed for all code-book blocks c_i, a general idea can be obtained about the colors present in the image. In this way, images containing drastically different color content can be separated from each other, or – if different from the query image regarding their colors - not given by the system as query results.

Image classifications and queries based on image-specific code books can be made even more specific if also the information related to the usage of code-book blocks for image reconstruction is taken into account. This information can easily be retrieved from the list of addresses for block correspondences. Then, by counting the numbers of times a_i, that a code-book block c_i is present in the address list, the image and an arbitrary image region can be represented as

$$image \Leftrightarrow [a_1 c_1, ..a_M c_M] \tag{17.4.1a}$$

$$image\ region \Leftrightarrow [a_m c_m, ..a_n c_n] \tag{17.4.1b}$$

After the average color of each block c_i has been computed, the expressions (17.4.1a-b) can provide a global idea about the color composition of the image (region), both in terms of which general colors are present there and in which amount. The higher the number a_i, the more important role plays the block c_i in the image (region), and thus also the average color of its pixels, that is

$$\max_i a_i = a_{dominant} \Rightarrow c_{dominant} \Rightarrow \text{dominant average color} \qquad (17.4.2)$$

As an example, we now estimate the computational complexity of obtaining the information on a general color composition of an image based on (17.4.1a) and on computing the average color of each block c_i. We estimate the complexity by determining the number of reading (O_{read}), adding (O_{add}), multiplying ($O_{multiply}$) and comparing ($O_{compare}$) operations which are to be performed.

With S addresses to be found in the last segment of the compressed image format in Figure 17.8, the number of operations required to obtain all the coefficients a_i can be estimated as follows:

$$C_{a_i} = S(O_{read} + O_{add}) \qquad (17.4.3)$$

Then, the average color is computed for each block of the code book with the following amount of operations:

$$C_{c_i} = 3(N^2 O_{add} + O_{multiply}) \qquad (17.4.4)$$

The entire complexity of obtaining the information on a general image color composition complying with (17.4.1a) is now given as

$$C_{color\ composition} = C_a + MC_{c_i} + MO_{multiply} \qquad (17.4.5)$$

As another example, histograms for image (regions) can easily be computed by collecting pixels of blocks c_i and taking into account the values a_i. Here, we compute the bins h of an 1-dimensional color histogram $H(h)$, where h can be the value of any pixel-color component K ($K=R$, G or B) and where only characteristic blocks c_i, used for reconstruction of an image region (17.4.1b), are considered:

$$H_{R,G,B}(h) = \sum_{i=m}^{n} a_i v_h(i) \qquad (17.4.6a)$$

with

$$v_h(i) = \begin{cases} l_{i,h}, & pixel(K = h) \in c_i \\ 0, & otherwise \end{cases}$$
(17.4.6b)

The function $v_h(i)$ indicates whether a pixel with its K color component corresponding to h is present in the block c_i, and in which amount $l_{i,h}$. Also by counting the number of operations required for the expression (17.4.6a), we estimate the complexity of obtaining an image region histogram directly from the compressed image in Figure 17.8 as

$$C_{hist} = C_{a_i} + (n-m)\left(N^2\left(O_{read} + O_{compare}\right) + \left(l_{i,h} + 1\right)O_{add} + O_{multiply}\right)$$
(17.4.7)

We now like to compare the C_{hist} from (17.4.7) with the complexity of computing the same histogram on the decoded image. This last complexity is given as:

$$C_{hist,decompressed} = C_{decompressed} + XY(O_{read} + O_{compare} + O_{add})$$
(17.4.8)

Since the size of the code book M is only a small fraction (average of 4.6% in our tests) of the total number of pixels in an image obtained by multiplying both image dimensions X and Y, and since $0 \le l_{i,h} \le N^2$ with $N=2$, the second summand in (17.4.8) can be considered considerably larger than the second summand in (17.4.7). Further, if, for instance, JPEG CODEC is used as alternative, the first summand $C_{decompressed}$ in (17.4.8) includes the total (high) number of operations required for JPEG decompression. As such, this summand can realistically be assumed far larger than the value C_{a_i} in (17.4.7). Therefore, it can be said that the histogram computation using (17.4.6) is computationally considerably less expensive than if performed after e.g. a JPEG-compressed image is decoded.

17.5 Conclusions

The image CODEC presented in this chapter was developed to suit emerging applications on large-scale image databases, where a fast and easy access to image content can considerably improve the efficiency of interacting with an image database. While the currently available CODECs are optimized with respect to the classical criteria (bit rate, image distortion and overall computational complexity), introducing an additional fourth criterion on content accessibility has the effect that existing CODECs are no longer optimal and, as discussed in Section 17.1, that the development of new CODECs is needed. It would be best if we could retain an excellent compression efficiency, low distortion of reconstructed images and nicely balanced and low computational complexity of JPEG in newly developed CODECs and still be able to easily perform any operation on image content. Such a perfect balance among the four optimization criteria needs indeed to be the guiding objective of research in this area. The development of the CODEC in this chapter can be understood as a first step in the process of reaching this objective. We deliberately left the powerful concept of transform-based CODECs in order to remain in the spatial domain and so to provide means for accessing the image content more easily. In this way we expected a priori a lowering of the compression efficiency and the quality of the reconstructed images, compared to JPEG. Also we took into account a possible misbalance and an increase of computational complexity at the encoder side. However, as a compensation, we are able to decompress an image much more quickly and to reach some of the characteristic image features directly in the compressed domain. Although we can say that in some way we found an acceptable trade-off between four optimization criteria, we are also aware of the fact that the developed CODEC is far from optimal. Nevertheless, we hope with our CODEC to provide a solid base for further research in this area.

Bibliography - Part III

[Aha96] Ahanger G., Little T.D.C.: *A Survey of Technologies for Parsing and Indexing Digital Video*, Jornal of Visual Communication and Image Representation, Vol. 7, No.1, pp.28-43, March 1996.

[Aku92] Akutsu A. et al.: *Video indexing using motion vectors*, Proceedings of VCIP'92, Boston 1992

[Ari96] Ariki Y., Saito Y.: *Extraction of TV News Articles based on Scene Cut Detection using DCT Clustering*, Proceedings of ICIP '96, Vol. 3, pp. 847-850, Lausanne CH, 1996

[Arm93a] Arman F., Hsu A., Chiu M.: *Feature Management for Large Video Databases*, Proceedings of IS&T/SPIE Storage and Retrieval for Image and Video Databases, February 1993.

[Avr98] Avrithis Y.S., Doulamis N.D., Doulamis A.D., Kollias S.D.: *Efficient Content Representation in MPEG Video Databases,* Proceedings of IEEE Workshop on Content-Based Access of Image and Video Databases, Santa Barbara CA, 1998

[Bor93] Bordwell D., Thompson K.: *Film Art: An Introduction*, McGraw-Hill, New York 1993

[Bor96] Boreczky J.S., Rowe L.: *Comparison of video shot boundary detection techniques*, Proceedings of IS&T/SPIE Storage and Retrieval for Still Image and Video Databases IV, Vol. 2670, February 1996.

[Boy99] Boykin S., Merlino A.: *Improving Broadcast News Segmentation Processing*, Proceedings of the IEEE International Conference on Multimedia Computing and Systems, Florence 1999

[Cha95a] Chang S.-F.: *Compressed-Domain Techniques for Image/Video Indexing and Manipulation*, Proceedings of ICIP'95, Washington DC, October 1995

[Cha95b] Chang S.-F.: *New Algorithms for Processing Images in the Transform-Compressed Domain*, Proceedings of IS&T/SPIE, Vol. 2501, February 1995

[Che97] Chen L., Faudemay P.: *Multi-Criteria Video Segmentation for TV News*, IEEE First Workshop on Multimedia Signal Processing, Princeton NJ, 1997

[Col76] Coll D.C., Choma G.K.: *Image Activity Characteristics in Broadcast Television*, IEEE Transactions on Communications, pp. 1201-1206, October 1976

[Cur65] Curran T.F., Ross M.: *Optimum Detection Thresholds in Optical Communications*, Proceedings of IEEE, pp. 1770-1771, November 1965

[Dav79] Davies D.L., Bouldin D.W.: *A Cluster Separation Measure*, IEEE Transactions on Pattern Analysis and Machine Intelligence, Vol. PAMI-1, No.2, pp. 224-227, April 1979

[DeM98] DeMenthon D., Kobla V., Doermann D.: *Video Summarization by Curve Simplification*, Proceedings of CVPR '98, Santa Barbara CA, 1998

[ETS94] ETS 300 421: *Digital broadcasting systems for television, sound and data services; framing structure, channel coding and modulation for 11/12 GHz staellite services*, EBU/ETSI JTC, December 1994

[ETS97] ETS 300 401 ed.2: *Radio broadcasting systems; Digital Audio Broadcasting (DAB) to mobile, portable and fixed receivers*, 1997

[Fis95] Fischer S., Lienhart R., Effelsberg W.: *Automatic Recognition of Film Genres*, Proceedings of ACM Multimedia '95, San Francisco 1995

[Fri67] Friedman H.P., Rubin J.: *On some Invariant Criteria for Grouping Data*, Journal Amer. Stat. Assoc., Vol.62, pp.1159-1178, 1967

[Fur95] Furht B., Smoliar S.W., Zhang H.: *Video and Image Processing in Multimedia Systems*, Kluwer Academic Publishers 1995

[Ger92] Gersho A., Gray R.M.:*Vector Quantization and Signal Compression*, Kluwer Academic Publishers, Boston 1992

[Gre97] Gresle P., Huang T.S.: *Video Sequence Segmentation and Key Frames Selection Using Temporal Averaging and Relative Activity Measure*, Proceedings of VISUAL '97, San Diego 1997

[Gon95] Gong Y. et al.: *Automatic Parsing of TV Soccer Programs*, Proceedings of the International Conference on Multimedia Computing and Systems, May 1995

[Gun98] Gunsel B., Tekalp. A.M.: *Content-Based Video Abstraction*, Proceedings of ICIP '98, Chicago USA, 1998

[Ham94] Hampapur A., Jain R., Weymouth T.: *Digital Video Segmentation*, Proceedings of ACM Multimedia'94, 1994

[Han97a]   Hanjalic A., Ceccarelli M., Lagendijk R.L., Biemond J.: *Automation of systems enabling search on stored video data*, Proceedings of IS&T/SPIE Storage and Retrieval for Image and Video Databases V, Vol. 3022, February 1997.

[Han97b]   Hanjalic A., Lagendijk R.L., Biemond J.: *A novel video parsing method with improved thresholding*, Proceedings of the Third Annual Conference of the Advanced School for Computing and Imaging, Heijen, The Netherlands, 1997

[Han97c]   Hanjalic A., Lagendijk R.L., Biemond J.: *A New Method for Key Frame based Video Content Representation*, in *Image Databases and Multi Media Search*, Eds. A.W.M. Smeulders, R. Jain, World Scientific, Singapore, 1997

[Han99b]   Hanjalic A., Lagendijk R.L., Biemond J.: *Automated High-Level Movie Segmentation for Advanced Video Retrieval Systems*, IEEE Transactions on Circuits and Systems for Video Technology, June 1999

[Han99c]   Hanjalic A., Lagendijk R.L., Biemond J.: *An Efficient Image CODEC with Reduced Content Access Work*, Proceedings of ICIP '99, Kobe 1999

[Han00]   Hanjalic A., Zhang H.: *An Integrated Scheme for Automated Video Abstraction based on Unsupervised Cluster-Validity Analysis*, IEEE Transactions on Circuits and Systems for Video Technology, December 1999

[Hua99a]   Huang S.: *Digital television: a new way to deliver information*, Proceedings of IS&T/SPIE Storage and Retrieval for Image and Video Databases VII, Vol. 3656, January 1999

[Hua99b]   Huang Q., Liu Z., Rosenberg A.: *Automated Semantic Structure Reconstruction and Representation Generation for Broadcast News*, Proceedings of IS&T/SPIE Storage and Retrieval for Image and Video Databases VII, Vol. 3656, January 1999

[ISO97]   ISO/IEC JTC1/SC29/WG11, *MPEG-7: Context and Objectives (v.4)*, July 1997.

[Jac92]   Jacquin A.E.: *Image Coding based on a Fractal Theory of Iterated Contractive Image Transformations*, IEEE Transactions on Image Processing, Vol.2, No.1, pp.18-30, January 1992

[Jai82]   Jain A.K., Chandrasekaran B.: *Dimensionality and sample size considerations in pattern recognition practice*, in Handbook of Statistics, Vol. 2 (P. Krishnaiah and L.N. Kanal eds.), North-Holland Publishing Company, Amsterdam 1982

[Jai88]   Jain A.K., Dubes R.C.: *Algorithms for Clustering Data*, Prentice Hall Advance Reference Series, 1988

[Ken98]   Kender J.R., Yeo B.: *Video Scene Segmentation Via Continuous Video Coherence*, Proceedings of IEEE Computer Society Conference on Computer Vision and Pattern Recognition, Santa Barbara, June 1998

[Kik92]   Kikukawa T., Kawafuchi S.: *Development of an automatic summary editing system for the audio visual resources*, Transactions of the Institute of Electronics, Information and Communication Engineers, Vol J75-A, No.2, 1992

[Lie99]   Lienhart R.: *Comparison of automatic shot boundary detection algorithms*, Proceedings of IS&T/SPIE Storage and Retrieval for Image and Video Databases VII, Vol. 3656, January 1999

[Liu98]   Liu Z., Huang Q.: *Classification of Audio Events in Broadcast News*, Proceedings of IEEE Workshop in Multimedia Signal Processing, December 1998

[Mai95]   Mai K., Miller J., Zabih R.: *A robust method for detecting cuts and dissolves in video sequences*, Proceedings of ACM Multimedia '95, San Francisco 1995

[Max60]   Max J.: *Quantization for Minimum Distortion*, IRE Transactions on Information Theory, pp. 7-12, 1960

[McG99]   McGee T., Dimitrova N.: *Parsing TV programs for identification and removal of nonstory segments*, Proceedings of IS&T/SPIE Storage and Retrieval for Image and Video Databases VII, Vol. 3656, January 1999

[Men95]   Meng J., Juan Y., Chang S.: *Scene Change Detection in a MPEG CompressedVideo Sequence*, Proceedings of IS&T/SPIE, Vol. 2419, February 1995

[Men96]   Meng J., Chang S.-F.: *Tools for Compressed-Domain Video Editing*, IS&T/SPIE Symposium Proceedings, Vol.2670, San Jose (CA, USA), February 1996

[Nag92]   Nagasaka A., Tanaka Y.: *Automatic video indexing and full-video search for object appearances*, in *Visual Database Systems II*, Eds. Knuth E. and Wegner L.M., volume A-7 of IFIP Transactions A: Computer Science and Technology, pages 113-127, North-Holland, Amsterdam 1992

[Oka93]   Okamoto H. et al.: *A Consumer Digital VCR for Advanced Television*, IEEE Transactions on Consumer Electronics, Vol. 39, No.3, pp. 199-204, August 1993.

[Ots93]   Otsuji K., Tonomura Y.: *Projection Detecting Filter for Video Cut Detection*, Proceedings of ACM Multimedia '93, 1993

[Ots91]   Otsuji K., Tonomura Y., Ohba Y.: *Video browsing using brightness data*, Proceedings of SPIE/IS&T VCIP'91, Vol.1606, 1991

[Paa97]   van Paasen R.: *Subjective Representation of Video: An Exploratory Study*, MSc. Thesis, Eindhoven University of Technology (Nl), 1997

[Pap84]   Papoulis A.: *Probability, Random Variables and Stochastic Processes*, McGraw-Hill, International Editions, 1984

[Pen93]   Pennebaker W.B., Mitchell J.L.: *The JPEG Still Image Data Compression Standard*, Van Nostrand Reinhold, New York 1993

[Pen94a]   Pentland A., Picard R., Davenport G., Haase K.: *Video and Image Semantics: Advances Tools for Tellecommunications*, IEEE MultiMedia, Summer 1994

[Pen94]   Pentland A., Picard R.W., Sclaroff S.: *Photobook: Tools for Content-Based Manipulation of Image Databases*, MIT Media Lab, Perceptual Computing Technical Report No. 255

[Pfe96] Pfeiffer S., Lienhart R., Fischer S., Effelsberg W.: *Abstracting Digital Movies Automatically*, Journal of Visual Communication and Image Representation, Vol. 7, No. 4, pp. 345-353, December 1996.

[Pic95b] Picard R.W.: *Content Access for Image/Video Coding: The Fourth Criterion*, Tech. Rep. 295, MIT Media Lab, Perceptual Computing, Cambridge MA, 1994 (also available as MPEG Doc. 127, Lausanne 1995)

[Sah88] Sahoo P.K., Soltani S., Wong A.K.C., Chen Y.C.: *A survey of thresholding techniques*, CVGIP, 41:233-260, 1988

[Sal73] Salt B.: *Statistical style analysis of motion pictures*, Film Quarterly, Vol 28, pp. 13-22, 1973

[Sau97] Saur D.D. et al.: *Automated Analysis and Annotation of Basketball Video*, Proceedings of IS&T/SPIE Vol. 3022, February 1997

[Sha95a] Shahraray B.: *Scene change detection and content-based sampling of video sequences*, Proceedings of IS&T/SPIE Vol. 2419, February 1995

[Sha95b] Shahraray B., Gibbon: *Automatic generation of pictorial transcripts of video programs*, Proceedings of IS&T/SPIE Digital Video Compression: Algorithms and Technologies, San Jose 1995

[SMA] SMASH project home page:
http://www.extra.research.philips.com/euprojects/smash/

[Sme97] Smeulders A.W.M., Jain R. (Eds.): *Image Databases and Multimedia Search*, Series on Software Engineering and Knowledge Engineering, Vol.8, World Scientific Singapore, 1997

[Smi94] Smith J.R., Chang S.-F.: *Quad-Tree Segmentation for Texture-Based Image Query*, Proceedings of ACM 2nd Multimedia Conference, San Francisco, October 1994

[Smi93] Smith B.C., Rowe L.A.: *Algorithms for Manipulating Compressed Images*, IEEE Computer Graphics & Applications, Septermber 1993

[Son98] Song S., Kwon T., Kim W.: *Detection of gradual scene changes for parsing of video data*, Proceedings of IS&T/SPIE Vol. 3312, 1998

[Sun97] Sun X., Kankanhalli M.S., Zhu Y., Wu J.: *Content-Based Representative Frame Extraction for Digital Video*, Proceedings of IEEE Multimedia Computing and Systems, Austin TX 1997

[TNO97] *TNO Magazine*, Vol.1, No.4, December 1997

[Ued91] Ueda H., Miyatake T., Yoshizawa S.: **IMPACT:** *An interactive natural-motion picture dedicated multimedia authoring system*, Proceedings of the CHI'91, 1991

[Vai98] Vailaya A., Jain A., Zhang H.: *On Image Classification: City vs. Landscape*, Proceedings of IEEE Workshop on Content-Based Access of Image and Video Libraries, Santa Barbara, June 1998

[Vai99]   Vailaya A., Jain A., Zhang H.: *A Bayesian Framework for Semantic Classification of Outdoor Vacation Images*, Proceedings of IS&T/SPIE Storage and Retrieval for Image and Video Databases VII, Vol. 3656, January 1999

[Vas98]   Vasconcelos N., Lippman A.: *A Bayesian Video Modeling Framework for Shot Segmentation and Content Characterization*, Proceedings of CVPR '98, Santa Barbara CA, 1998

[dWi92]   de With P.H.N.: *Data Compression Systems for Home-Use Digital Video Recording*, IEEE Journal on Selected Areas in Communication, Vol 10, No. 1, pp. 97-121, Januar 1992.

[dWi93]   de With P.H.N., Rijckaert A.M.A., Keesen H.-W., Kaaden J., Opelt C.: *An Experimental Digital Consumer HDTV Recorder using MC-DCT Video Compression*, IEEE Transactions on Consumer Electronics, Vol. 39, No.4, pp. 711-722, November 1993.

[Wol96]   Wolf W.: *Key frame selection by motion analysis*, in Proceedings of IEEE ICASSP'96, 1996

[Yan93]   Yanagihara N., Siu C., Kanota K., Kubota Y.: *A Video Coding Scheme with a High Compression Ratio for Consumer Digital VCRs*, IEEE Transactions on Consumer Electronics, Vol. 39, No. 3, pp 192-198, August 1993.

[Yeo95a] Yeo B.-L., Liu B.: *Rapid Scene Analysis on Compressed Video*, IEEE Transactions on Circuits and Systems for Video Technology, Vol.5, No.6, December 1995

[Yeo95b] Yeo B.-L., Liu B.: *On the Extraction of DC Sequence from MPEG compressed Video*, Proceedings of IEEE ICIP'95, October 1995

[Yeu95a] Yeung M.M., Liu B.: *Efficient Matching and Clustering of Video Shots*, Proceedings of IEEE ICIP '95, Vol. I, October 1995

[Yeu97]   Yeung M., Yeo B.-L.: *Video Visualisation for Compact Presentation and Fast Browsing of Pictorial Content*, IEEE Transactions of Circuits and Systems for Video Technology, Special Issue on Multimedia Technology, Systems and Applications, October 1997.

[Zha93]   Zhang H., Kankanhalli A., Smoliar S.W., *Automatic partitioning of full-motion video*, Multimedia Systems, Vol.1, pp. 10-28, 1993

[Zha95a] Zhang H., Low C.Y., Smoliar S.W.: *Video Parsing and Browsing using Compressed Data*, *Multimedia Tools and Applications*, vol. 1, pp. 89-111, Kluwer Academic Publishers, 1995.

[Zha97a] Zhang H., Wang J.Y.A., Altunbasak Y.: *Content-Based Retrieval and Compression: A Unified Solution*, Proceedings of ICIP '97, Santa Barbara CA, 1997

[Zha97b] Zhang H., Wu J., Zhong D., Smoliar S.W.: *An Integrated System for Content-Based Video Retrieval and Browsing*, Pattern Recognition, Vol. 30, No.4, pp.643-658, 1997

[Zhu98]   Zhuang Y., Rui Y., Huang T.S., Mehrotra S.: *Key Frame Extraction Using Unsupervised Clustering*, Proceedings of ICIP '98, Chicago 1998

Index

Analysis
    cluster-validity 371
    high-level video 379
    news programs 387
    sport programs 390

Benchmarking
    methods 274
    the DEW algorithm 273, 285

Blotch correction techniques 63

Blotch detection
    and correction 51
        system for 51
        techniques 54
    improved 66
    with increased temporal
        aperture 76

Clustering 369

Compression trends 413

Coring 93, 95, 98
    Heuristic 97
    I, P and B frames 113

In wavelet and subband
    domains 98

CODEC
    alternative (concept) 418
    based on simplified VQ 422
    evaluation 427
        classical criteria 428
        content accessibility 431

Detection approaches (shots) 333

Detector
    abrupt shot-boundaries 343

DEW
    benchmarking 273, 285
    concept for MPEG/JPEG
        encoded video 229
    detailed algorithm
        description 232
    evaluation for MPEG
        video data 240
    extension for EZW
        coded images 247
    finding optimal
        parameters 251

introduction 227
modeling for JPEG
  compressed video 253

Drift 221, 245

Evaluation
  alternative image CODEC 427
  anchorperson detection 407
  DEW algorithm
    for MPEG video 286
    for still images 291
  key-frame extraction 363, 375
  LSU-boundary detection 400
  shot-boundary detection 347

Features 343

Film
  genre recognition 386

Image
  restoration 20
    background 3
    evaluation 123
      assessment 123
      experiments 128
    modeling for 9

Information
  retrieval: an introduction 313

Intensity flicker
  a model for 28
  correction 27
  estimating 30
  experiments on 43

Interpolation
  of missing data 79

Key frames
  extraction based on cluster-
    validity analysis 368

extraction based on visual-
  content variations 359
from clusters 375
previous work 355

LSU
  definition 393
  detection 391, 396

Metrics 343

Movies
  extracting most characteristic
    segments 386
  scene-boundary detection 386
  segmentation into LSUs 391

Motion
  incorporating 34
  vector repair 60

News
  analysis 387, 403
  anchorperson detection
    387, 403

Noise
  reduction by coring 93
  techniques 94

Probability function
  a priori 344
  conditional 346

Robustness 222, 246

Shot-boundary
  detection 323
    previous work 328
    statistical
      framework 339

Simoncelli pyramid 100

INDEX

Sport
    program analysis   390

Video
    abstraction using
        key frames   351
    high-level analysis   379
    related work   384

Watermarking
    correlation-based   172
        in spatial domain   172
        in the coefficient domain   210
    extended-correlation based   191
    fractal-based   201
    history   162
    MPEG video bit streams   207
    non-correlation-based   198
    re-labeling resistant   223
    requirements   159
    state-of-the-art   171
    the need for   157

Watermark
    attacks   276
    differential energy   227
    energy adaptation   187
    estimation   278
    payload of the   217, 240
    visual impact of the   218, 241